T0142927

Conceptual Exploration

Bernhard Ganter • Sergei Obiedkov

Conceptual Exploration

Bernhard Ganter
Mathematik
Technische Universität Dresden
Dresden, Germany

Sergei Obiedkov
Higher School of Economics
National Research University
Moskva, Russia

ISBN 978-3-662-56999-3 ISBN 978-3-662-49291-8 (eBook)
DOI 10.1007/978-3-662-49291-8

Softcover reprint of the hardcover 1st edition 2016

Printed on acid-free paper

This Springer imprint is published by Springer Nature
The registered company is Springer-Verlag GmbH Berlin Heidelberg

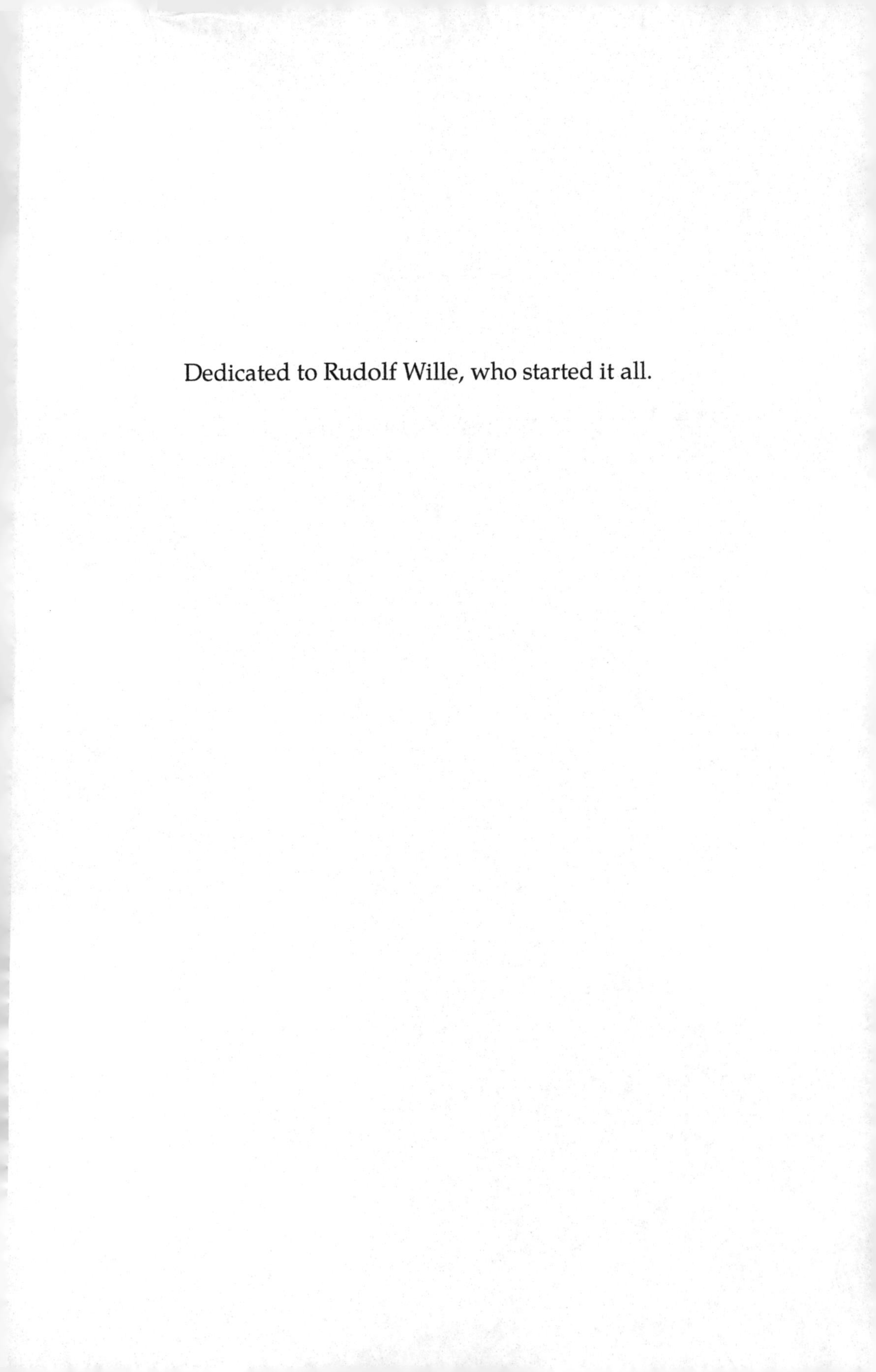

Dedicated to Rudolf Wille, who started it all.

Contents

About this book

This book describes a method called "attribute exploration", its applications, its theory, and some of the many possible generalizations.

The basic task is very simple: for a given list of *attributes* we want to find all meaningful *attribute combinations*. This should be done in the simplest possible manner, avoiding all work that is not absolutely necessary and letting the computer do the technicalities. There is indeed a method to achieve this. It makes clever use of a mathematical theory called *Formal Concept Analysis*. The book offers a self-contained introduction to those aspects of this theory that are used for attribute exploration.

As an appetizer, we present a toy example. Suppose you are a researcher investigating nano-particles that are embedded into paint for enhancing the color. You wonder how the particles are distributed in the paint. You study colored surfaces under a microscope. Figure 1 shows a simplified example of what you might see.

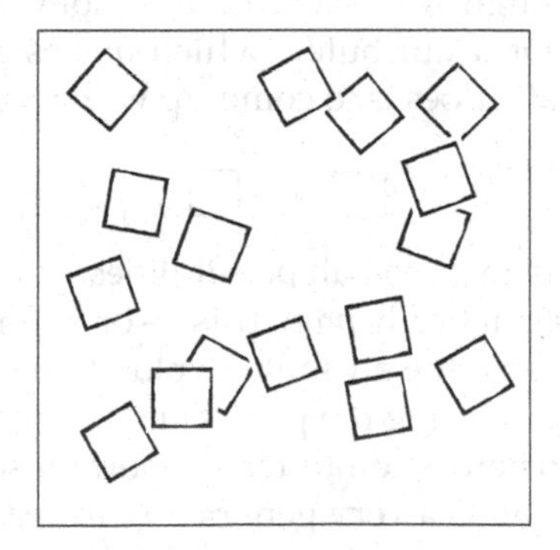

Figure 1: Particles under a microscope (schematic)

Apparently, these particles are squares of equal size. There is no obvious pattern in their distribution, and you decide that you should study much larger samples using computerized scans. The computer program has difficulties recognizing overlapping squares, and it is necessary to adjust the program code. For that, you need to understand how two squares can be situated relative to each other. This is a theme from elementary geometry: discussing *pairs of squares*. Here are two possible such pairs:

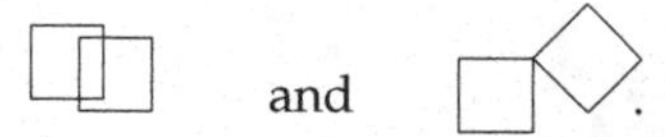

The two squares on the left *overlap*, because they have an interior point in common. They are also *parallel* in the sense that each side of one square is parallel to some side of the other. The two squares in the second diagram do not have these properties. Instead, they have a *common vertex*, but no *common edge*, not even a *common segment* of an edge. They are not *disjoint*, since they share a point.

We have collected a small list consisting of six attributes:

overlap, parallel, disjoint, common vertex, common edge, common segment,

and we have seen an object (a pair of squares) with the attribute combination

overlap, parallel,

and another one with the attribute "combination"

common vertex.

You may wonder which other attribute combinations are possible. Some, like

overlap, parallel, disjoint, common vertex,

are obviously not.

The idea of going through all possible cases is not very inviting. You would have to check 2^n cases for n attributes, which makes already 64 for our toy example. What one usually does is to come up with more examples, like[1]

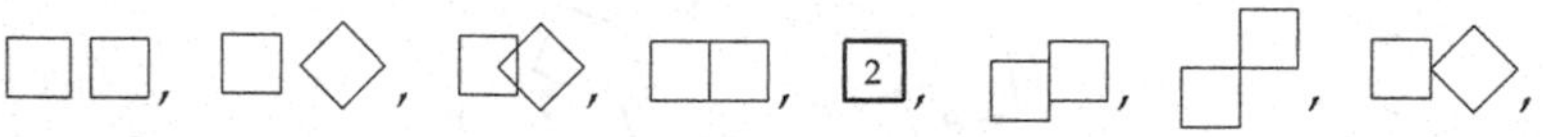

$\boxed{2}$

hoping to have intuitively included all possibilities.

But how can you systematically *prove* this — or find more examples in case you missed some? This is where attribute exploration comes into play, giving you computer support for solving the problem.

What we do *not* mean here is computer support for solving geometry problems. Attribute exploration is a very general *knowledge acquisition method*, applicable to all kinds of situations and not restricted to mathematical questions. What we expect from the method is help with the classification procedure!

The basic idea is to let the computer discover common properties of all examples considered so far. For example, the computer may find that in all listed examples where the two squares have a *common vertex* and a *common segment* they also have a *common edge*. It will then ask you to check if this implication

[1]The symbol $\boxed{2}$ stands for two squares at the same position.

if *common vertex* and *common segment*, then *common edge*

must be true for all possible examples, i.e., if it can be proved for pairs of squares in general. If not, then there must be a counterexample, an example that is not yet in your list!

However, the above implication is indeed true in general (for squares of equal size). Therefore, this implication does not lead to a new example. But there may be other ones that do. We stop the procedure here and postpone this exploration until Section 4.2.2, which gives the complete answer.

The general strategy is this: we let the computer find a simple logical rule holding in all known examples and try to prove this rule in general. If it fails, then there must be a yet unknown example, an example violating the rule. If not, we confirm the rule and let the computer find another one. After finitely many steps, we are done. Really? Well, we need some theory to answer this precisely.

Then, once we have a successful method and a mathematical theory behind it, we want to generalize. There are indeed many possibilities of doing so, but these require more and more theoretical background.

Thus, before we come to applications, we present the fundamentals of the theory, beginning with concept lattices in Chapter 1. For most of the readers, it will not be necessary to read all this. If you are only interested in understanding what attribute exploration is, then have a look at the example below. If you find the method interesting, we suggest that you start with Chapter 4 (p. 125) and study the more ambitious examples given there. Try it out, make your own explorations using the available software tools (see Section 1.4.4)!

Attribute exploration: a very small example

We present another example, now showing a complete exploration. More precisely, we completely explore a subset of the attributes under consideration and leave some open questions to the reader. Before doing so, we point out what to expect: the typical result of an attribute exploration is a classification (of admissible attribute combinations). So do not expect a result expressed as a number or as a theorem, expect a conceptual classification!

The first step is to delimit the topic. We shall explore a theme from decorative arts (in architecture, ceramics, textiles) called **frieze patterns**. Figure 2 shows an example. Such a frieze is an ornamental strip with a repeating decoration. Classifying frieze types is of interest, since different cultures have developed different kinds of friezes, but it is also important for applications in crystallography.

Usually, one investigates the *symmetries* which a frieze shows. This is simplified by some abstraction: the frieze is assumed to extend *ad infinitum* in both

Figure 2: A frieze from the Alhambra palace, Granada

directions and to exactly repeat its pattern infinitely often. There are only few geometric congruence mappings that such a frieze may have. One type is *translations*, i.e., pattern preserving shifts. Each frieze is assumed to admit such shifts, and there must be one of minimal positive length, making the pattern repetition discrete and not continuous. Apart from translations, there are only four other types:

- a *vertical* mirror reflection at some axis perpendicular to the strip[2],
- the *horizontal* mirror reflection at the middle line of the strip,
- a *turn*, i.e., a 180° rotation at some point on the middle line, and
- a *glide* reflection, a combination of a translation and a horizontal reflection.

It is common to use these four symmetry types as classification attributes. The frieze in Figure 2 has only one of these, a *vertical* reflection (and, of course, translations). But it also has another, more sophisticated type of symmetry. The black area is congruent to the white (we tacitly exclude the points on the color borders). This frieze can be rotated 180° in such a way that all black areas are mapped onto white ones and vice versa. In other words, it admits a *color-changing turn*, abbreviated $\overline{\textit{turn}}$. A closer look reveals that the frieze also admits a $\overline{\textit{glide}}$, a color-changing glide reflection. We shall see an example of another frieze with $\overline{\textit{horizontal}}$ and $\overline{\textit{vertical}}$ reflections below.

So this is what we choose as the universe of this exploration: black and white friezes, to be classified by their color-preserving and their color-changing self-map types

$\overline{\textit{turn}}$, $\overline{\textit{horizontal}}$, $\overline{\textit{vertical}}$, $\overline{\textit{glide}}$, *turn*, *horizontal*, *vertical*, and *glide*.

The interactive exploration algorithm will ask questions about possible combinations of these attributes, which we, the "domain experts", are supposed to answer. We start with our only example so far:

[2]The reflection *axis* is vertical (assuming that the frieze extends horizontally). Each point is mapped to a point on the same horizontal line.

	$\overline{turn}$	$\overline{horizontal}$	$\overline{vertical}$	$\overline{glide}$	*turn*	*horizontal*	*vertical*	*glide*
	×			×			×	

,

to which the algorithm reacts by asking

Do all friezes have symmetries $\overline{turn}$, $\overline{glide}$, and *vertical*?

It is easy to come up with a counterexample, and we extend our example list as follows:

	$\overline{turn}$	$\overline{horizontal}$	$\overline{vertical}$	$\overline{glide}$	*turn*	*horizontal*	*vertical*	*glide*
	×			×			×	
				×		×		×

.

We decide to explore the color-preserving symmetries first. Then the next question asks if all friezes with a *glide* symmetry also have a *horizontal* reflection axis, abbreviated to

glide $\rightarrow$ *horizontal* ?

It is refuted by the third example

	$\overline{turn}$	$\overline{horizontal}$	$\overline{vertical}$	$\overline{glide}$	*turn*	*horizontal*	*vertical*	*glide*
	×			×			×	
				×		×		×
	×	×	×	×	×		×	×

.

The next two queries

vertical, glide $\rightarrow$ *turn* ? and *horizontal* $\rightarrow$ *glide* ?

must be approved, while

turn $\rightarrow$ *vertical, glide* ?

is refuted by yet another example

	$\overline{turn}$	$\overline{horizontal}$	$\overline{vertical}$	$\overline{glide}$	*turn*	*horizontal*	*vertical*	*glide*
	×			×			×	
				×		×		×
	×	×	×	×	×		×	×
		×	×	×	×			

Two more queries

$$turn, glide \quad \rightarrow \quad vertical\,? \quad \text{and} \quad turn, vertical \quad \rightarrow \quad glide\,?$$

must be confirmed before the exploration of the four color-preserving symmetries (only!) is completed. From the exploration result, we can build a *concept lattice*. This will be explained in detail later. It has seven elements and is shown in Figure 3. It leads us to three more frieze types, since the so-called "reducible" (see Section 1.4.3) attribute combinations turn out to be admissible as well. We have added examples for these too, see Figure 4.

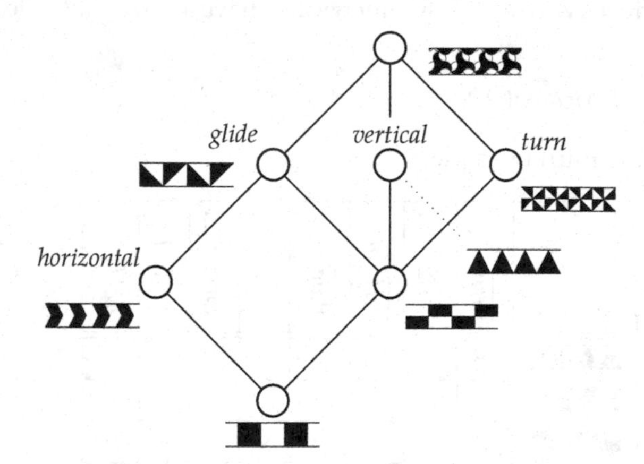

Figure 3: The result for the exploration of the color-preserving symmetries as a concept lattice. Examples of friezes for the join-reducible elements were added

This result that there are precisely seven "frieze groups" is well known. It can be found, for example, in Wikipedia. In contrast, we know of no reference for the full exploration including all eight attributes for black and white

	$\overline{turn}$	$\overline{horizontal}$	$\overline{vertical}$	$\overline{glide}$	*turn*	*horizontal*	*vertical*	*glide*
	×		×	×	×	×	×	×
	×		×					×
	×							

Figure 4: Three more friezes

friezes.[3] We shall not reveal the result here. The reader is invited to finish the exploration. Standard software like ConExp [229] can be used for generating the queries. Some of the techniques that we develop later in this book may ease the workload. The details, as well as supplementary information about the theme of this book, can be found on `www.attribute-exploration.org`.

Acknowledgements

Writing this book took way too long. We started with lecture notes in 2002, when much of the methodology was already available. A few years later, Gerd Stumme used and extended these notes for his own lectures and Sebastian Rudolph defended his PhD thesis on Relational Exploration. This inspired us to make plans for a book. Well, plans. Other projects seemed more urgent and got priority. When we revived our project, again some years later, we found that meanwhile there were more research results than we could handle. The Description Logic group around Franz Baader in Dresden had made remarkable progress, and "Fuzzy Concept Analysis" had been developed to a theory of its own by Radim Bělohlávek (Olomouc) and his colleagues. We had to draw a line. These achievements are only mentioned in this book, but not thoroughly treated. It still was a long way to the present version.

We were lucky to get encouragement and advice from many colleagues. Stumme and Rudolph gave the permission to use their materials. Ivan Arzhantsev, Radim Bělohlávek, Daniel Borchmann, Çynthia V. Glodeanu, Robert Jäschke, Michal Krupka, Sergei O. Kuznetsov, Rokia Missaoui, Amedeo Napoli, Manuel Ojeda-Aciego, L. John Old, Uta Priss, Barış Sertkaya, and Francisco Valverde-Albacete suggested many improvements. Special thanks go to the AIFB at the Karlsruhe Institute of Technology for hosting an incredible number of revisions of our project in their SVN repository.

[3] Note that the problem treated in the beautiful book by Conway et al. [67] is different.

Chapter 1

Concept lattices

This book is about **attribute exploration**, a simple but useful knowledge acquisition technique from **Formal Concept Analysis** (FCA). The latter is a mathematical theory of *concepts* and *conceptual hierarchies*, called *concept lattices*. FCA offers several highly practical methods for working with concrete qualitative data. A solid knowledge of FCA is not a prerequisite for attribute exploration, but knowing the basic ideas will help. We therefore start with an introduction.

Formal Concept Analysis studies how *objects* can hierarchically be grouped together according to their common *attributes*. Thus one of the aspects of FCA is *attribute logic*, the study of possible attribute combinations. Most of the time this will be very elementary. Readers with a background in mathematical logic might say that attribute logic is just propositional calculus, and thus Boolean logic, or even a fragment of this. G. Boole himself used the intuition of attributes ("signs") rather than of propositions. So in fact, attribute logic goes back to Boole.

But our style is different from that of logicians. Our logic is *contextual*, which means that we are interested in the logical structure of concrete data (of the *context*). Of course, the general rules of mathematical logic are important for this and will be utilized.

In the subsequent sections, it will be explained in detail what a concept lattice is, how it can be represented by a diagram, and how to read such diagrams.[1]

1.1 Examples of concept lattices

Probably, many readers are familiar with lattice diagrams or have at least seen examples such as the one in Figure 1.1. It shows the divisors of the number 200, ordered by divisibility, and is quite intuitive to read. By simply following the ascending paths one discovers that 4 is a divisor of 40 and of 100, but not of 50.

[1]Detailed reading instructions will be given in Section 1.3.1 (p. 13).

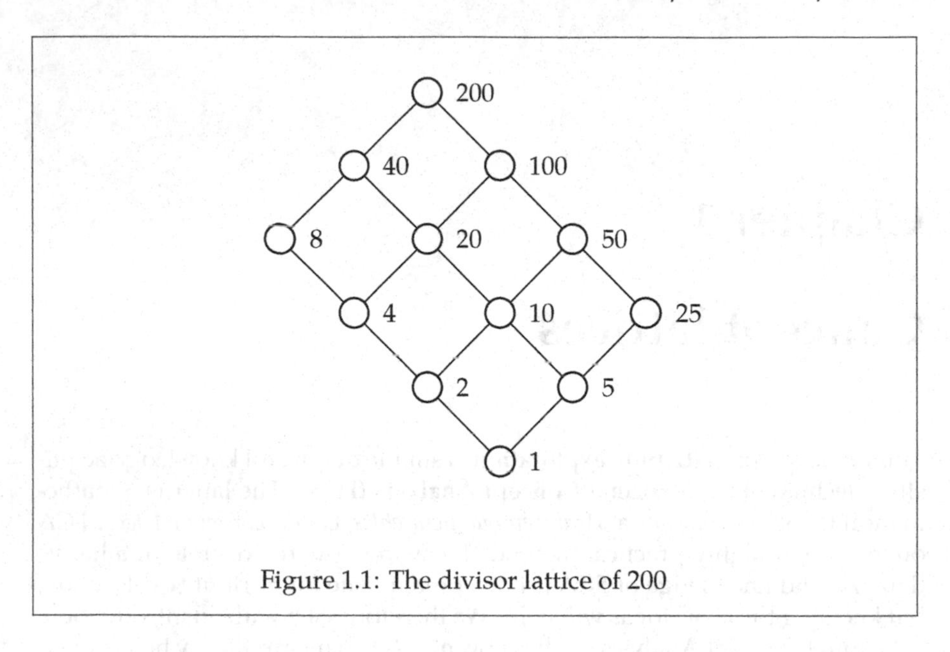

Figure 1.1: The divisor lattice of 200

The diagram in Figure 1.1 is not exactly a *concept lattice* diagram, but it differs only in the way in which it is labeled (compare Figure 1.18 (p. 29)).

A concept lattice with the same structure is depicted in Figure 1.2. It shows the recommended serving temperatures for selected red wines. The recommendation for Trollinger is 15–16 °C, for Beaujolais 15–17 °C, and for Barbera it is 16–18 °C.

The simplicity of this picture is partially due to the fact that we have used only an excerpt of the data (extensive wine temperature charts can be found on the Internet). However, the *planarity* is caused by the interval nature of the attribute combinations. This explains why the slightly more complex recommendations for white wines in Figure 1.3 can also be expressed by a planar diagram.

Often, concept lattice diagrams are not that simple, and it is advisable to visualize only a part of the data. Figure 1.4 displays some paragraphs from a regional building law ("Bauordnung Nordrheinwestfalen") and to which parts of a building these apply. This data is from an information system that used concept lattices for browsing a large number of such regulations and various aspects of buildings [83].

Figure 1.5 is from a linguistic investigation by Priss and Old [180] on the *conceptual exploration of semantic mirrors*. Some English words in the neighborhood of the word *good* and some neighbors of the Norwegian word *god* are displayed according to their relationship in a dictionary. Legend: the English

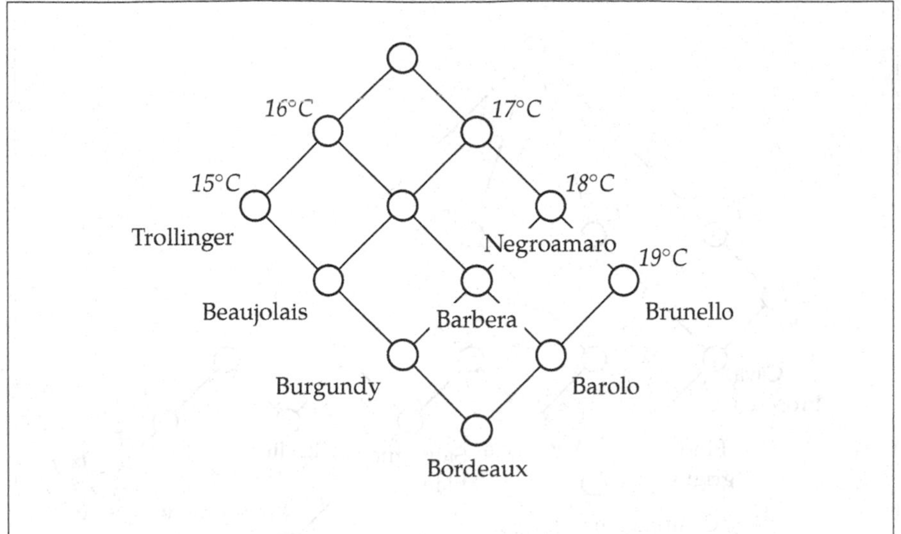

Figure 1.2: Recommended serving temperatures of some red wines

word *clever* may be translated into *god, flink,* or *skarp,* but not into *skjonn,* the latter translating into *beautiful* and *pretty.*

Our final introductory example displays data of some airlines belonging to the "Star Alliance" and the areas where they operated in 2002. The concept lattice in Figure 1.6 is based on the data in Figure 1.7.

1.2 Basic notions

One of the strengths of Formal Concept Analysis is that it is a mathematical theory and can draw on a rich stock of scientific results. The main reference for the mathematical foundations, the book by Ganter and Wille [110], goes far beyond what is needed for our topic here. We restrict ourselves to an introduction to the basic mathematical concepts of the theory.

Concept lattice diagrams, like the ones in Figures 1.2 to 1.6, are not merely *illustrations* of some data, but are faithful representations in a certain, well-defined sense. The data is required to be in a standard form, called a *formal context*. This is defined next.

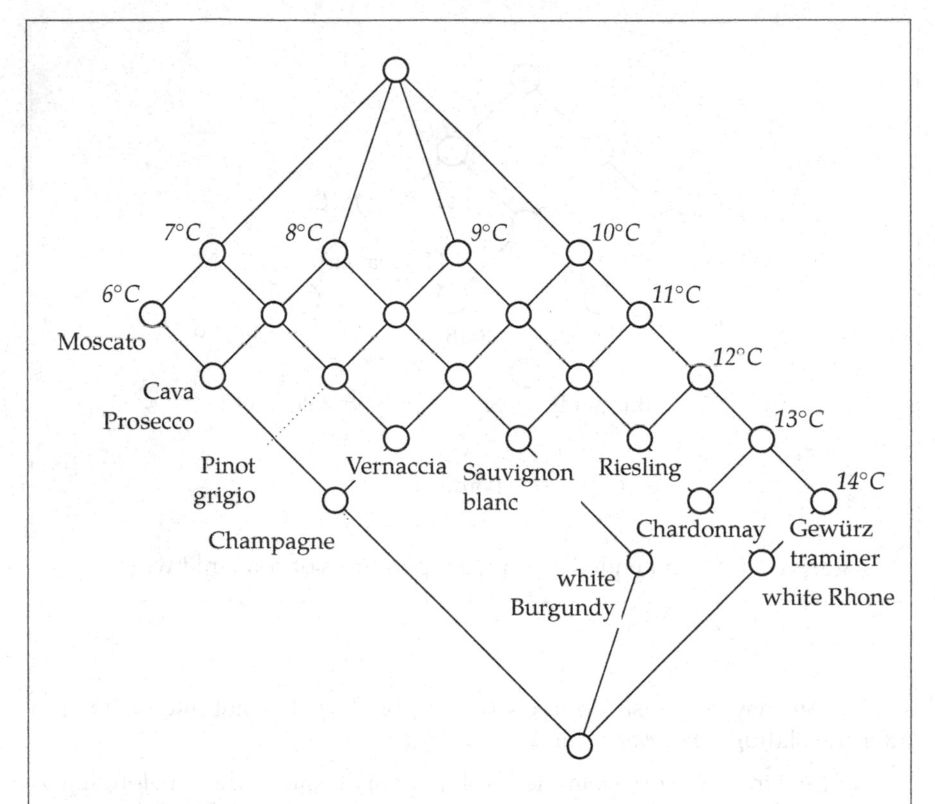

Figure 1.3: Recommended serving temperatures for white wines. Both Cava and Prosecco should be served at 6–8 °C, while Chardonnay is better at 9–13 °C

1.2.1 Formal contexts and cross-tables

The simplest format for writing down a formal context is a **cross-table**[2], as shown in Figure 1.7: we write a rectangular table with one row for each object and one column for each attribute, having a cross in the intersection of row g with column m iff object g has attribute m. The simplest data type for computer storage is that of a bit matrix.[3]

Definition 1 A **formal context** (G, M, I) consists of two sets, G and M, and a

[2]Not to be confused with a *contingency table*.

[3]It is not easy to say which is the *most efficient* data type for formal contexts. This depends, of course, on the operations we want to perform with formal contexts. The most important ones are the derivation operators, to be defined in the next subsection.

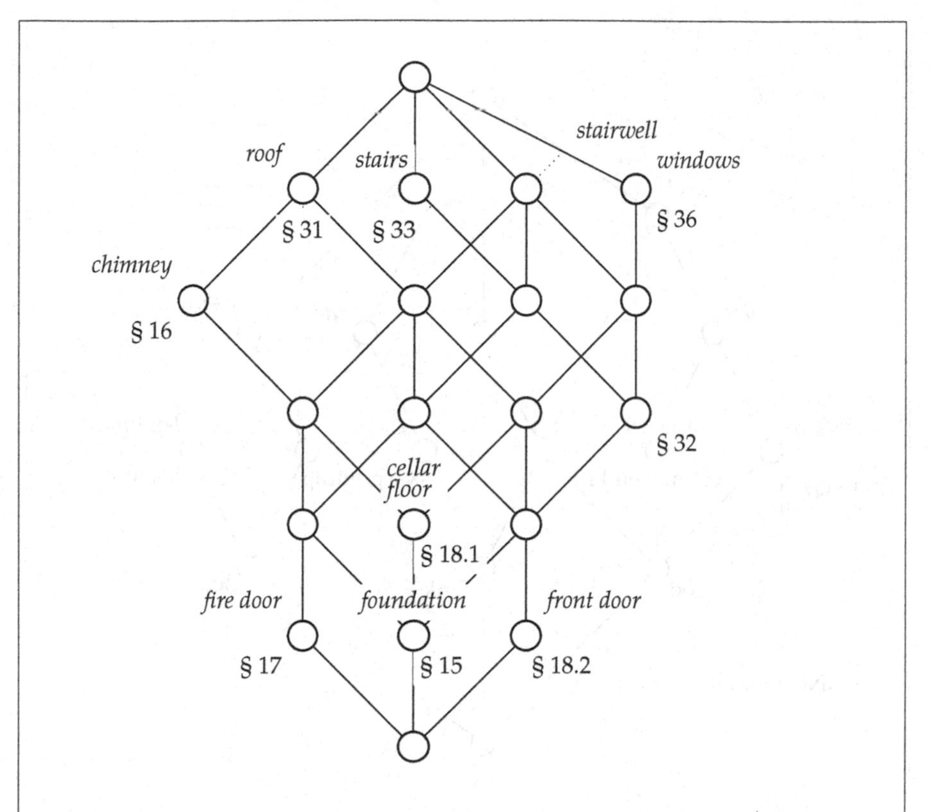

Figure 1.4: Snapshot from an information system on bye-law regulations for buildings. § 32 applies to stairs, stairwells, and windows, while for chimneys one should refer to §§ 15, 16, 17, and 18.1

binary relation $I \subseteq G \times M$. The elements of G are called the **objects** and those of M the **attributes** of (G, M, I). If $g \in G$ and $m \in M$ are in relation I, we write $(g, m) \in I$ or $g\ I\ m$ and read this as *"the object g* **has** *the attribute m"*. $\Diamond$

Note that the definition of a formal context is very general. There are no restrictions on the nature of objects and attributes. Anything that is a *set* in the mathematical sense may be taken as the set of objects or of attributes of some formal context. We may interchange the roles of objects and attributes: if (G, M, I) is a formal context, then so is the **dual context** (M, G, I^{-1}) (with $(m, g) \in I^{-1} :\iff (g, m) \in I$). Also, it is not required that G and M are disjoint. They need not even be different, as the example in Figure 1.8 shows.

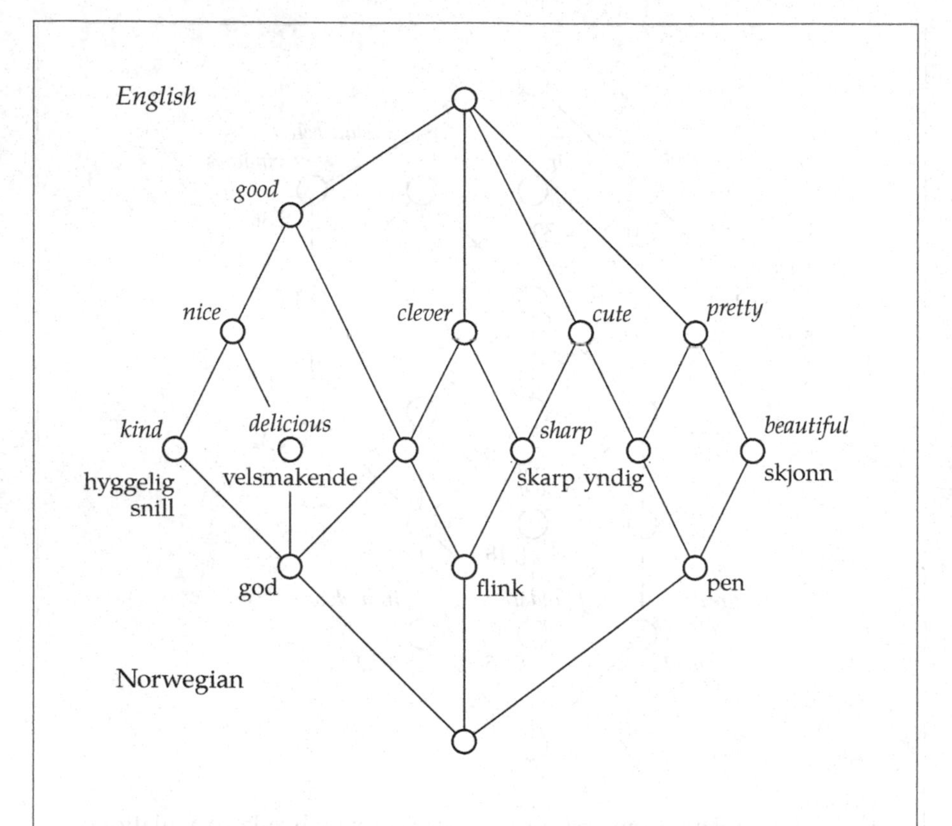

Figure 1.5: A neighborhood lattice for English/Norwegian: "good/god" displaying a result of the Semantic Mirror method (Step 1)

On the other hand, the definition of a formal context is rather restrictive when applied to real-world phenomena. Language phrases like "all human beings" or "all chairs" are too vague if used to define sets: there is no "set of all chairs", because the decision if something is a chair is not a matter of fact, but of subjective interpretation. The notion of a "formal concept", which we base on the definition of a "formal context", is much narrower than what is commonly understood as a concept of human cognition. There are actually many different views of what "human concepts" may be, but none of them assumes that a concept may only be built from pre-selected sets of objects and attributes. The step from "context" to "formal context" is quite an incisive one. It is the step from "real world" to "data". Later on, when we get tired of saying "formal concepts of a formal context", we will sometimes omit the word "formal". But

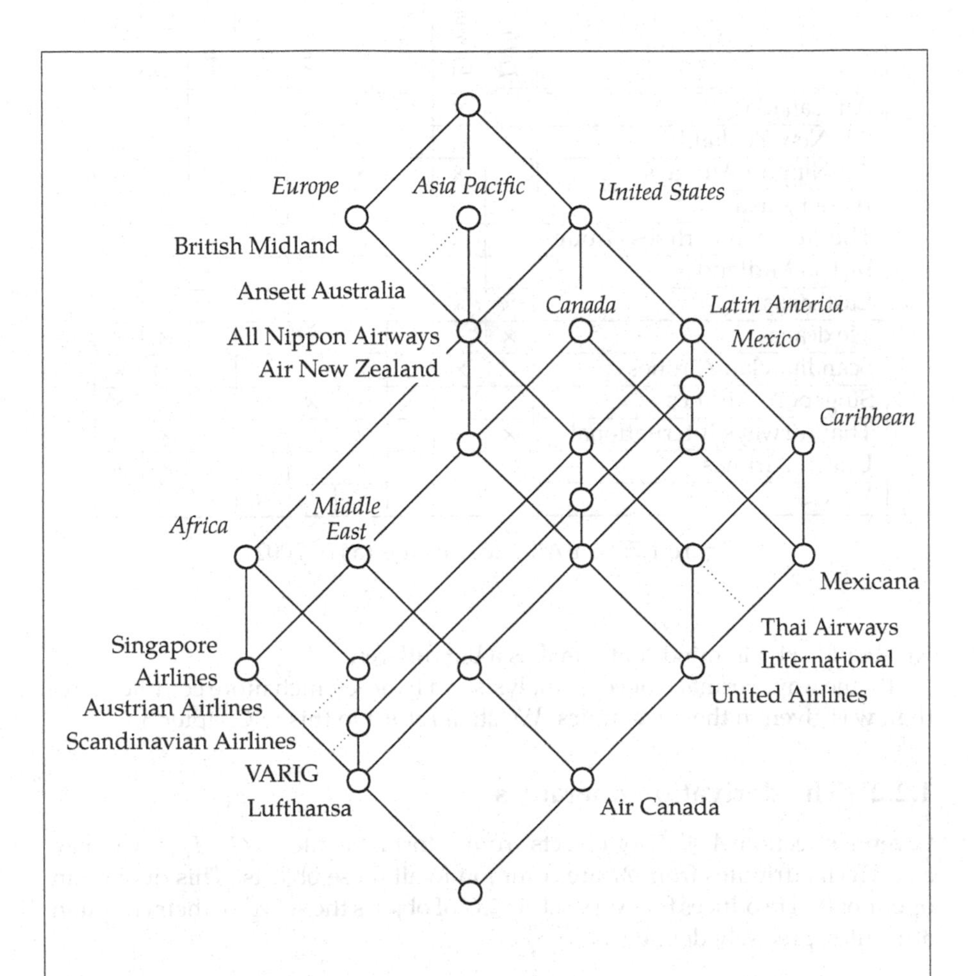

Figure 1.6: Star Alliance airlines and their destinations (as of 2002). Mexicana flies to the Caribbean, Mexico, Latin America, Canada, and the United States. Other airlines flying to the Caribbean are Thai Airways, United Airlines, and Air Canada

	Latin America	*Europe*	*Canada*	*Asia Pacific*	*Middle East*	*Africa*	*Mexico*	*Caribbean*	*United States*
Air Canada	×	×	×	×	×		×	×	×
Air New Zealand		×		×					×
All Nippon Airways		×		×					×
Ansett Australia				×					
The Austrian Airlines Group		×	×	×	×	×			×
British Midland		×							
Lufthansa	×	×	×	×	×	×	×		×
Mexicana	×		×				×	×	×
Scandinavian Airlines	×	×		×		×			×
Singapore Airlines		×	×	×	×	×			×
Thai Airways International	×	×		×				×	×
United Airlines	×	×	×	×			×	×	×
VARIG	×	×		×		×	×		×

Figure 1.7: Star Alliance airlines (as of 2002)

we should keep in mind that it makes a big difference.

By the way: Formal Concept Analysis can handle much more complex data than was given in these examples. We shall return to this in Chapter 5.

1.2.2 The derivation operators

Given a selection $A \subseteq G$ of objects from a formal context (G, M, I), we may ask which attributes from M are common to all these objects. This defines an operator that produces for every set $A \subseteq G$ of objects the set $A^{\uparrow}$ of their common attributes, precisely defined as

$$A^{\uparrow} := \{m \in M \mid g\ I\ m \text{ for all } g \in A\} \quad (\text{for } A \subseteq G).$$

Dually, we may introduce for selected attributes $B \subseteq M$ the set $B^{\downarrow}$ by

$$B^{\downarrow} := \{g \in G \mid g\ I\ m \text{ for all } m \in B\} \quad (\text{for } B \subseteq M).$$

Then $B^{\downarrow}$ denotes the set consisting of those objects in G that have all the attributes from B.

These two operators are the **derivation operators** for (G, M, I). Usually, we do not distinguish them in writing and use the notation A', B' instead (and

	1	2	4	5	10	20	25	50	100
1	×	×	×	×	×	×	×	×	×
2		×	×		×	×		×	×
4			×			×			×
5				×	×	×	×	×	×
10					×	×		×	×
20						×			×
25							×	×	×
50								×	×
100									×

Figure 1.8: This cross-table represents a formal context (G, M, I) the objects and attributes of which are the divisors of 100. We thus have

$$G = M = \{1, 2, 4, 5, 10, 20, 25, 50, 100\}.$$

In this context, $g\ I\ m$ if g is a divisor of m. In the dual context, "g has m" means that g is a multiple of m

even write g', m' instead of $\{g\}'$ and $\{m\}'$). This is convenient, but not always sufficient. Sometimes, when we need to differentiate between the derivation operators of two contexts, say, (G, M, I) and (H, N, J), we use the symbol designating the incidence relation instead of $'$ and write, e.g., A^I and A^J.

If A is a set of objects, then A' is a set of attributes, to which we can apply the second derivation operator to obtain A'' (which is short for $(A^{\uparrow})^{\downarrow}$), a set of objects. Dually, starting with a set B of attributes, we may form the set B'' (short for $(B^{\downarrow})^{\uparrow}$), which is again a set of attributes. We have the following simple facts:

Proposition 1 *For subsets $A, A_1, A_2 \subseteq G$, we have*

1. $A_1 \subseteq A_2 \Rightarrow A_2' \subseteq A_1'$,
2. $A \subseteq A''$,
3. $A' = A'''$.

Dually, for subsets $B, B_1, B_2 \subseteq M$ we have

1'. $B_1 \subseteq B_2 \Rightarrow B_2' \subseteq B_1'$,

2'. $B \subseteq B''$,

3'. $B' = B'''$.

The elementary proof is omitted here. The reader may refer to [110] for mathematical details. The mathematically interested reader may have noticed that the derivation operators constitute a **Galois connection** between (the power sets of) the sets G and M.

The not so mathematically oriented reader should try to express the statements of Proposition 1 in common language. We give an example: Statement 1.) says that *if a selection of objects is enlarged, then the attributes in common to all objects of the larger selection are among the common attributes of the smaller selection.* Try to formulate 2.) and 3.) in a similar manner!

Example 1 Consider the formal context in Figure 1.7. Let A_1 = {All Nippon Airways, Singapore Airlines} and A_2 = {All Nippon Airways, Singapore Airlines, Mexicana}. We have $A_1 \subseteq A_2$ and $A_2' \subseteq A_1'$. Indeed, A_1' = {*Europe, Asia Pacific, United States*} is the set of all areas served by both All Nippon Airways and Singapore Airlines. Out of these three areas, Mexicana serves only the *United States*; therefore, A_2', the set of areas served by all the three airlines from A_2, is equal to {*United States*}.

A_2'' is then the set of all airlines flying to A_2' = {*United States*}. Thus A_2'' includes all the airlines except *Ansett Australia* and *British Midland*. Note that A_2'' necessarily includes all the airlines from A_2, since they all fly to A_2'. This illustrates the $A \subseteq A''$ statement of Proposition 1.

If we apply the derivation operator once again, we obtain A_2''' = {*United States*} = A_2'. The reason for this is that A_2''' is the set of all areas served by all airlines from A_2'', but these include all the airlines of A_2, whose common destinations are given by A_2'. Therefore, applying the derivation operator to A_2'' does not allow us to go beyond A_2', and so $A_2' = A_2'''$.

1.2.3 Formal concepts, extents and intents

In what follows, (G, M, I) always denotes a formal context.

Definition 2 (A, B) is a **formal concept** of (G, M, I) iff

$$A \subseteq G, \quad B \subseteq M, \quad A' = B, \quad \text{and} \quad A = B'.$$

The set A is called the **extent** of the formal concept (A, B), and the set B is called its **intent**.

◊

According to this definition, a formal concept has two parts: its extent and its intent. The extent consists of precisely those objects that have all attributes of the intent. And the intent contains precisely those attributes that all objects in the extent have in common. This follows an old tradition in philosophical concept logic, as expressed in the *Logic of Port Royal, 1662* [16]; see also [78].

Given a context, the description of a concept by its extent and intent is redundant, because each of the two parts determines the other (since $B = A'$ and $A = B'$). But for many reasons, this redundant description is very convenient.

When a formal context is written as a cross-table, every formal concept (A, B) corresponds to a rectangular subtable with row set A and column set B (see Example 2). To make this more precise, note that in the definition of a formal context there is no *order* on the sets G or M. Permuting the rows or the columns of a cross-table therefore does not change the formal context it represents. A *rectangular subtable* may, in this sense, omit some rows or columns; it must be rectangular after an appropriate rearrangement of the rows and the columns. It is then easy to characterize the rectangular subtables that correspond to formal contexts: they are full of crosses and maximal with respect to this property.

A formal context may have many formal concepts. In fact, it is not difficult to come up with examples where the number of formal concepts is exponential in the size of the formal context (see Section 2.2.4 (p. 49)). The set of all formal concepts of (G, M, I) is denoted

$$\mathfrak{B}(G, M, I).$$

Later on, we shall discuss algorithms and software for computing all formal concepts of a given (G, M, I).

Example 2 Consider again the formal context in Figure 1.8. Let $A := \{1, 4, 25\}$ be a subset of objects of this context. Since the objects and the attributes of this context are the same numbers, it makes sense to be explicit and use $(\cdot)^{\uparrow}$ and $(\cdot)^{\downarrow}$ instead of $(\cdot)'$. Here, $A^{\uparrow}$ is the set of all common multiples of the three numbers in A (but only among those that are divisors of 100 and are therefore in the context). There is only one such number, 100; thus, $A^{\uparrow} = \{100\}$. Now, $A^{\uparrow\downarrow}$ is the set of all common divisors of all the multiples (again, mind the context) of all the numbers in A or, simply put, the set of all divisors of 100, which coincides with the object set G of the context. This is one example of a formal concept: $(G, \{100\})$. Figure 1.9 shows the corresponding rectangle.

Next, consider the attribute subset $B := \{25, 50\}$. The set $B^{\downarrow} = \{1, 5, 25\}$ consists of all the divisors of both 25 and 50. In this case, $B^{\downarrow\uparrow} = \{25, 50, 100\}$, since, besides 25 and 50, the only multiple of 1, 5, and 25 is 100. The concept rectangle corresponding to the resulting concept $(\{1, 5, 25\}, \{25, 50, 100\})$ is shown in Figure 1.10. In this case, the rectangle has gaps, which can be eliminated by rearranging the rows of the context. Note also that this rectangle (as any rectangle corresponding to a formal concept) is maximal in the following sense: if one adds a row or a column, the enlarged rectangle will contain at least one cell without a cross.

More generally, every number in the extent of a formal concept of the context in Figure 1.8 is a divisor of every number in the intent of this concept.

	1	2	4	5	10	20	25	50	100
1	×	×	×	×	×	×	×	×	×
2		×	×		×	×		×	×
4			×			×			×
5				×	×	×	×	×	×
10					×	×		×	×
20						×			×
25							×	×	×
50								×	×
100									×

Figure 1.9: The formal concept $(\{1,2,4,5,10,20,25,50,100\},\{100\})$

	1	2	4	5	10	20	25	50	100
1	×	×	×	×	×	×	×	×	×
2		×	×		×	×		×	×
4			×			×			×
5				×	×	×	×	×	×
10					×	×		×	×
20						×			×
25							×	×	×
50								×	×
100									×

Figure 1.10: The formal concept $(\{1,5,25\},\{25,50,100\})$

Obviously, every number in the intent is then a multiple of every number in the extent. More than that, the extent and the intent contain all such divisors and multiples, respectively, among the numbers from the object and attribute sets of the context.

1.2.4 Conceptual hierarchy

Formal concepts can be (partially) ordered in a natural way. Again, the definition is inspired by the way humans usually order concepts in a *subconcept–superconcept hierarchy*: "pig" is a subconcept of "mammal", because every pig is a mammal. Transferring this to formal concepts, the natural definition is as follows:

Definition 3 Let (A_1, B_1) and (A_2, B_2) be formal concepts of (G, M, I). We say that (A_1, B_1) is a **subconcept** of (A_2, B_2) and, equivalently, that (A_2, B_2) is a **superconcept** of (A_1, B_1) iff $A_1 \subseteq A_2$. We use the $\leq$-sign to express this relation and thus have

$$(A_1, B_1) \leq (A_2, B_2) :\Longleftrightarrow A_1 \subseteq A_2.$$

The set of all formal concepts of (G, M, I), ordered by this relation, is denoted

$$\underline{\mathfrak{B}}(G, M, I)$$

and is called the **concept lattice** of the formal context (G, M, I). ◇

This definition is natural, but irritatingly asymmetric. What about the intents? A look at Proposition 1 reveals that for concepts (A_1, B_1) and (A_2, B_2)

$$A_1 \subseteq A_2 \quad \text{is equivalent to} \quad B_2 \subseteq B_1.$$

Therefore

$$(A_1, B_1) \leq (A_2, B_2) :\Longleftrightarrow A_1 \subseteq A_2 \quad (\Longleftrightarrow B_2 \subseteq B_1).$$

1.3 The algebra of concepts

The concept lattice of any formal context is not only a graphically vivid unfolding of the data, it is also an algebraic structure called a *complete lattice*. Of the extensive mathematical **lattice theory**, however, you need only a small fraction in order to understand this text.[4]

1.3.1 Reading a concept lattice diagram

The concept lattice of (G, M, I) is the set of all formal concepts of (G, M, I), ordered by the subconcept–superconcept relation. Ordered sets of moderate size can conveniently be displayed as **order diagrams**. Many readers will be familiar with such diagrams. Nevertheless, we explain how to read a concept lattice diagram by means of the example given in Figure 1.11. Later on we shall discuss how to *draw* such diagrams.

Figure 1.11 refers to the "warmup" problem that was mentioned in the introduction (page xi). Think of two squares of equal size drawn on paper. There are different ways to arrange two squares: they may be disjoint (= have no point in common); they may overlap (= have a common interior point); they may share a vertex, an edge, or a line segment of the boundary (of length > 0); and they may be parallel or not.

[4]Standard texts on lattice theory are the books by Birkhoff [47] and Grätzer [118]. The book by Davey & Priestley [70] is a popular introduction.

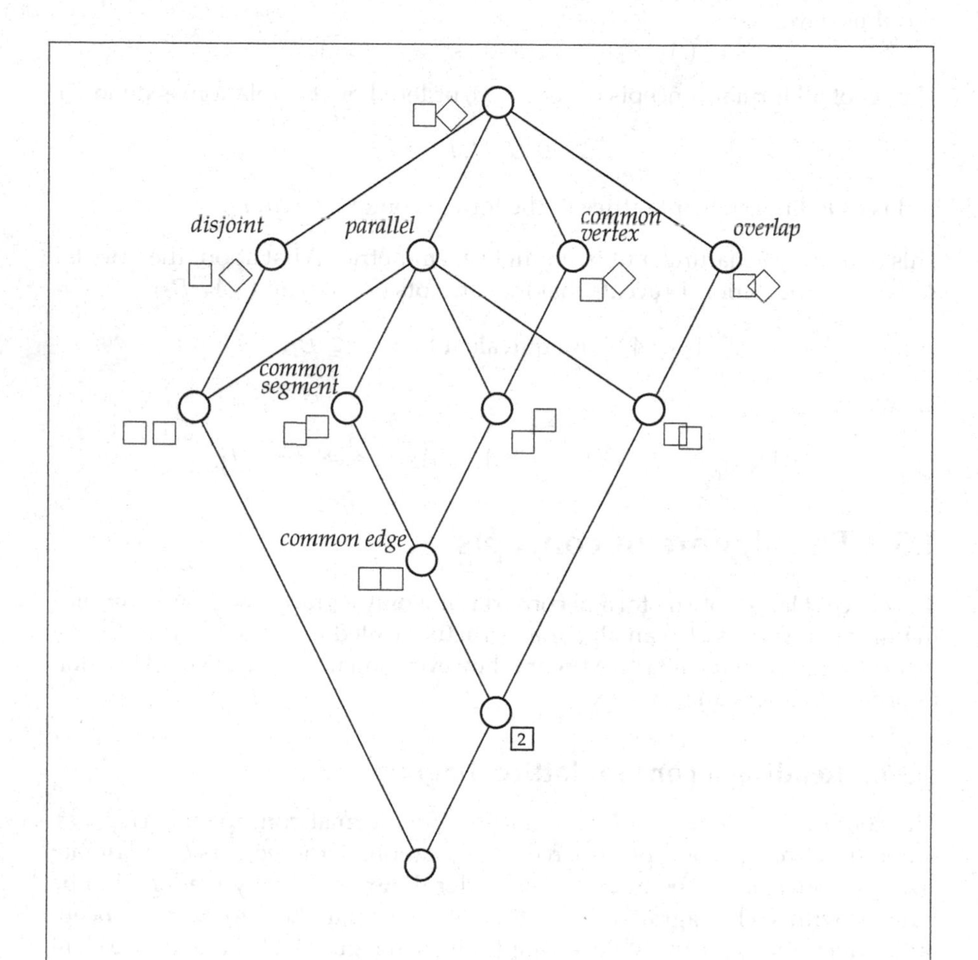

Figure 1.11: A concept lattice diagram. The objects are pairs of unit squares. The attributes describe their mutual position. The topic will be investigated more closely in Section 4.2.2 (p. 139)

Figure 1.11 shows a concept lattice unfolding these possibilities. It consists of twelve formal concepts, these being represented by the twelve small circles in the diagram. The names of the six attributes are given. Each name is attached to one of the formal concepts and is written slightly above the respective circle. The ten objects are represented by little pictures, each showing a pair of unit squares. Again, each object is attached to exactly one formal concept; the picture representing the object is drawn slightly below the circle representing the object concept.

Some of the circles are connected by edges. These express the concept order. With the help of the edges, we can read from the diagram which concepts are subconcepts of which other concepts and which objects have which attributes. To do so, one has to follow *ascending paths* in the diagram.

For example, consider the object □□. From the corresponding circle we can reach, via ascending paths, four attributes: "common edge", "common segment", "common vertex", and "parallel". □□ does in fact have these properties and does not have the other ones: the two squares are neither "disjoint" nor do they "overlap".

Similarly, we can find all objects that have a given attribute by following all descending paths. For example, to find all objects that "overlap", we start at the attribute concept labelled "overlap" and follow the edges downward. We can reach three objects (namely □◇, □□, and [2], the latter symbolizing two squares at the same position).

With the same method, we can read the intent and the extent of every formal concept in the diagram. For example, consider the concept circle labeled □□. Its extent consists of all objects that can be reached from that circle on a descending path. The extent therefore is {□□, [2] }. Similarly, we find by an inspection of the ascending paths that the intent of this formal concept is {overlap, parallel}.

The diagram contains all necessary information. We can read off the objects, the attributes, and the incidence relation I. Thus we can perfectly reconstruct the formal context from the diagram ("the original data"). Moreover, for each formal concept we can easily determine its extent and intent from the diagram.

So in a certain sense, concept lattice diagrams are perfect. But there are, of course, limitations. Take another look at Figure 1.11. Is it *correct*? Is it *complete*? The answer is that, since a concept lattice faithfully unfolds the formal context, the information displayed in the lattice diagram can be only as correct and complete as the formal context is. In our specific example, it is easy to check that the given examples in fact do have the properties as indicated. But a more difficult problem is if our selection of objects is representative. Are there possible combinations of two squares that lead to an attribute combination that does not occur in our sample?

We shall come back to this question later.

1.3.2 Supremum and infimum

Can we compute with formal concepts? Yes, we can. The concept operations are however quite different from addition and multiplication of numbers. They resemble more of the operations the *greatest common divisor* and the *least common multiple*, which we know from integers. We do not use the symbols $\cdot$ and $+$ for lattice operations (as some authors do), but use $\wedge$ and $\vee$ instead.

Definition 4 For any two formal concepts (A_1, B_1) and (A_2, B_2) of some formal context, we define

- the **greatest common subconcept** of (A_1, B_1) and (A_2, B_2) as

$$(A_1, B_1) \wedge (A_2, B_2) := (A_1 \cap A_2, (B_1 \cup B_2)''),$$

- the **least common superconcept** of (A_1, B_1) and (A_2, B_2) as

$$(A_1, B_1) \vee (A_2, B_2) := ((A_1 \cup A_2)'', B_1 \cap B_2).$$

◇

It is not difficult to prove that what is suggested by this definition is true: $(A_1, B_1) \wedge (A_2, B_2)$ and $(A_1, B_1) \vee (A_2, B_2)$ are in fact formal concepts (of the same context). More precisely: $(A_1, B_1) \wedge (A_2, B_2)$ is a subconcept of both (A_1, B_1) and (A_2, B_2), and any other common subconcept of (A_1, B_1) and (A_2, B_2) is also a subconcept of $(A_1, B_1) \wedge (A_2, B_2)$. Similarly, $(A_1, B_1) \vee (A_2, B_2)$ is a superconcept of (A_1, B_1) and of (A_2, B_2) and is a subconcept of any common superconcept of these two formal concepts.

Instead of "least common superconcept" we usually say **supremum**, and instead of "greatest common subconcept" we use the word **infimum**.

With some practice, one can read off the infimum and supremum from the lattice diagram. Choose any two concepts from Figure 1.11 and follow the descending paths from the corresponding nodes in the diagram. There is always a unique highest point where these paths meet, that is, the highest concept that is below both, namely, the infimum. Any other concept below both can be reached from the highest one on a descending path. Similarly, for any two formal concepts there is always the lowest node (the supremum of the two) that can be reached from both concepts via ascending paths. And any common superconcept of the two is on an ascending path from their supremum.

It may be that the supremum of two concepts is one of them; then, the other is their infimum. This happens if one concept is more general than the other one. Take, for example, the concepts corresponding to the nodes labeled by and in Figure 1.11. The first is the concept of squares sharing a common vertex and is the supremum of the two concepts. The second concept is one of *parallel* squares sharing a common vertex and is the infimum. On the other hand, considering this second concept together with the concept of parallel overlapping squares (labeled by), we note that their supremum is the more general concept of parallel squares and their infimum is the less general concept that is depicted by the node labeled with [2]. Since it is less general, its intent includes all the attributes of the other two concepts: "overlap", "parallel", and "common vertex". But there is only one object, [2], with these three attributes in our context and it has additional attributes, "common segment" and "common edge". Thus, in our context, the infimum of the concept of parallel overlapping squares and that of parallel squares sharing a common vertex is the concept of squares that are at the same position.

1.3.3 Complete lattices

The operations for computing with formal concepts, infimum and supremum, are not as weird as one might suspect. In fact, each concept lattice automatically carries an algebraic structure called a "lattice", and such structures occur frequently in mathematics and computer science.[5] "Lattice theory" is an active field of research in mathematics. A **lattice** is an algebraic structure with two operations (called "meet" and "join" or "infimum" and "supremum") that satisfy certain natural conditions. We shall not extensively discuss the algebraic theory of lattices in this book. Many universities offer courses in lattice theory, and there are excellent textbooks available on the subject, see the footnote on page 13.

The concept lattices have an additional nice property: they are **complete lattices**. This means that the operations of infimum and supremum do not only work for an input consisting of two elements, but for arbitrarily many elements. In other words, each collection of formal concepts has the greatest common subconcept and the least common superconcept. This is even true for infinite sets of concepts. The operations "infimum" and "supremum" are not necessarily binary, they work for any input size. This is very useful, but will make essentially no difference for our considerations, because we shall mainly be concerned with finite formal contexts and finite concept lattices. Well, this is not completely true. In fact, although the concept lattice in Figure 1.11 is finite, its ten objects are representatives for *all* possible combinations of two unit squares. Of course, there are infinitely many such possibilities. It is true that we shall consider finite concept lattices, but our examples may be taken from an infinite reservoir.

[5] Unfortunately, the word "lattice" is used with different meanings in mathematics. It also refers to generalized grids.

1.3.4 The basic theorem

We conclude the section by giving a mathematically precise formulation of the algebraic properties of concept lattices. The theorem below is not difficult, but a foundation for many other results. Its formulation contains some technical terms that we have not mentioned so far. The notion of "isomorphism" is probably familiar to most readers and will not be explained.

To understand the notion of a "supremum-dense" set, think of the positive integers with the least common multiple (**lcm**) in the role of the supremum operation. Some numbers can be written as an **lcm** of other integers. For example, $12 = \mathbf{lcm}\{3, 4\}$. Therefore, 12 is called *lcm-reducible*. Prime numbers cannot be written as an **lcm** of other numbers. Therefore, prime numbers are **lcm**-irreducible. But prime numbers are not the only **lcm**-irreducible numbers. For example, 25 is not the least common multiple of other numbers, and thus it is **lcm**-irreducible. It is easy to see that the **lcm**-irreducible integers are precisely the *prime powers*.[6] The set of prime powers is *lcm-dense*, which means that every positive integer can be written as an **lcm** of prime powers. Any larger set, that is, any set containing all prime powers, is also **lcm**-dense, of course.

There are no **gcd**-irreducible numbers (where **gcd** stands for the "greatest common divisor"). Every positive integer can be written as the **gcd** of other numbers. But there are **gcd**-dense sets of numbers. For example, the set of all integers greater than 1000 is **gcd**-dense.

In a complete lattice, an element is called **supremum-irreducible** or **join-irreducible** if it cannot be written as the supremum of other elements and **infimum-irreducible** or **meet-irreducible** if it can not be expressed as the infimum of other elements. It is very easy to locate the irreducible elements in a lattice diagram: the supremum-irreducible elements are precisely those from which there is exactly one edge going downward. An element is infimum-irreducible if and only if it is the start of exactly one upward edge. In Figure 1.11, there are precisely nine supremum-irreducible concepts and precisely five infimum-irreducible concepts. Exactly four concepts have both properties, they are **doubly irreducible**.

A set of elements of a complete lattice is called **supremum-dense** if every lattice element is a supremum of elements from this set. Dually, a set is called **infimum-dense** if the infima that can be computed from this set exhaust all lattice elements.

Theorem 1 (The basic theorem on concept lattices [215]) *The concept lattice of any formal context (G, M, I) is a complete lattice. For an arbitrary set*

$$\{(A_t, B_t) \mid t \in T\} \subseteq \mathfrak{B}(G, M, I)$$

[6]The number 1 requires special treatment. We shall not go into this.

of formal concepts, the supremum is given by

$$\bigvee_{t\in T}(A_t, B_t) = \left((\bigcup_{t\in T} A_t)'', \bigcap_{t\in T} B_t\right)$$

and the infimum is given by

$$\bigwedge_{t\in T}(A_t, B_t) = \left(\bigcap_{t\in T} A_t, (\bigcup_{t\in T} B_t)''\right).$$

A complete lattice $\underline{L}$ *is isomorphic to* $\mathfrak{B}(G, M, I)$ *iff there are mappings* $\tilde{\gamma} : G \to L$ *and* $\tilde{\mu} : M \to L$ *such that* $\tilde{\gamma}(G)$ *is supremum-dense and* $\tilde{\mu}(M)$ *is infimum-dense in* $\underline{L}$ *and*

$$g\, I\, m \iff \tilde{\gamma}(g) \leq \tilde{\mu}(m).$$

In particular, $\underline{L} \cong \mathfrak{B}(L, L, \leq)$.

The theorem is less complicated than it may first seem. We give some explanations below, but readers in a hurry may continue with the next section.

The first part of the theorem gives the precise formulation for the infimum and supremum of arbitrary sets of formal concepts. In a similar way to which we use the symbol $\sum$ instead of + for arbitrary sums, we use the large symbols $\bigvee$ and $\bigwedge$ for arbitrary suprema and infima.

The second part of the theorem gives (among other information) an answer to the question whether or not concept lattices have any special properties. The answer is "no": every complete lattice is (isomorphic to) a concept lattice. This means that for every complete lattice we must be able to find a set G of objects, a set M of attributes, and a suitable relation I, such that the given lattice $\underline{L} := (L, \leq)$ is isomorphic to $\mathfrak{B}(G, M, I)$. The theorem not only says how this can be done, it also describes *all* possible ways to achieve this.

In Figure 1.11, every object is attached to a unique concept, the corresponding *object concept*. Similarly, for each attribute there is an *attribute concept*. These can be defined as follows:

Definition 5 Let (G, M, I) be some formal context. Then

- for each object $g \in G$ the corresponding **object concept** is

$$\gamma g := (\{g\}'', \{g\}'),$$

- and for each attribute $m \in M$ the **attribute concept** is given by

$$\mu m := (\{m\}', \{m\}'').$$

◊

Using Definition 2 and Proposition 1, it is easy to check that these expressions indeed define formal concepts of (G, M, I). The set of all object concepts of (G, M, I) is denoted γG, the set of all attribute concepts is μM.

We have that $\gamma g \leq (A, B) \iff g \in A$. A look at the first part of the Basic Theorem shows that each formal concept is the supremum of all the object concepts below it. Therefore, the set γG of all object concepts is supremum-dense. Dually, the attribute concepts form an infimum-dense set in $\underline{\mathfrak{B}}(G, M, I)$. The Basic Theorem says that, conversely, any supremum-dense set in a complete lattice $\underline{L}$ can be taken as the set of objects and any infimum-dense set can be taken as the set of attributes for a formal context with concept lattice isomorphic to $\underline{L}$.

Let us illustrate this with a small example.

Example 3 Consider the lattice $\underline{L}$ in the left part of Figure 1.12. Apart from the top element, 1, and the bottom element, 0, it contains seven elements: a, b, c, d, e, f, and g. All but c and f are supremum-irreducible; thus, the set G = {a, b, d, e, g} is supremum-dense in $\underline{L}$. The set M = {a, e, f, g} is infimum-dense. The Basic Theorem tells us that the concept lattice of the context (G, M, I), where $g\ I\ m$ whenever $g \leq m$ in $\underline{L}$, is isomorphic to $\underline{L}$. This context is shown in the lower part of Figure 1.12. We leave it as an easy exercise to check that its concept lattice is indeed isomorphic to $\underline{L}$.

We conclude with a simple observation that often helps finding errors in concept lattice diagrams. The fact that the object concepts form a supremum-dense set implies that every supremum-irreducible concept must be an object concept (the converse is not true). Dually, every infimum-irreducible concept must be an attribute concept. This yields the following rule for concept lattice diagrams:

Proposition 2 *Suppose that a formal context (G, M, I) is given together with a finite order diagram labeled by the objects from G and the attributes from M. For $g \in G$, let $\tilde{\gamma}(g)$ denote the element of the diagram that is labeled with g, and, for $m \in M$, let $\tilde{\mu}(m)$ denote the element labeled with m. Then the given diagram is a correctly labeled diagram of $\underline{\mathfrak{B}}(G, M, I)$ if and only if it fulfills the following conditions:*

1. *The diagram is a correct lattice diagram,*
2. *every supremum-irreducible element is labeled by some object,*
3. *every infimum-irreducible element is labeled by some attribute,*
4. $g\ I\ m \iff \tilde{\gamma}(g) \leq \tilde{\mu}(m)$.

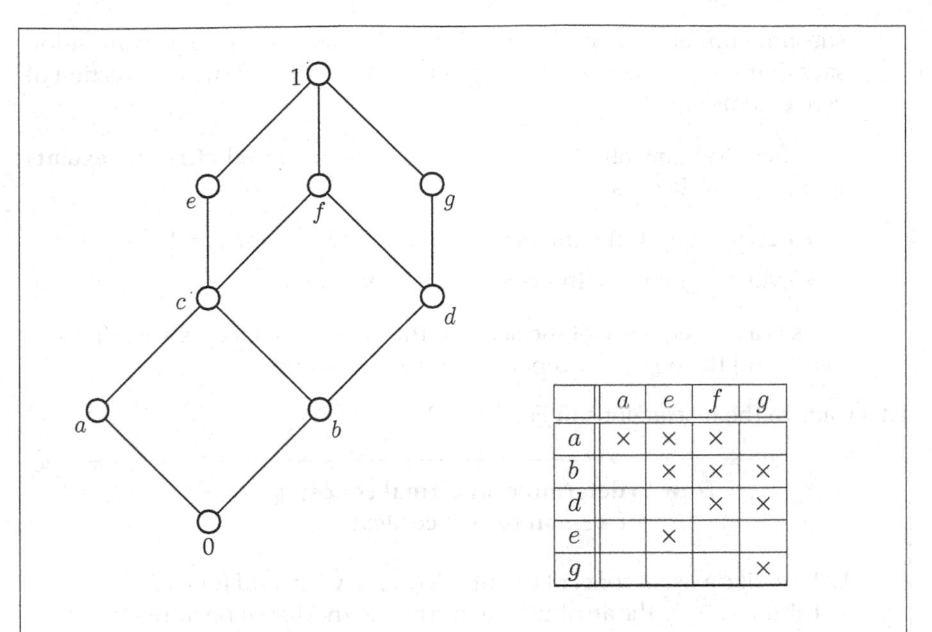

	a	e	f	g
a	×	×	×	
b		×	×	×
d			×	×
e		×		
g				×

Figure 1.12: A lattice and a formal context with an isomorphic concept lattice

1.4 How to draw a concept lattice diagram

There are computer programs that help finding and drawing concept lattices. We shall discuss some of them in Section 1.4.4, but we find it instructive to start with some small examples that can be drawn by hand.

1.4.1 Finding the concepts

The first step is to determine all concepts. Later we shall investigate a fast algorithm for this problem. Again, we start with a naive method. It is based on the following observations:

1. It suffices to determine all concept extents (or all intents) of (G, M, I)

   (since we can always determine the other part of the formal concept with the help of the derivation operators).

2. The intersection of arbitrarily many extents is an extent, and the intersection of arbitrarily many intents is an intent.

   (This follows easily from the formulas given in the Basic Theorem. Incidentally, a convention that may seem absurd at first glance allows one to subsume under "arbitrarily many" also the case "zero". The convention says that the intersection of zero intents equals M and the intersection of zero extents equals G.)

3. It suffices to know all **object intents** $\{g\}'$, $g \in G$, or all **attribute extents** $\{m\}'$, $m \in M$, because

   - every extent is the intersection of attribute extents, and
   - every intent is the intersection of object intents.

   (This is a consequence of the fact that the attribute concepts are infimum-dense and the object concepts are supremum-dense.)

This leads to the instructions in Figure 1.13.

How to determine all formal concepts of a small formal context

1. Initialize a list of concept extents. To begin with, add for each attribute $m \in M$ the attribute extent $\{m\}'$ to this list (if not already present).
2. For any two sets in this list, compute their intersection. If the result is a set that is not yet in the list, then extend the list by this set. With the extended list, continue to build all pairwise intersections.
3. If for any two sets of the list their intersection is also in the list, then extend the list by the set G (provided it is not yet contained in the list). The list now contains all concept extents (and nothing else).
4. For every concept extent A in the list compute the corresponding intent A' to obtain a list of all formal concepts (A, A') of (G, M, I).

Figure 1.13: Instructions for finding all formal concepts by hand

1.4.2 An example

We illustrate the method by means of an example from elementary geometry. The objects of our example are seven triangles. The attributes are five standard properties that triangles may or may not have, see Figure 1.14.

We follow the instructions in Figure 1.13 and proceed step by step:

Triangles				
symbol	coordinates			diagram
T_1	(0,0)	(6,0)	(3,1)	
T_2	(0,0)	(1,0)	(1,1)	
T_3	(0,0)	(4,0)	(1,2)	
T_4	(0,0)	(2,0)	$(1,\sqrt{3})$	
T_5	(0,0)	(2,0)	(5,1)	
T_6	(0,0)	(2,0)	(1,3)	
T_7	(0,0)	(2,0)	(0,1)	

Attributes	
symbol	property
a	*equilateral*
b	*isosceles*
c	*acute-angled*
d	*obtuse-angled*
e	*right-angled*

	a	b	c	d	e
T_1		×		×	
T_2		×			×
T_3			×		
T_4	×	×	×		
T_5				×	
T_6		×	×		
T_7					×

Figure 1.14: A formal context of triangles and their attributes. The pictures of the triangles have different scales

1. Write the attribute extents to a list.

No.		extent	found as
e_1	:=	$\{T_4\}$	$\{a\}'$
e_2	:=	$\{T_1, T_2, T_4, T_6\}$	$\{b\}'$
e_3	:=	$\{T_3, T_4, T_6\}$	$\{c\}'$
e_4	:=	$\{T_1, T_5\}$	$\{d\}'$
e_5	:=	$\{T_2, T_7\}$	$\{e\}'$

2. Compute all pairwise intersections, then 3. add G.

No.		extent	found as
e_1	:=	$\{T_4\}$	$\{a\}'$
e_2	:=	$\{T_1, T_2, T_4, T_6\}$	$\{b\}'$
e_3	:=	$\{T_3, T_4, T_6\}$	$\{c\}'$
e_4	:=	$\{T_1, T_5\}$	$\{d\}'$
e_5	:=	$\{T_2, T_7\}$	$\{e\}'$
e_6	:=	$\emptyset$	$e_1 \cap e_4$
e_7	:=	$\{T_4, T_6\}$	$e_2 \cap e_3$
e_8	:=	$\{T_1\}$	$e_2 \cap e_4$
e_9	:=	$\{T_2\}$	$e_2 \cap e_5$
e_{10}	:=	$\{T_1, T_2, T_3, T_4, T_5, T_6, T_7\}$	step 3

4. Compute the intents

Concept No.	(extent	,	intent)
1	$(\{T_4\}$	,	$\{a, b, c\})$
2	$(\{T_1, T_2, T_4, T_6\}$	,	$\{b\})$
3	$(\{T_3, T_4, T_6\}$	,	$\{c\})$
4	$(\{T_1, T_5\}$	,	$\{d\})$
5	$(\{T_2, T_7\}$	,	$\{e\})$
6	$(\emptyset$	,	$\{a, b, c, d, e\})$
7	$(\{T_4, T_6\}$	,	$\{b, c\})$
8	$(\{T_1\}$	,	$\{b, d\})$
9	$(\{T_2\}$	,	$\{b, e\})$
10	$(\{T_1, T_2, T_3, T_4, T_5, T_6, T_7\}$	,	$\emptyset)$

We have now computed all ten formal concepts of the context of triangles. With the help of this list, we can start drawing a diagram. Before doing so, we give two simple definitions.

Definition 6 Let (A_1, B_1) and (A_2, B_2) be formal concepts of (G, M, I).

We say that (A_1, B_1) is a **proper subconcept** of (A_2, B_2), if $(A_1, B_1) \leq (A_2, B_2)$ and, in addition, $(A_1, B_1) \neq (A_2, B_2)$ holds. As an abbreviation, we write $(A_1, B_1) < (A_2, B_2)$.

We say that (A_1, B_1) is a **lower neighbor** of (A_2, B_2), if $(A_1, B_1) < (A_2, B_2)$, but $(A_1, B_1) < (A, B) < (A_2, B_2)$ holds for no formal concept (A, B) of (G, M, I). The abbreviation for this is $(A_1, B_1) \prec (A_2, B_2)$. ◊

With these definitions we are prepared to follow the instructions in Figure 1.15.

Instruction for drawing a line diagram of a small concept lattice

5. Take a sheet of paper and draw a small circle for every formal concept in the following manner: a circle for a concept is always positioned higher than all the circles for its proper subconcepts.
6. Connect each circle with the circles of the lower neighbors.
7. Label with attribute names: attach each attribute m to the circle representing the concept $(\{m\}', \{m\}'')$.
8. Label with object names: attach each object g to the circle representing the concept $(\{g\}'', \{g\}')$.

Figure 1.15: How to draw a line diagram

5. Draw a circle for each of the formal concepts:

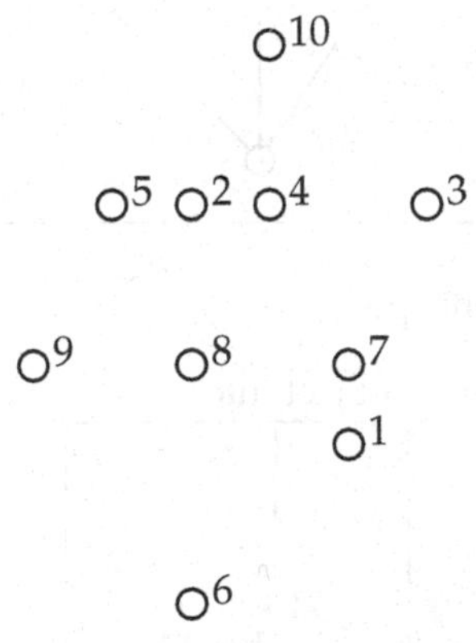

○10

○5 ○2 ○4 ○3

○9 ○8 ○7

○1

○6

6. Connect circles with their lower neighbors:

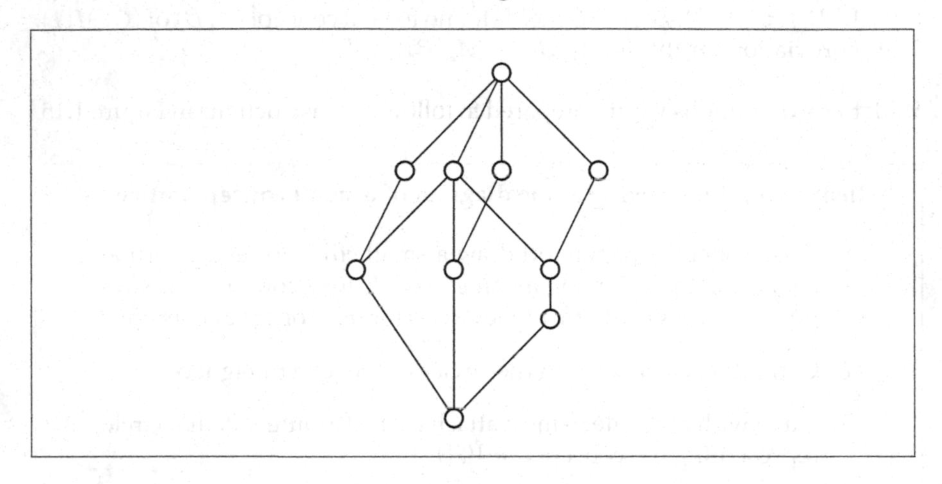

7. Write attribute names:

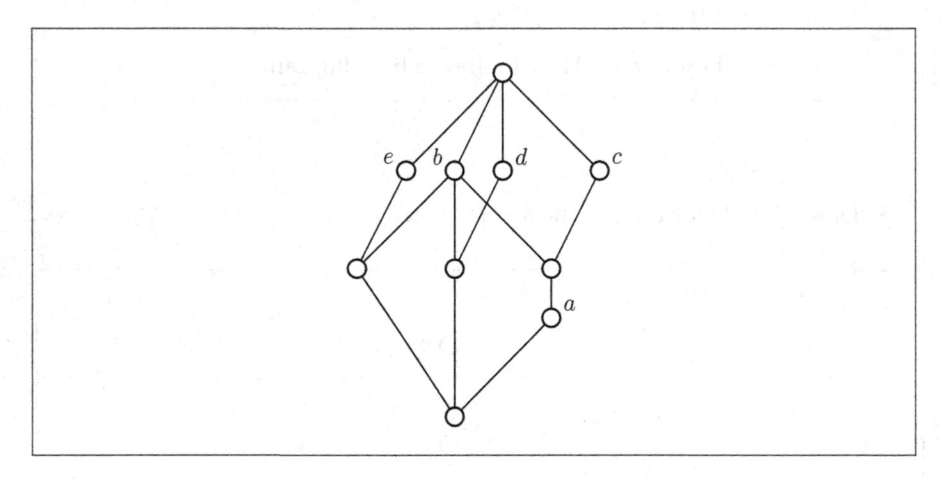

8. Determine object concepts:

Object g	object intent $\{g\}'$	No. of concept
T_1	$\{b, d\}$	8
T_2	$\{b, e\}$	9
T_3	$\{c\}$	3
T_4	$\{a, b, c\}$	1
T_5	$\{d\}$	4
T_6	$\{b, c\}$	7
T_7	$\{e\}$	5

and write object names to the diagram:

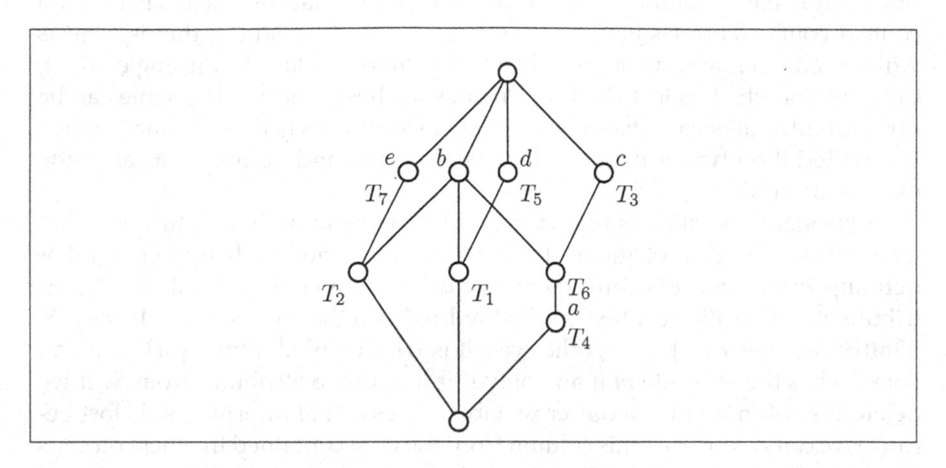

Ready! It usually takes several attempts before a nice, readable diagram is achieved. Finally we avoid abbreviations and increase the readability. The final result is shown in Figure 1.16.

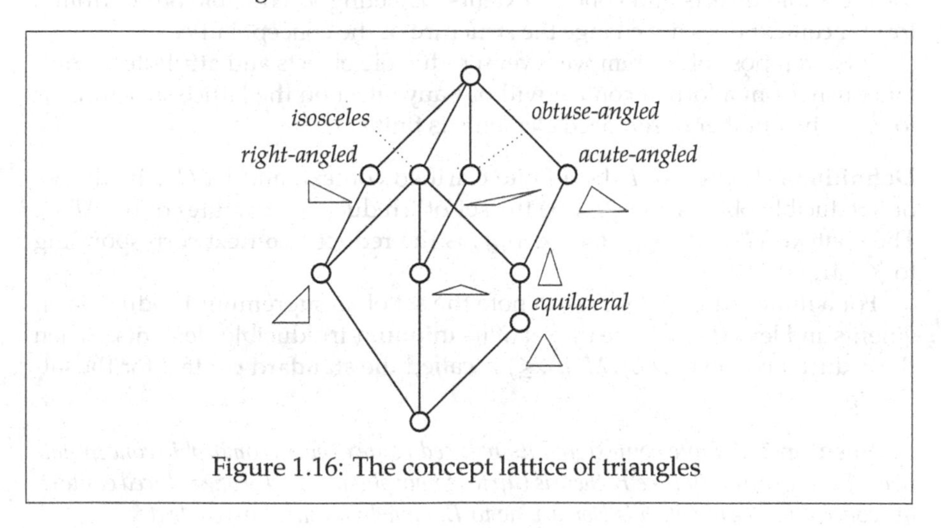

Figure 1.16: The concept lattice of triangles

1.4.3 Clarifying and reducing a formal context

There are context manipulations available that simplify a formal context without changing the diagram except the labelling. It is usually advisable to make these manipulations first, before starting serious computations.

The simplest operation is **clarification**, which refers to identifying "equal rows" and "equal columns" of a formal context. What this means is that if a context contains objects $g_1, g_2, \ldots$ with $\{g_i\}' = \{g_j\}'$ for all i, j, that is, objects with exactly the same attributes, then these can be replaced by a single object, the name of which is just the list of names of these objects. The same can be done for attributes with the same attribute extent. We say that a formal context is **clarified** if no two of its object intents are equal and no two of its attribute extents are equal.

A stronger operation is **reduction**, which refers to omitting attributes that are equivalent to combinations of other attributes (and dually for objects). For defining reduction, it is convenient to work with a clarified context. An attribute m of a clarified context is called **reducible** if there is a set $S \subseteq M, m \notin S$, of attributes with $\{m\}' = S'$, otherwise it is **irreducible**.[7] When $\{m\}' = S'$, an object g has the attribute m if and only if it has all the attributes from S. If we delete the column m from our cross-table, no essential information is lost because we can reconstruct this column from the data contained in other columns (those of S). Moreover, deleting that column does not change the number of concepts, nor the concept hierarchy, because $\{m\}' = S'$ implies that m is in the intent of a concept if and only if S is contained in that intent. The same is true for reducible objects and concept extents. Deleting a reducible object from a formal context does not change the structure of the concept lattice.

It is even possible to remove several reducible objects and attributes simultaneously from a formal context without any effect on the lattice structure, as long as the number of removed elements is finite.

Definition 7 Let (G, M, I) be a finite clarified context, and let G_{irr} be the set of irreducible objects and M_{irr} be the set of irreducible attributes of (G, M, I). The context $(G_{irr}, M_{irr}, I \cap G_{irr} \times M_{irr})$ is the **reduced context** corresponding to (G, M, I).

For a finite lattice L, let $J(L)$ denote the set of its supremum-irreducible elements and let $M(L)$ denote the set of its infimum-irreducible elements.[8] Then the reduced context $(J(L), M(L), \leq)$ is called the **standard context** for the lattice L.

◊

Proposition 3 *A finite context and its reduced context have isomorphic concept lattices. For every finite lattice L, there is (up to isomorphism) exactly one reduced context, the concept lattice of which is isomorphic to L, namely its standard context.*

Example 4 Figure 1.17 shows the reduced context corresponding to the context of divisors of 100 from Figure 1.8. Recall that the relation of this context

[7] An attribute m is irreducible if and only if the attribute concept $\mu(m)$ is infimum-irreducible, as it was defined in Section 1.3.4.

[8] Other words for supremum and infimum are **join** and **meet**. This explains the abbreviations $J(L)$ and $M(L)$.

	4	*20*	*25*	*50*
2	×	×		×
4	×	×		
5		×	×	×
25			×	×

Figure 1.17: The reduced context derived from Figure 1.8 (p. 9)

holds between an object and an attribute if the object is a divisor of the attribute. Here, reducible objects (with the exception of 1) are numbers that can be represented as products of other different numbers in the context. Thus, the irreducible objects are exactly the (positive) powers of prime numbers. Every reducible attribute (with the exception of 100) is the greatest common divisor of a subset of the remaining numbers in the context. The diagrams of the isomorphic concept lattices of the initial and reduced contexts are shown in Figure 1.18.

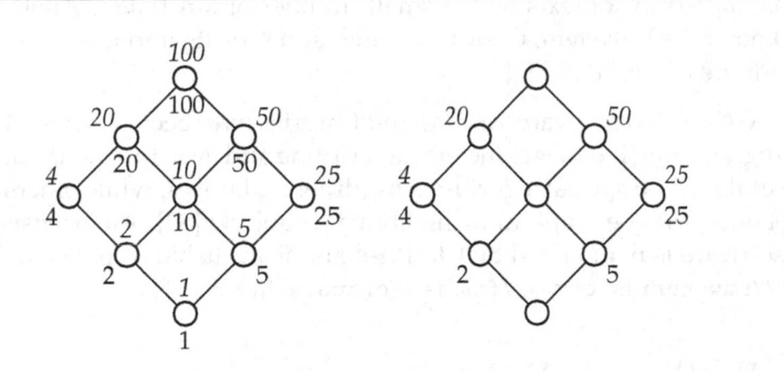

Figure 1.18: Diagrams of the concept lattices of the context in Figure 1.8 (left) and the reduced context in Figure 1.17 (right). The attribute labels are in italics for better readability

1.4.4 Computer programs for concept generation and lattice drawing

Later on we shall discuss algorithms for computer-supported concept lattice generation. There are quite a few programs available.

- One of the most widely-used FCA tools is Concept Explorer (ConExp) [229], which implements most of the essential FCA techniques: context editing, lattice construction and drawing, generation of implications and association rules (see Chapter 3), and a basic form of attribute exploration, which is the main topic of this book. It is an open-source project written in Java.

- ToscanaJ is a somewhat more advanced software suite for conceptual data processing [34] (also written in Java). It includes support for multi-level data analysis using so-called nested line diagrams [33] and provides an interface to database systems.

- Another open-source Java project is Galicia [210]. Its features include operations for many-valued contexts (see Section 5.4), as well as support for multiple data formats.

- The first computer programs for Formal Concept Analysis were written for a *Commodore Pet* Computer with 8 kB memory, and on a *VAX 11* UNIX computer. Among older DOS programs, Glad [77] implements sophisticated and efficient algorithms for lattice decomposition and drawing, but is limited to contexts with a small number of attributes. Another well-known DOS system, Conimp, is interesting for its implementation of attribute exploration [57].

Other FCA-related software systems and libraries have been developed in various programming languages including scripting and even functional languages. Some of them use special algorithms for drawing lattices, while others rely on the general-purpose graph-drawing library Graphviz [82]. An extensive list of FCA software is maintained by Uta Priss and is available from her web page: `http://www.upriss.org.uk/fca/fcasoftware.html`.

1.5 Further reading

The mathematical construction of building a complete lattice from an arbitrary binary relation is due to G. Birkhoff. It is already contained in the first edition of his "Lattice Theory" [46] and subsequently was discussed by many other authors. It was R. Wille who suggested the conceptual interpretation of this construction in 1979 (first publication [215] in 1982). He introduced the notions of a "formal context", a "formal concept", and a "concept lattice" and systematically worked out a version of lattice theory from this point of view, which he called Formal Concept Analysis.

There is very extensive literature on Formal Concept Analysis. Many of the early publications are documented in the already mentioned book by Ganter and Wille [110]. Numerous publications can be found in the Proceedings of the conference series CLA (*"Concept Lattices and their Applications"*), ICFCA (*"International Conference on Formal Concept Analysis"*), and ICCS (*"International Conference on Conceptual Structures"*).

Applications are discussed in the books [111], [224], [207], and [107] by Ganter, Stumme, Wille, Wolff, and Zickwolff. The book by Carpineto and Romano [61] is geared towards computer science practitioners, and it focuses on potential applications of concept lattices, in particular, in data mining and information retrieval.

Many computer programs offer automated diagram drawing, see Section 1.4.4. This is useful, since diagram design is often tedious, time-consuming, and error-prone. However, none of the diagrams in this book is machine-made. We use decent computer generated diagrams as feedstock for our hand-made versions. But automated drawing of good concept lattice diagrams is one of the unresolved issues in Formal Concept Analysis. The main difficulty is that it remains unclear what constitutes a "good" diagram. Optimizing single geometric parameters, such as minimizing the number of edge crossings, leads to interesting effects, but is insufficient. One may actually have doubts that the problem is a purely geometric one. It may well be that the same concept lattice has different "good" diagrams depending on its intended interpretation.

An early collection of exemplary concept lattice diagrams was published by Wille [216]. He has discussed automated drawing in several papers (with coauthors, e.g. [150, 157]). See also Cole et al [63]. Eklund et al [81] investigate readability of diagrams. Skorsky [198] promotes using parallelograms in lattice drawings, a technique which in a more general form is described as *additive* line diagram drawing in [110]. Ganter [97] suggests an optimization parameter for such additive diagrams, which Zschalig [231] combined with a "force-directed placement" heuristics to draw reasonably good diagrams of all 13596 lattices having exactly five meet-irreducible elements (i.e., of all concept lattices with five irreducible attributes).

1.6 Exercises

1.1 Consider the concept lattice diagram in Figure 1.19.

(a) What are the attributes of "Classic"?

(b) Which knives have a toothpick and a corkscrew?

(c) What is the intent of the smallest formal concept having "Outdoorsman" and "New Tinker" in its extent?

(d) What is the extent of that formal concept?

1.2 Draw diagrams for each of the small formal contexts in Figure 1.20.

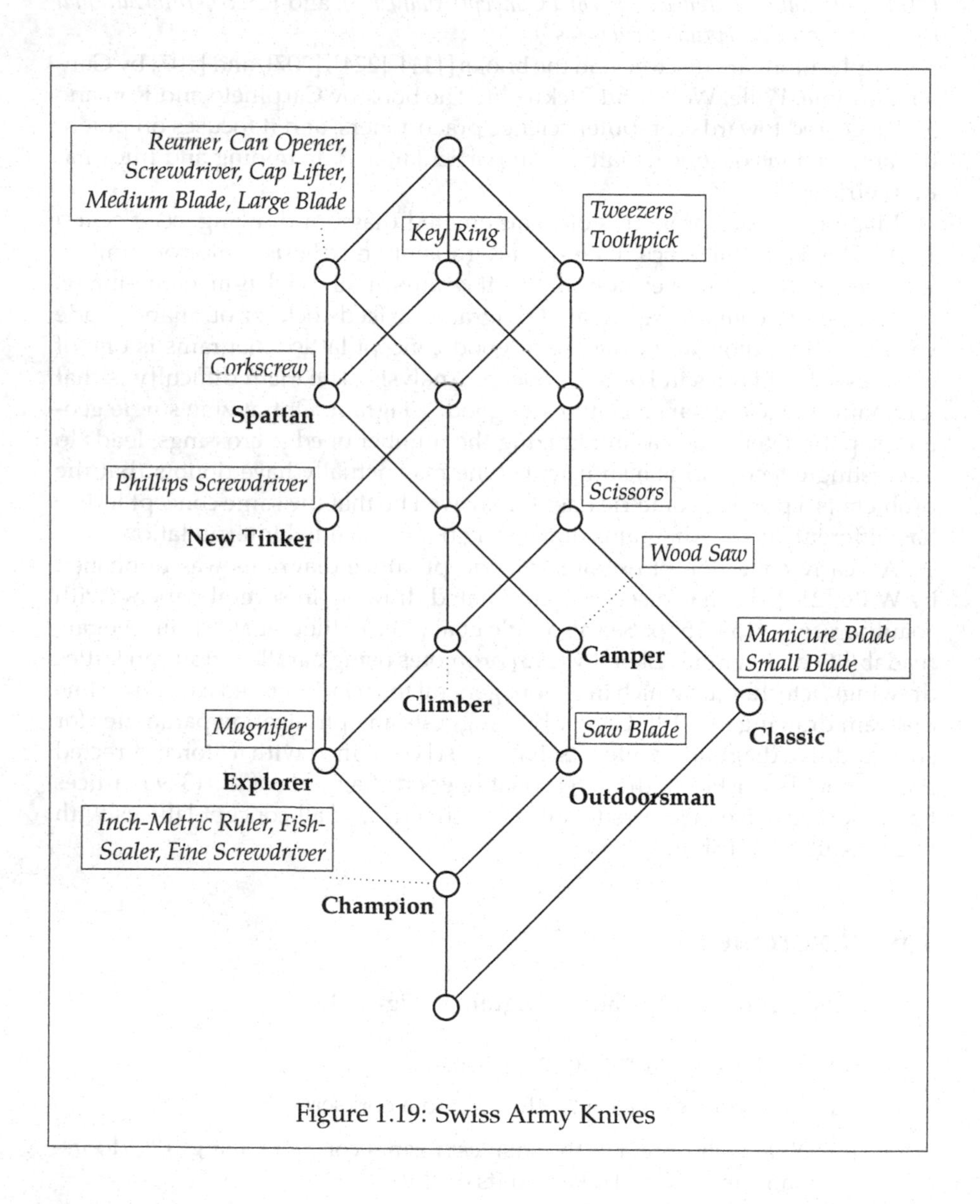

Figure 1.19: Swiss Army Knives

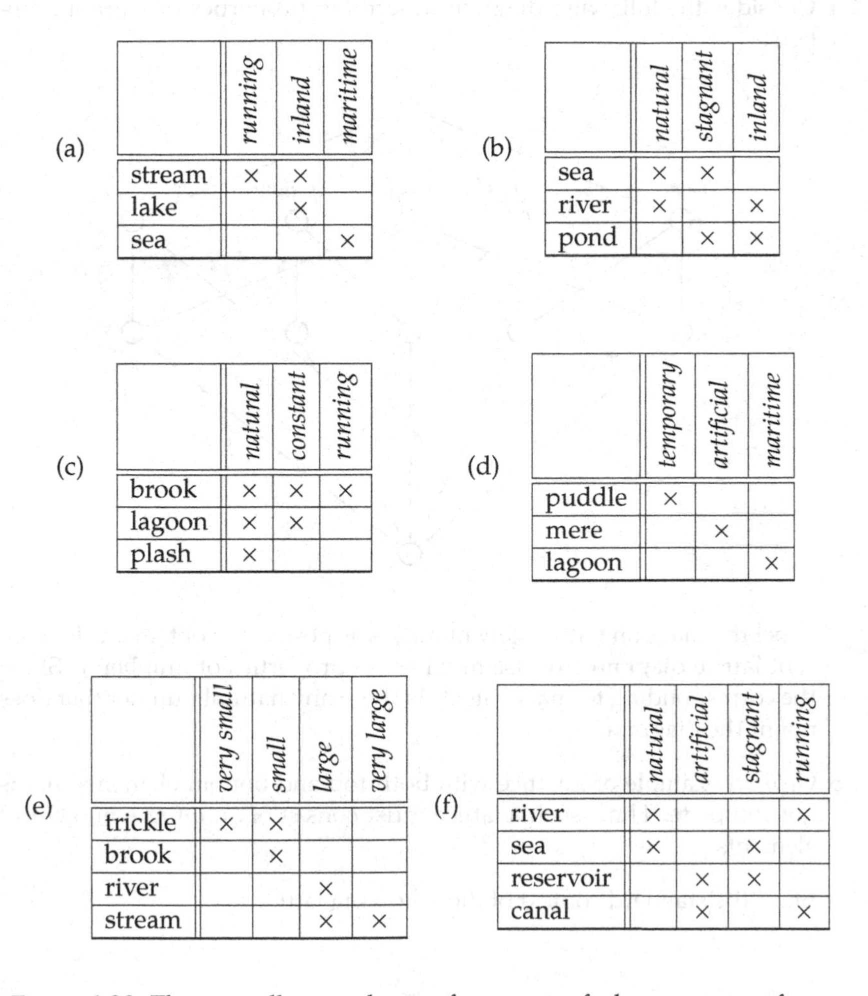

(a)

	running	*inland*	*maritime*
stream	×	×	
lake		×	
sea			×

(b)

	natural	*stagnant*	*inland*
sea	×	×	
river	×		×
pond		×	×

(c)

	natural	*constant*	*running*
brook	×	×	×
lagoon	×	×	
plash	×		

(d)

	temporary	*artificial*	*maritime*
puddle	×		
mere		×	
lagoon			×

(e)

	very small	*small*	*large*	*very large*
trickle	×	×		
brook		×		
river			×	
stream			×	×

(f)

	natural	*artificial*	*stagnant*	*running*
river	×			×
sea	×		×	
reservoir		×	×	
canal		×		×

Figure 1.20: These small examples are fragments of a larger context from [216]

1.3 Take a look at the formal context and the line diagram of its concept lattice in Figure 1.21. Label the diagram.

Do the same for Figures 1.22 and 1.23.

1.4 Consider the following diagram describing properties of natural numbers:

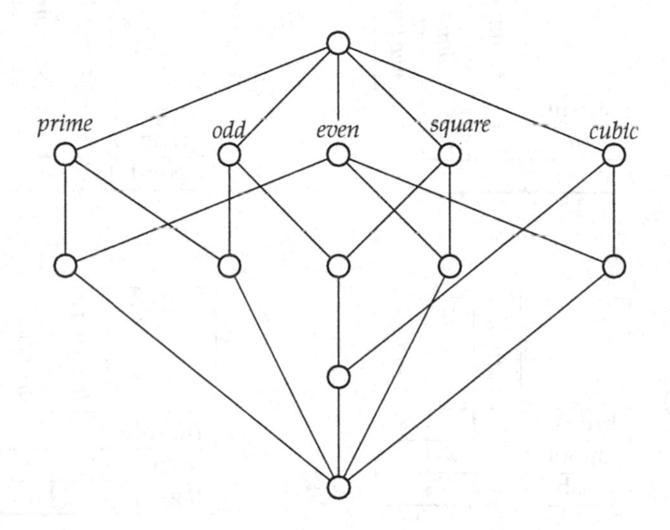

Label the diagram with as few numbers as possible to obtain a valid concept lattice diagram (consistent with the properties of numbers). Show the corresponding formal context. Is there any natural number that does not fit the diagram?

1.5 Give an example of a lattice with both top and bottom elements that is not complete. Hint: such a lattice must consist of an infinite number of elements.

1.6 Build the standard context of the following lattice:

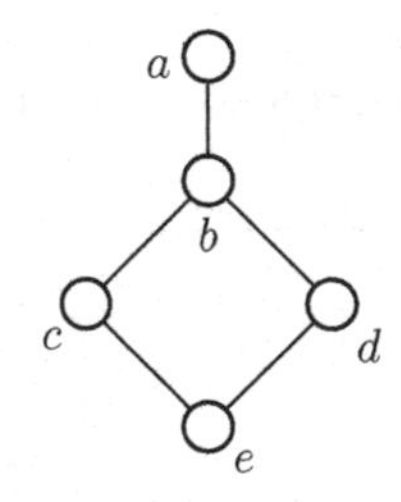

	alcoholic	*non-alcoholic*	*caffeine*	*hot*	*sparkling*
vodka	×				
champagne	×				×
coffee		×	×	×	
cola		×	×		×
cacao		×		×	
Fanta		×			×

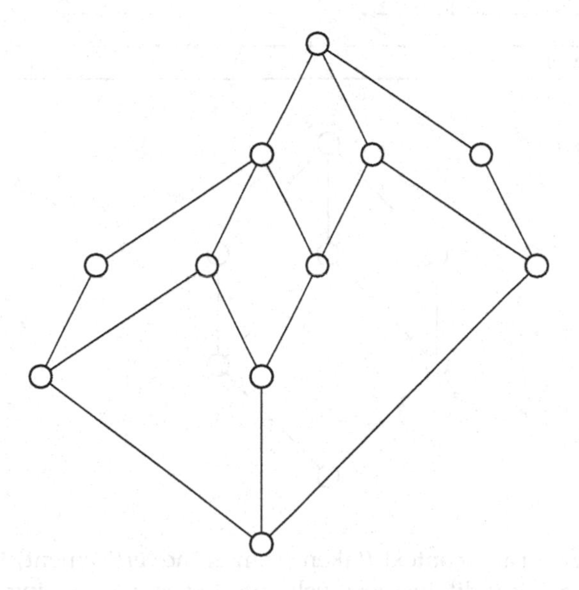

Figure 1.21: A formal context of various drinks and its unlabeled concept lattice diagram

	sugar level	*body weight*	*cholesterol level*
raspberry			
cherry		×	×
strawberry			
blueberry	×		
cranberry			
cowberry	×		
orange		×	
apple and celery		×	×
tomato and celery		×	×
beetroot	×		×
pumpkin	×		×
carrot			
bell pepper			

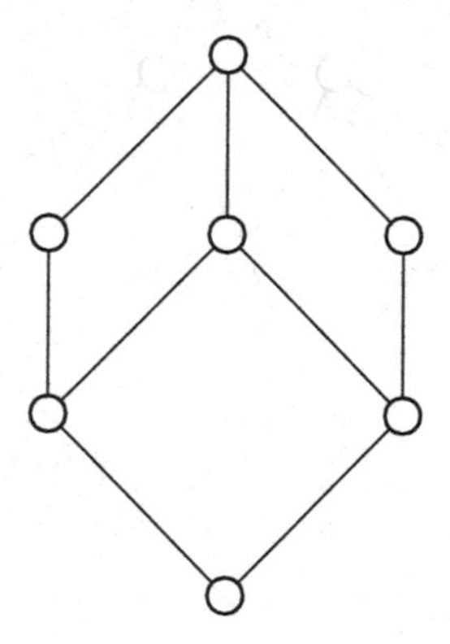

Figure 1.22: A formal context (taken from an advertisement) that describes a person's conditions positively affected by various ingredients in candies and an unlabeled concept lattice diagram of this context

	antioxidant	*burns fat*	*detoxifies the body*	*suppresses inflammation*
ginger and lemon			×	×
thyme and black tea	×		×	
licorice and orange	×			×
cowberry leaves and juice	×		×	
melissa and strawberry				
pineapple	×		×	
grapefruit	×	×	×	
jasmine and hibiscus			×	
wild rose and cranberry	×		×	×

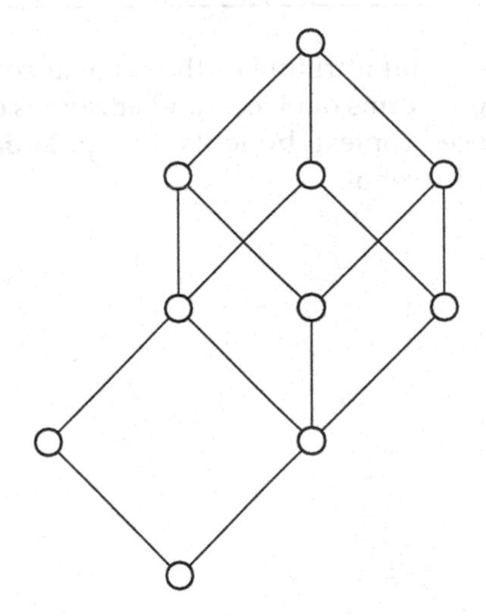

Figure 1.23: Yet another formal context of ingredients in candies and the concept lattice diagram of this context

1.7 Clarify and reduce the following formal context:

	nervous system	*brain*	*blood pressure*	*digestion*	*metabolism*
raspberry					
cherry	×				×
strawberry					
blueberry					
cranberry	×				
cowberry			×		
orange					×
apple and celery		×		×	
tomato and celery	×	×			
beetroot					
pumpkin					×
carrot	×			×	
bell pepper					

For each reducible object and attribute of the original context, indicate how it can be expressed in terms of objects and attributes of the reduced context. From the reduced context, build its concept lattice, but label it with original objects and attributes.

Chapter 2

An algorithm for closure systems

The knowledge acquisition algorithm that is the central theme of our book can most naturally be phrased in the language of Formal Concept Analysis. It essentially relies on a single property of concept lattices, namely, that the set of concept intents is closed under intersection. The technique works for an arbitrary intersection-closed family of sets, that is, for a *closure system*. Readers who are familiar with closure systems but not with Formal Concept Analysis may prefer this approach.

But note that this means no generalization. We shall show that closure systems are not more general than systems of concept intents.

2.1 Definition and examples

Closure systems occur frequently in mathematics and computer science. Their definition is very simple, but not very intuitive when encountered for the first time. The reason is the higher level of abstraction: closure systems are *sets of sets* with certain properties.

Let us recall some elementary notions how to work with sets of sets. For clarity, we shall normally use small Latin letters for elements, capital Latin letters for sets, and calligraphic letters for sets of sets. Given a set $\mathcal{S}$ of sets, we may ask:

- Which elements occur in these sets? The answer is given by the **union** of $\mathcal{S}$, denoted by
$$\bigcup \mathcal{S} := \{x \mid \exists_{S \in \mathcal{S}}\ x \in S\}.$$
- Which elements occur in each of these sets? The answer is given by the **intersection** of $\mathcal{S}$, denoted by
$$\bigcap \mathcal{S} := \{x \mid \forall_{S \in \mathcal{S}}\ x \in S\},$$

at least when $\mathcal{S}$ is nonempty. Some confusion with this definition is caused by the fact that a set of sets may (of course) be empty. Applying the above definition to the case $\mathcal{S} = \emptyset$ gives a problem for the intersection, because then the condition $\forall_{S \in \mathcal{S}} \; x \in S$ is fulfilled by *all* x (because there is nothing to fulfill). But there is no *set of all* x, the assumption of such a set leads to contradictions.

For the case $\mathcal{S} = \emptyset$, the intersection is defined only with respect to some base set M. If we work with subsets of some specified set M (as we often do, for example, with the set of all attributes of some formal context), then we define

$$\bigcup \emptyset := \emptyset \quad \text{and} \quad \bigcap \emptyset := M.$$

A set M with, say, n elements, has 2^n subsets. The set of all subsets of a set M is denoted $\mathfrak{P}(M)$ and is called the **power set** of the set M. To indicate that $\mathcal{S}$ is a set of subsets of M, we may therefore simply write $\mathcal{S} \subseteq \mathfrak{P}(M)$.

2.1.1 Closure systems

A closure system on a set M is a set of subsets that contains M and is closed under intersections. More formally,

Definition 8 A **closure system** on a set M is a set $\mathcal{C} \subseteq \mathfrak{P}(M)$ satisfying

- $M \in \mathcal{C}$, and
- if $\mathcal{D} \subseteq \mathcal{C}$, then $\bigcap \mathcal{D} \in \mathcal{C}$.

(Technically, the first condition may be subsumed under the second as the special case for $\mathcal{D} = \emptyset$, as explained above.) ◇

Closure systems are everywhere:

- The *subtrees* of any tree form a closure system, because the intersection of subtrees is always a subtree.

  That this is true can be seen as follows: recall that in a tree any two vertices are connected by a unique path. A set of vertices induces a subtree if and only if it contains, with any two of its vertices, all vertices on the path between these two. Now consider any selection of subtrees and let S be the set of vertices common to all these subtrees (i.e., their intersection). Let v, w be two vertices in S. The path between v and w belongs to every subtree containing v and w and therefore to each of the selected subtrees. Consequently, it is also contained in their intersection. Thus, S is the vertex set of a subtree.

- Take any algebraic structure, for example, a group, and take the set of its *subalgebras* (*subgroups*). This is a closure system, because the intersection of arbitrary subgroups is again a subgroup, and, more generally, the intersection of subalgebras is a subalgebra.

  Similarly, let M be the set A^* of all (finite) words over some alphabet A. In other words, consider the free monoid over A. The set of its *submonoids* is a closure system, because any intersection of submonoids is closed under multiplication and contains the empty word.

- Or take the set of *subintervals* of the real interval $[0, 1]$, including the empty interval. Since any intersection of intervals is again an interval, we have a closure system.

  This example can be generalized to n dimensions in several ways. One of them yields the closure system of all *convex sets*.

- The set of downsets of any ordered set $(M, \leq)$ is a closure system on M. A **downset** (or **order ideal**) of $(M, \leq)$ is a subset $D \subseteq M$ such that, whenever $d \in D$ and $m \in M$ with $m \leq d$, we have $m \in D$. Dually, the set of all **order filters** (or **upsets**, respectively) is a closure system.

- Consider all **preorders** on some fixed base set S, that is, all transitive reflexive relations $R \subseteq S \times S$. Any intersection of transitive reflexive relations is again transitive and reflexive. Therefore, *preorders* form a closure system.

2.1.2 Closure operators

Definition 9 A **closure operator** φ on M is a map assigning a **closure** $\varphi X \subseteq M$ to each set $X \subseteq M$ which is

monotone: $X \subseteq Y \Rightarrow \varphi X \subseteq \varphi Y$

extensive: $X \subseteq \varphi X$, and

idempotent: $\varphi\varphi X = \varphi X$.

(Conditions to be fulfilled for all $X, Y \subseteq M$.) $\Diamond$

For every example in the above list of closure systems, we can also give a closure operator. In each of the following examples, φ is a closure operator on the set M.

- Let (M, E) be a tree with vertex set M and φ be the mapping taking each set X of vertices to the vertex set of the smallest subtree containing X.

- For a group with carrier set M, define φX to be the subgroup generated by X.

  For a set X of words over an alphabet A, let $\varphi X := X^*$ be the submonoid of $M := A^*$ generated by X.
- For any $X \subseteq M := [0, 1]$ let φX be the smallest interval containing X (i.e., the convex closure of X).
- If $(M, \leq)$ is an ordered set, then, for any $X \subseteq M$, let

$$\varphi X := \{m \in M \mid m \leq x \text{ for some } x \in X\}$$

  be the downset generated by X.
- For any relation R on a set S, in other words, for any subset $R \subseteq S \times S =: M$, let φR denote the reflexive transitive closure of R.

The examples indicate why closure operators are so frequently met: their axioms describe the natural properties of a *generating process*. We start with some generating set X, apply the generating process and obtain the generated set, φX, the closure of X. Such generating processes occur in fact in many different variants in mathematics and computer science.

Closure systems and closure operators are closely related. In fact, there is a natural way to obtain from each closure operator a closure system and conversely. It works as follows: for any closure operator, the set of all closures is a closure system. Conversely, given any closure system $\mathcal{C}$ on M, there is, for each subset X of M, a unique smallest set $C \in \mathcal{C}$ containing X, namely,

$$\bigcap_{X \subseteq D \in \mathcal{C}} D.$$

Taking this as the closure of X defines a closure operator. The two transformations are inverse to each other.

2.1.3 The closure systems of intents and of extents

Thus, closure systems and closure operators are essentially the same. We can add to this: a closure system $\mathcal{C}$ on M can be considered as an ordered set, ordered by set inclusion $\subseteq$. This ordered set has a special property: it is a complete lattice. The infimum of any subfamily $\mathcal{D} \subseteq \mathcal{C}$ is obviously equal to $\bigcap \mathcal{D}$, and the supremum is just the closure of $\bigcup \mathcal{D}$. Conversely, we can find for any complete lattice L a closure system that is isomorphic to L. So closure systems and complete lattices are also very closely related.

It comes at no surprise that concept lattices can be subsumed under this relationship. It follows from the Basic Theorem (Theorem 1 (p. 18)) that the set of all concept intents of a formal context is closed under intersection and thus is a closure system on M. Dually, the set of all concept extents is always a closure system on G. The corresponding closure operators are just the two operators

$$X \mapsto X''$$

on M and G, respectively.

Conversely, given any closure system $\mathcal{C}$ on a set M, we can construct a formal context such that $\mathcal{C}$ is the set of concept intents. It can be concluded from the Basic Theorem that, for example, $(\mathcal{C}, M, \ni)$ is such a context. In particular, whenever a closure operator on some set M is considered, we may assume that it is the closure operator

$$A \mapsto A''$$

on the attribute set of some formal context (G, M, I).

Thus, closure systems and closure operators, complete lattices, systems of concept intents or of concept extents, all these are very closely related. It is not appropriate to say that they are "essentially the same", but it is true that all these structures have the same degree of generality; none of them is a generalization of another. A substantial result proved for one of these structures can usually be transferred to the others without much effort.

2.1.4 The closure system of closure systems

A closure system on M is a set of subsets of M which is closed under arbitrary intersections. There are many closure systems, even when the base set M is of modest size. Their numbers are known up to $|M| = 7$ [121, 64]:

Size of the base set M	0	1	2	3	4	5
Number of closure systems	1	2	7	61	2 480	1 385 552

$\|M\|$	6	7
# closure syst.	75 973 751 474	14 087 648 235 707 352 472

The intersection of arbitrary many closure systems (on the same base set) is again a closure system. This is easy to verify. Therefore, the family of closure systems of M is itself a closure system. There is a closure operator associated with the closure system of closure systems. This operator assigns to an arbitrary family $\mathcal{F} \subseteq \mathfrak{P}(M)$ of subsets of M the smallest closure system containing $\mathcal{F}$, which we call the **closure system generated by** $\mathcal{F}$. It is obtained by adding M to $\mathcal{F}$ together with all those subsets of M that can be obtained as intersections of $\mathcal{F}$-sets. Formally, the closure system generated by $\mathcal{F}$ is

$$\left\{ \bigcap \mathcal{D} \mid \mathcal{D} \subseteq \mathcal{F} \right\}.$$

All closure systems can be represented as closure systems of extents of some formal context, and so can the closure system of closure systems. We shall describe this context in Section 3.4.4 (p. 117). An example for the case $M = \{a, b, c\}$ is shown in Figure 3.9 (p. 118).

2.2 The NEXT CLOSURE algorithm

We present a simple algorithm that solves the following task: for a given closure operator on a finite set M, it computes all closed sets. There are many ways to achieve this. Our algorithm is particularly simple. We shall discuss its efficiency below. It also allows many useful modifications, some of which will be used in our more advanced applications. We start with the simplest version.

2.2.1 Representing sets by bit vectors

We begin by giving our finite base set M an arbitrary linear order, so that

$$M = \{m_1 < m_2 < \cdots < m_k\},$$

where k is the number of elements of M. Then every subset $S \subseteq M$ can conveniently be described by its **characteristic vector** $\varepsilon_S : \{1, 2, \ldots, k\} \to \{0, 1\}$, given by

$$\varepsilon_S(i) := \begin{cases} 1 & \text{if } m_i \in S \\ 0 & \text{if } m_i \notin S \end{cases}.$$

For example, if the base set is $M := \{a < b < c < d < e < f < g\}$, then the characteristic vector of the subset $S := \{a, c, d, f\}$ is 1011010. In concrete examples, we prefer to write a cross instead of a 1 and a blank or a dot instead of a 0, as in the cross tables representing formal contexts. The characteristic vector of the subset $S := \{a, c, d, f\}$ will therefore be written as

×	·	×	×	·	×	·

.

In this notation, it is easy to see if a given set is a subset of another given set.

The set $\mathfrak{P}(M)$ of all subsets of the base set M is naturally ordered by the subset–order $\subseteq$. This is a complete lattice order, and $(\mathfrak{P}(M), \subseteq)$ is called the **power set lattice** of M. The subset-order is a *partial* order. We can also introduce a *linear* or *total* order of the subsets, for example, the **lectic order** $\leq$, defined as follows. Let $A, B \subseteq M$ be two distinct subsets. We say that A is *lectically smaller* than B if the smallest element in which A and B differ belongs to B. Formally,

$$A < B \quad :\Longleftrightarrow \quad \exists_i \, (i \in B,\ i \notin A,\ \forall_{j<i} \, (j \in A \iff j \in B)).$$

For example, $\{a, c, e, f\} < \{a, c, d, f\}$, because the smallest element in which the two sets differ is d and this element belongs to the larger set. This becomes even more apparent when we write the sets as vectors and interpret these as natural numbers in binary encoding:

×	·	×	·	×	×	·

$$\updownarrow$$

×	·	×	×	·	×	·

Note that the lectic order extends the subset-order, i.e.,

$$A \subseteq B \Rightarrow A \leq B.$$

The following notation is helpful:

$$A <_m B \quad :\Longleftrightarrow \quad (m \in B,\ m \notin A,\ \forall_{n<m}\ (n \in A \iff n \in B)).$$

In words: $A <_m B$ iff m is the smallest element in which A and B differ and $m \in B$.

Proposition 4 *For $A, B \subseteq M$:*

1. *$A < B$ if and only if $A <_m B$ for some $m \in M$.*
2. *If $A <_m B$ and $A <_n C$ with $m < n$, then $C <_m B$.*

Proof 1. is immediate from the definition of the lectic order. To show 2., observe that the three sets A, B, and C contain the same elements $< m$. B contains m, but C does not; therefore, $C < B$. □

2.2.2 Closures in lectic order

We consider a closure operator

$$A \mapsto A''$$

on the base set M. It associates each subset $A \subseteq M$ with its closure $A'' \subseteq M$.[1] Our task is to find a list of all these closures. In principle, we might just follow the definition, compute for each subset $A \subseteq M$ its closure A'' and include it in the list. The problem is that different subsets may have identical closures. So if we want a list that contains each closure *exactly once*, we will have to check many times if a computed closure already exists in the list. Moreover, the number of subsets is exponential: a set with n elements has 2^n subsets.

The naive algorithm

[1]For our algorithm, it is not important *how* the closure is computed.

> **for all** $A \subseteq M$ **do**
>   compute A'' and check if the result is already listed

therefore requires an exponential number of lookups in a list that may have exponential size.

A better idea is to generate the closures in some predefined order, thereby guaranteeing that every closure is generated only once. The reader may already have guessed that we shall generate the closures in lectic order. We will show how to compute, given a closed set, the *lectically next* one. Then no lookups are necessary. Actually, it will not even be necessary to store the list. For many applications, it will suffice to generate the list elements on demand. Therefore, we do not have to store exponentially many closed sets. Instead, we shall store just *one*!

To find the next closure, we define for $A \subseteq M$ and $m_i \in M$

$$A \oplus m_i := ((A \cap \{m_1, \ldots, m_{i-1}\}) \cup \{m_i\})''.$$

We illustrate this definition by an example. Let $A := \{a, c, d, f\}$ and $m_i := e$.

				↓		
×	·	×	×	·	×	·

We first remove all elements that are greater than or equal to m_i from A:

				↓		
×	·	×	×	·	·	·

Then we insert m_i

				↓		
×	·	×	×	×	·	·

and form the closure. Since we have not yet specified a closure operator, the example stops here with

$$A \oplus e = \{a, c, d, e\}''.$$

The following theorem characterizes the "next" closure:

Theorem 2 *The smallest closed set larger than a given set $A \subset M$ with respect to the lectic order is*

$$A \oplus m,$$

m being the largest element of M with $A <_m A \oplus m$.

A *proof* will be given for a more general version in Theorem 5 below. Theorem 2 suggests an algorithm (see Algorithm 1) for finding all closed sets: start with the lectically first one (see Algorithm 2) and then repeatedly compute the next closure (see Algorithm 3) until all are computed.

Algorithm 1 ALL CLOSURES(M, $''$): Generating all closed sets

Input: A closure operator $X \mapsto X''$ on a finite linearly ordered set M.
Output: All closed sets in lectic order.

```
A := FIRST CLOSURE('')                    {see Algorithm 2}
while A ≠ ⊥ do
    output A
    A := NEXT CLOSURE(A, M, '')           {see Algorithm 3}
```

Algorithm 2 FIRST CLOSURE($''$)

Input: A closure operator $X \mapsto X''$ on a finite linearly ordered set.
Output: The lectically first closed set, i.e., the closure of the empty set.

```
return ∅''
```

Algorithm 3 NEXT CLOSURE(A, M, $''$)

Input: A closure operator $X \mapsto X''$ on a finite linearly ordered set M and a subset $A \subseteq M$.
Output: The lectically next closed set after A if it exists; $\perp$, otherwise.

```
for all m ∈ M in reverse order do
    if m ∈ A then
        A := A \ {m}
    else
        B := (A ∪ {m})''
        if B \ A contains no element < m then
            return B
return ⊥
```

2.2.3 Finding formal concepts

Recall that, for any formal context (G, M, I), the set of extents is a closure system on G and the set of intents is a closure system on M. We can therefore apply the NEXT CLOSURE algorithm to compute all extents or all intents – and thereby all formal concepts – of a formal context.

When we use the NEXT CLOSURE algorithm for extents, we call it NEXT EXTENT, and if we use it for concept intents, we speak of the NEXT INTENT algorithm. These algorithms therefore are not new, but merely specializations of the general algorithm to distinct closure systems.

The NEXT CLOSURE algorithm requires some linear order of the base set. In the case that a formal context is given as a cross-table, both G and M come with such an order. As a default, we shall use that order when we use the NEXT EXTENT or the NEXT INTENT algorithm.

We demonstrate NEXT INTENT on a small example given in Figure 2.1. The

		a	b	c	d	e	f	g	h
		brilliant white	*fine white*	*white*	*for high-performance copiers*	*for copiers*	*for liquid toner copiers*	*for typewriters*	*for double-sided copying*
1	Copy-Lux	×			×	×		×	×
2	Copy-X		×		×	×		×	×
3	Copy			×	×	×		×	×
4	Liquid-Copy			×		×	×		×
5	Office			×		×		×	
6	Offset			×				×	

Figure 2.1: Paper types for copiers

formal context (copied from an advertisement) shows which types of paper of the respective company serve which copying purposes.

Figure 2.2 shows in detail how the sixteen intents of this formal context are found by the algorithm. Each row corresponds to one iteration of the main loop of NEXT CLOSURE, but we omit rows corresponding to cases when $m \in A$.

The intents can be read off from the column with the header "last generated closure".

Figure 2.3 gives the list of all formal concepts of the context in Figure 2.1. These could easily be derived from the intents: for each intent B computed in Figure 2.2, we have the formal concept (B', B), and these are all.

How the algorithm works is perhaps better understood from Figure 2.4, which displays the same data as Figure 2.3, but in a different format.

2.2.4 The complexity of the NEXT CLOSURE algorithm

Given an input of size n, how many steps does it take to produce all formal concepts? In other words, what is the computational complexity of ALL CLOSURES (Algorithm 1)?

This has an easy answer. For certain inputs, the computation time is exponential for a simple reason: the output size may be exponential. Consider, for example, the formal contexts

$$\mathbb{N}_n^c := (\{1, 2, \ldots, n\}, \{1, 2, \ldots, n\}, \neq).$$

Such contexts are called **contranominal scales**. Every subset of $\{1, 2, \ldots, n\}$ is an extent of $\mathbb{N}_n^c$, and the concept lattice is isomorphic to the power-set lattice of $\{1, 2, \ldots, n\}$. There are 2^n formal concepts, while the input context has size only $n \times n$. No matter how fast our algorithm computes each single concept, we need exponential time just for the output. In general, it is difficult to predict the size of the concept lattice given a formal context. Computing the size of a concept lattice was shown to by #$\mathcal{P}$-complete by Kuznetsov [140, 142]. The contranominal scales are not the only formal contexts with a huge concept lattice, but every context with many formal concepts must contain a large contranominal scale, as was shown by Albano and Chornomaz [3].

Thus, asking how much time in terms of the input size the algorithm needs to generate all closures is not particularly useful. A more interesting question is how much time the algorithm needs *per closed set*. Similarly, we may ask for the total time needed to compute all formal concepts from a given input context divided by the number of concepts. This is the *average* time to compute one closed set.

A look at Algorithm 1 reveals that the first closure is generated by Algorithm 2, appropriately named FIRST CLOSURE, while all other closures are generated by repetitive application of the NEXT CLOSURE procedure (Algorithm 3). All that FIRST CLOSURE does is computing the closure of the empty set. The complexity of this operation depends on how the closure operator is specified. For now, assume that computing a closure takes time γ. The main loop of the NEXT CLOSURE procedure is repeated at most $|M|$ times, at each iteration performing at most $\gamma + O(|M|)$ work.

last generated closure	A	m	$B = (A \cup \{m\})''$	smallest new element	success?
	$\emptyset$	$= \emptyset''$	(first intent)		yes
$\emptyset$	$\emptyset$	h	$\{e, h\}$	e	no
	$\emptyset$	g	$\{g\}$	g	yes
$\{g\}$	$\{g\}$	h	$\{d, e, g, h\}$	d	no
	$\emptyset$	f	$\{c, e, f, h\}$	c	no
	$\emptyset$	e	$\{e\}$	e	yes
$\{e\}$	$\{e\}$	h	$\{e, h\}$	h	yes
$\{e, h\}$	$\{e\}$	g	$\{e, g\}$	g	yes
$\{e, g\}$	$\{e, g\}$	h	$\{d, e, g, h\}$	d	no
	$\{e\}$	f	$\{c, e, f, h\}$	c	no
	$\emptyset$	d	$\{d, e, g, h\}$	d	yes
$\{d, e, g, h\}$	$\{d, e\}$	f	$\{a, b, c, d, e, f, g, h\}$	a	no
	$\emptyset$	c	$\{c\}$	c	yes
$\{c\}$	$\{c\}$	h	$\{c, e, h\}$	e	no
	$\{c\}$	g	$\{c, g\}$	g	yes
$\{c, g\}$	$\{c, g\}$	h	$\{c, d, e, g, h\}$	d	no
	$\{c\}$	f	$\{c, e, f, h\}$	e	no
	$\{c\}$	e	$\{c, e\}$	e	yes
$\{c, e\}$	$\{c, e\}$	h	$\{c, e, h\}$	h	yes
$\{c, e, h\}$	$\{c, e\}$	g	$\{c, e, g\}$	g	yes
$\{c, e, g\}$	$\{c, e, g\}$	h	$\{c, d, e, g, h\}$	d	no
	$\{c, e\}$	f	$\{c, e, f, h\}$	f	yes
$\{c, e, f, h\}$	$\{c, e, f\}$	g	$\{a, b, c, d, e, f, g, h\}$	a	no
	$\{c\}$	d	$\{c, d, e, g, h\}$	d	yes
$\{c, d, e, g, h\}$	$\{c, d, e\}$	f	$\{a, b, c, d, e, f, g, h\}$	a	no
	$\emptyset$	b	$\{b, d, e, g, h\}$	b	yes
$\{b, d, e, g, h\}$	$\{b, d, e\}$	f	$\{a, b, c, d, e, f, g, h\}$	a	no
	$\{b\}$	c	$\{a, b, c, d, e, f, g, h\}$	a	no
	$\emptyset$	a	$\{a, d, e, g, h\}$	a	yes
$\{a, d, e, g, h\}$	$\{a, d, e\}$	f	$\{a, b, c, d, e, f, g, h\}$	b	no
	$\{a\}$	c	$\{a, b, c, d, e, f, g, h\}$	b	no
	$\{a\}$	b	$\{a, b, c, d, e, f, g, h\}$	b	yes
M	M	$= \{a, b, c, d, e, f, g, h\}$	(last intent)		

Figure 2.2: The NEXT INTENT algorithm, applied to the formal context in Figure 2.1

$(G, \emptyset)$	$(\{1,2,3,5,6\},\{g\})$	$(\{1,2,3,4,5\},\{e\})$
$(\{1,2,3,4\},\{e,h\})$	$(\{1,2,3,5\},\{e,g\})$	$(\{1,2,3\},\{d,e,g,h\})$
$(\{3,4,5,6\},\{c\})$	$(\{3,5,6\},\{c,g\})$	$(\{3,4,5\},\{c,e\})$
$(\{3,4\},\{c,e,h\})$	$(\{3,5\},\{c,e,g\})$	$(\{4\},\{c,e,f,h\})$
$(\{3\},\{c,d,e,g,h\})$	$(\{2\},\{b,d,e,g,h\})$	$(\{1\},\{a,d,e,g,h\})$
$(\emptyset, M)$		

Figure 2.3: The formal concepts of the "copy"–context in Figure 2.1

extents						intents							
1	2	3	4	5	6	a	b	c	d	e	f	g	h
×	×	×	×	×	×								
×	×	×		×	×							×	
×	×	×	×	×						×			
×	×	×	×							×			×
×	×	×		×						×		×	
×	×	×							×	×		×	×
		×	×	×	×			×					
		×		×	×			×				×	
		×	×	×				×		×			
		×	×					×		×			×
		×		×				×		×		×	
			×					×		×	×		×
		×						×	×	×		×	×
	×						×		×	×		×	×
×						×			×	×		×	×
						×	×	×	×	×	×	×	×

Figure 2.4: The same as Figure 2.3, but in tabular form, making apparent that the intents are in lectic order

If the closure operator is specified by a formal context (G, M, I), the closure of an attribute set A can be computed in time $O(|G||M|)$ by identifying all objects g such that $A \subseteq \{g\}'$ and intersecting all such $\{g\}'$ (see Algorithm 4). Provided that object intents, $\{g\}'$, are precomputed, the complexity of FIRST CLOSURE in this case is $O(|G||M|)$ and the complexity of NEXT CLOSURE is $O(|G||M|^2)$. This gives us an upper bound on the time that passes before each next concept is generated or the algorithm stops. Therefore, we can say that the algorithm ALL CLOSURES has a *polynomial delay* [129] of $O(|G||M|^2)$. This is true for the NEXT INTENT version of the algorithm; for NEXT EXTENT, this will be $O(|G|^2|M|)$. For algorithms with necessarily exponential complexity, a polynomial delay is a nice property: it means that the algorithm keeps producing stuff with relatively short intervals. Obviously, a polynomial delay is also an upper bound on the average time to compute one concept.

Algorithm 4 Computing A'' for $A \subseteq M$ in (G, M, I)

Input: A formal context (G, M, I) and an attribute set $A \subseteq M$.
Output: The closure of A in (G, M, I).

$B := M$
for all $g \in G$ **do**
  if $A \subseteq \{g\}'$ **then**
    $B := B \cap \{g\}'$
return B

There are faster algorithms, and they will be briefly discussed in Section 2.4.4 below, but NEXT CLOSURE is simple and fast enough for many applications. For example, Behrisch & Ganter used the algorithm to count all closure systems on a 6-element set and found that there are 75 973 751 474 of them, verifying the result by Habib and Nourine [121] mentioned in Section 2.1.4 (p. 43). The algorithm was executed in 2006 on a standard PC in less than a day. The algorithm produced on average more than a million concepts per second.

2.3 Computing only certain closed sets

2.3.1 Closures that contain and avoid prescribed elements

In some applications, it is desired not to produce all closed sets, but only some. For example, one might wish to generate only closed sets that contain certain fixed elements $C = \{c_1, \ldots, c_j\}$ and avoid certain others, say, $A = \{a_1, \ldots, a_i\}$.

This can easily be achieved using the fact that the NEXT CLOSURE algorithm produces the closed sets in lectic order. In fact, if

$$M = a_1 < \ldots < a_i < c_1 < \ldots < c_j < b_1 < \ldots < b_k,$$

then, starting from C'', NEXT CLOSURE will produce these closed sets in a row. As a first example, observe in Figure 2.4 that the intents containing c but not a or b were produced consecutively. This is automatically so whenever the elements to be avoided form an initial segment in the linear order of M and the elements that must be contained form an initial segment among the remaining elements. Since NEXT CLOSURE makes no assumptions on the order of M, we can freely choose it. For given A and C, we can arrange M so that attributes from A precede those from C, which precede all the other attributes.

Note that, if we wish to generate all closed sets containing at least *some* attributes from C instead of *all* attributes from C, we can proceed as above, but start with $\{c_j\}$ (or the lectically next closed set after $\{c_j\}$ if $\{c_j\}$ is not closed) instead of $\{c_1, \ldots, c_j\}''$.

A more interesting example is derived from the ordered set in Figure 2.5. It depicts precedence constraints for 14 steps (or "tasks") of launching a new product, from the initial plan to the actual production. Each step can be begun only after its predecessors are completed. For example, before a prototype is built, drawings and specifications should be ready, but that may happen before or after a market study is made.

Since the constraint diagram is not linear, there are different possibilities for processing it. A **state** is a situation where some, but not necessarily all of the steps have been completed, and the precedence constraints were respected. For example,

$$\{\textit{approve plan},\ \textit{drawings},\ \textit{write specs},\ \textit{prototype},\ \textit{materials},\ \textit{study market}\}$$

is a possible state.

It is easy to see that the states are precisely the **downsets** (also called **order ideals**) of the precedence order. These are the sets that contain, with any given element, also all elements below it; see Section 2.1.1 (p. 40). And quite obviously, these sets form a closure system. The closure operator adds to each set A all elements that are below some element in A.

An application of the NEXT CLOSURE algorithm shows that, in this example, there are precisely 41 states. Actually, there is a well-known construction of a formal context, called a **contraordinal scale**, the extents of which are precisely the downsets. It is shown in Figure 2.6.

We shall not list all 41 states here. The interested reader may use the formal context in Figure 2.6 as input for one of the programs mentioned in Section 1.4.4 (p. 29) to compute them.

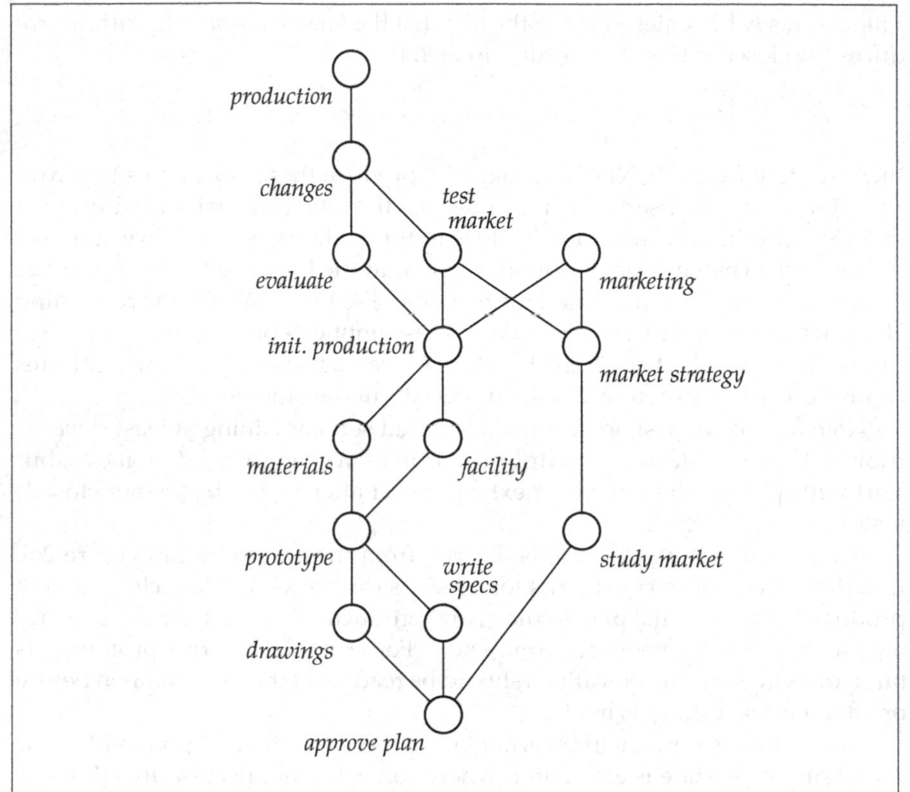

Figure 2.5: Precedence constraints for bringing an idea to the market. This is a modified version of an example given in the SAS manual, where it is represented as a Gantt diagram [193]

We shall use the NEXT CLOSURE algorithm (in its NEXT EXTENT version) to generate precisely those states for which the "*materials*" have been decided, but not the "*market strategy*". According to what was said in the beginning of this subsection, we let

$$A := \{\textit{market strategy}\},$$
$$C := \{\textit{materials}\},$$

and linearly order the set of tasks so that "*market strategy*" comes first, "*materials*" comes second, and then come the remaining steps in some order. Then we repeat the NEXT EXTENT algorithm, beginning with the closure of C, as long as it produces closed sets disjoint from A. The eight solutions are shown in Figure 2.7. They can easily be verified by hand.

$(P, P, \not\geq)$	*approve plan*	*drawings*	*write specs*	*prototype*	*study market*	*materials*	*facility*	*init. production*	*market strategy*	*evaluate*	*test market*	*marketing*	*changes*	*production*
approve plan		×	×	×	×	×	×	×	×	×	×	×	×	×
drawings			×	×	×	×	×	×	×	×	×	×	×	×
write specs		×		×	×	×	×	×	×	×	×	×	×	×
prototype					×	×	×	×	×	×	×	×	×	×
study market		×	×	×		×	×	×	×	×	×	×	×	×
materials					×		×	×	×	×	×	×	×	×
facility					×	×		×	×	×	×	×	×	×
init. production					×				×	×	×	×	×	×
market strategy		×	×	×		×	×	×		×	×	×	×	×
evaluate					×				×		×	×	×	×
test market										×		×	×	×
marketing										×	×		×	×
changes												×		×
production												×		

Figure 2.6: The contraordinal scale for the precedence order $(P, \leq)$ in Figure 2.5

market strategy	*materials*	*approve plan*	*drawings*	*write specs*	*prototype*	*study market*	*facility*	*init. production*	*evaluate*	*test market*	*marketing*	*changes*	*production*	
	×													init
	×	×	×	×	×									1.
	×	×	×	×	×		×							2.
	×	×	×	×	×		×	×						3.
	×	×	×	×	×		×	×	×					4.
	×	×	×	×	×	×								5.
	×	×	×	×	×	×	×							6.
	×	×	×	×	×	×	×	×						7.
	×	×	×	×	×	×	×	×	×					8.
×		×				×								stop

Figure 2.7: Generating states that contain *"materials"*, but not *"market strategy"*

2.3.2 Small closures only and large closures only

Another common task is to generate all *small* closed sets. For example, we might want to generate all closed sets of a certain size $\leq k$. But this is only a special case. All we need for the modified algorithm is that subsets of small sets are small as well. Therefore, let $\mathcal{F}$ be an order ideal (downset) of closed sets, i.e., a family of closed sets satisfying

$$\text{if } C \text{ and } D \text{ are both closed, } C \subseteq D, \text{ and } D \in \mathcal{F}, \text{ then } C \in \mathcal{F},$$

and call a closed set "small" iff it belongs to $\mathcal{F}$.

The NEXT CLOSURE algorithm can easily be modified to generate only those closed sets that are contained in a given order ideal $\mathcal{F}$. The modification is based on the following result:

Theorem 3 *Let $A \mapsto A''$ be a closure operator on M, and let $\mathcal{F}$ be an order ideal in the set of all closed sets. The smallest closed set in $\mathcal{F}$ that is larger than a given set $A \subset M$, provided such a set exists, is*

$$A \oplus m,$$

m being the largest element of M satisfying both $A <_m A \oplus m$ and $A \oplus m \in \mathcal{F}$.

So all we have to do is to change in Algorithm 3 (p. 47) the line

if $B \setminus A$ contains no element $< m$ **then**

to

if $B \setminus A$ contains no element $< m$ **and** $B \in \mathcal{F}$ **then**

This will then produce exactly the closed sets in $\mathcal{F}$, as will be proved in Section 2.4.1. See also Algorithm 5.

Again, we can illustrate the algorithm with the scheduling example from Figure 2.5. Suppose that we want to list all admissible states in which at most eight steps have already been completed. To find these, we may invoke NEXT SMALL CLOSURE with characteristic function

$$small : B \mapsto \begin{cases} \textbf{true}, & \text{if } |B| \leq 8; \\ \textbf{false}, & \text{otherwise.} \end{cases}$$

The steps of the algorithm are shown in Figure 2.8. As a variation, we have used yet another ordering of the base set, a special one that makes the computation particularly easy. But this is not necessary for the desired result.

A popular instance of finding only small closures occurs in "market basket analysis" (see, e.g., Agrawal *et al.* [1]), where "**transactions**" and "**items**" play the role of objects and attributes. Each item set B has a **support** (or **frequency**), which reflects the number $|B'|$ of transactions where it occurs. Often

production	*changes*	*marketing*	*test market*	*evaluate*	*market strategy*	*init. production*	*facility*	*materials*	*study market*	*prototype*	*write specs*	*drawings*	*approve plan*	
														1.
													×	2.
												×	×	3.
											×		×	4.
											×	×	×	5.
										×	×	×	×	6.
									×				×	7.
									×			×	×	8.
									×		×		×	9.
									×		×	×	×	10.
									×	×	×	×	×	11.
								×		×	×	×	×	12.
								×	×	×	×	×	×	13.
							×			×	×	×	×	14.
							×		×	×	×	×	×	15.
							×	×		×	×	×	×	16.
							×	×	×	×	×	×	×	17.
						×	×	×		×	×	×	×	18.
						×	×	×	×	×	×	×	×	19.
					×				×				×	20.
					×				×			×	×	21.
					×				×		×		×	22.
					×				×		×	×	×	23.
					×				×	×	×	×	×	24.
					×			×	×	×	×	×	×	25.
					×		×		×	×	×	×	×	26.
					×		×	×	×	×	×	×	×	27.
				×		×	×	×		×	×	×	×	28.

Figure 2.8: The states having not more than eight elements, as produced by the NEXT SMALL CLOSURE algorithm

Algorithm 5 NEXT SMALL CLOSURE(A, M, $''$, *small*)

Input: A closure operator $X \mapsto X''$ on a finite linearly ordered set M, the characteristic function *small* of an order ideal of closed sets, and a subset $A \subseteq M$.
Output: The lectically next small closed set after A if it exists; $\perp$, otherwise.

```
for all m ∈ M in reverse order do
    if m ∈ A then
        A := A \ {m}
    else
        B := (A ∪ {m})''
        if B \ A contains no element < m and small(B) then
            return B
return ⊥
```

the relative support $|B'|/|G|$ is used as a parameter, where G is the set of all transactions. Among all item sets comparable to B that have the same support as B, there is always the largest one, namely, B'', and such maximal sets are naturally called **closed item sets**.

When a **minimal support** is chosen as a number `minsupp` between 0 and 1, an item set B is called **frequent** if $|B'|/|G| \geq \texttt{minsupp}$.

A standard task is to find, for a given matrix of transactions and items, all **frequent closed item sets**, i.e., all **frequent intents**. It is easy to see that subsets of frequent sets are also frequent. Therefore, frequent intents form an order ideal in the set of all intents and they can be found by repetitive application of Algorithm 5 NEXT SMALL CLOSURE[2] in combination with characteristic function

$$small : B \mapsto \begin{cases} \textbf{true}, & \text{if } |B'|/|G| \geq \texttt{minsupp}; \\ \textbf{false}, & \text{otherwise.} \end{cases}$$

This will produce exactly the frequent closed item sets.

Presenting in detail a realistic example of computing frequent intents is somewhat ambitious, because such computation is typically done in application to very large data sets, which do not fit well in a book. We present as an example Rudolf Wille's data set about handicapped children in Paris hospitals from 1984. It is shown in Figure 2.9. Each of the 88 objects has a multiplicity, reflecting the number of cases with this attribute combination.

[2]For large data sets, other algorithms, such as *Titanic* to be discussed in the next subsection, may be a better choice.

	#	a	b	c	d	e	f	g	h
1	1212	×							
2	227								×
3	261		×						
4	45				×				
5	29					×			
6	136			×					
7	83						×		
8	138							×	
9	11		×					×	
10	11				×			×	
11	123	×				×			
12	220	×							×
13	15		×			×			
14	27			×		×			
15	64			×					×
16	9					×			×
17	23				×				×
18	104	×		×					
19	18	×			×				
20	15			×			×		
21	36		×						×
22	39	×					×		
23	8		×	×					
24	12						×		×
25	117	×						×	
26	7		×				×		
27	16							×	×
28	3					×	×		
29	4						×	×	
30	1				×		×		
31	2		×		×				
32	1					×		×	
33	20	×				×			×
34	31	×		×		×			
35	1			×			×	×	
36	5		×	×		×			
37	9			×			×		×
38	2			×		×		×	
39	1		×			×		×	
40	13			×		×			×
41	3		×					×	×
42	2		×				×		×
43	36	×		×					×
44	17	×					×		×
45	9		×	×					×
46	27	×						×	×
47	1		×	×			×		
48	4			×				×	×
49	2		×			×			×
50	1		×	×				×	
51	1						×	×	×
52	6	×					×	×	
53	6	×				×		×	
54	2	×		×				×	
55	15	×			×				×
56	6	×		×			×		
57	3	×				×	×		
58	1				×	×			×
59	1					×		×	×
60	1		×			×	×		
61	1			×		×	×		
62	1	×			×		×		
63	2		×		×				×
64	1					×	×		×
65	2				×			×	×
66	1	×			×			×	
67	5			×		×		×	×
68	1			×		×	×	×	
69	3	×					×	×	×
70	5		×	×		×			×
71	4	×		×		×	×		
72	9			×		×	×		×
73	1		×			×	×		×
74	1	×			×			×	×
75	12	×		×		×			×
76	2	×		×				×	×
77	1			×			×	×	×
78	2	×		×		×		×	
79	1	×		×			×		×
80	1	×		×			×	×	
81	1	×				×	×		×
82	1	×				×	×	×	
83	3	×		×			×	×	×
84	2	×		×		×	×		×
85	2	×		×		×		×	×
86	2	×		×		×	×	×	
87	4			×		×	×	×	×
88	2	×		×		×	×	×	×

Figure 2.9: Context of some research on handicapped children; a: *emotionally handicapped*, b: *mentally ill*, c: *handicapped by brain injury*, d: *disabled*, e: *epileptic*, f: *visually handicapped*, g: *deaf*, h: *suffering from somatic heart disease*. The #-column gives the number of cases. From [216]

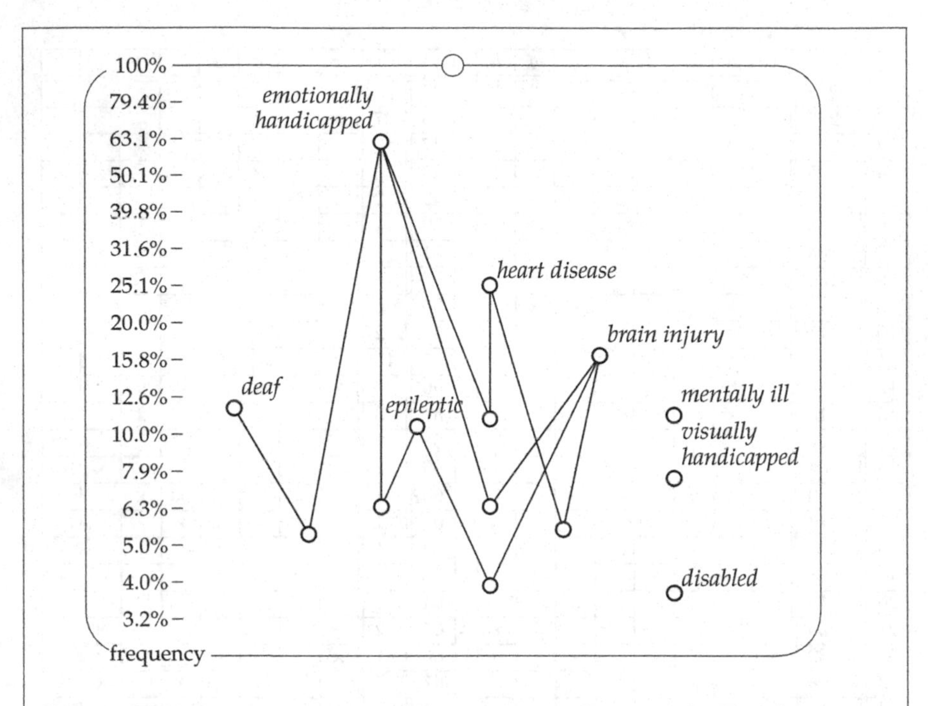

Figure 2.10: Frequent intents of the context in Figure 2.9. The diagram shows the 15 intents for which the respective extents comprise at least 3% of all cases. The top element ("no symptoms specified") is symbolized by the circle in the upper boundary. Its adjacent edges are omitted

For `minsupp` := 0.03, fifteen of the 92 intents are "frequent", meaning that their extents contain at least 3% of all cases listed in the data set. They can easily be computed using the algorithm described above. These frequent intents are shown in Figure 2.10. For better readability, we have omitted the edges that connect to the top element (which corresponds to the empty intent). The height in the diagram reflects the frequency, i.e., the relative size of the corresponding extent. A logarithmic scale was chosen for the height. Reading example: 11.6% of the children in the data set are deaf (386 out of 3316), while 5.5% combine a heart disease and a brain injury (183 of 3316). The attribute combination {deaf, brain injury} is not shown, because it is not frequent: only 1% of all cases (35 of 3316) show both attributes.

A slightly more complicated task is to generate only large closures, that is, to generate only the closed sets of a given order filter of closed sets. This is easy when the closure operator is represented by a formal context, because we can use the duality between extents and intents: given an order filter of intents, the corresponding extents form an order ideal, and vice versa. Thus, to generate an order filter of *extents*, we may use Algorithm 6, which is a modification of the NEXT SMALL CLOSURE algorithm. In particular, Algorithm 6 can be used to enumerate sets of transactions associated with frequent closed item sets. Of course, an order filter of *intents* can be enumerated dually.

Algorithm 6 NEXT LARGE EXTENT(C, (G, M, I), *large*)

Input: A formal context (G, M, I) over finite object set G and finite linearly ordered attribute set M, the characteristic function *large* of an order filter of extents, and a subset $C \subseteq G$.
Output: The large extent with lectically next intent after C' if it exists; $\perp$, otherwise.

```
A := C'
for all m ∈ M in reverse order do
    if m ∈ M then
        A := A \ {m}
    else
        B := (A ∪ {m})''
        if B \ A contains no element < m and large(B') then
            return B'
return ⊥
```

Every order filter of formal concepts, including the filter of concepts with frequent intents, is closed under suprema. If one adds the (dummy) smallest element, a complete lattice is obtained. This is called an **iceberg lattice** of the given lattice, because it may be thought of as the "very top" part of the original lattice (see Section 3.4.2 for a precise definition and for examples). Icebergs may be attacked with the TITANIC algorithm below.

2.3.3 Small generating sets and the TITANIC algorithm

A much different problem is that of finding closed sets with small generating sets. A **generating set** of a closed set A is a subset $S \subseteq A$ such that $A = S''$, and, obviously, a **minimal generating set** of A is a subset of A minimal with respect to this property. Every finite closed set has at least one minimal generating set, often it has many. A well-known observation is that the family of all minimal generating sets of a closure system is an order ideal in the power set, as the following proposition states:

Proposition 5 *Let $X \mapsto X''$ be a closure operator on M and set $S \subseteq M$ be a minimal generating set of a closed set. Then every subset $T \subseteq S$ of S is a minimal generating set of its closure T''.*

The **proof** is straightforward.

Algorithm 7, which is based on Proposition 5, can be used to iterate through all minimal generating sets of size less than or equal to a given threshold in lectic order. The test

$$\forall b \in B \quad (B \setminus \{b\})'' \neq B''$$

ensures that B is indeed a minimal generating set by checking that each of its proper subsets is a generating set of a different closure. One way to make this test more time-efficient is to store all computed minimal generating sets together with their closures in a prefix tree. Then, instead of computing $(B \setminus \{b\})''$, we could search for $B \setminus \{b\}$ in the prefix tree (in $O(|M|)$ time). If $B \setminus \{b\}$ is not found, then it is not a minimal generating set and, therefore, B cannot be one, either. Otherwise, we take the associated closure from the tree and use it in the above test.

Algorithm 7 NEXT MINIMAL GENERATING SET($A, M, '', k$)

Input: A closure operator $X \mapsto X''$ on a finite linearly ordered set M, a subset $A \subseteq M$, and a number k.
Output: The lectically next minimal generating set after A of size $\leq k$ if it exists; $\perp$, otherwise.

```
for all m ∈ M in reverse order do
    if m ∈ A then
        A := A \ {m}
    else if |A| < k then
        B := A ∪ {m}
        if (B \ {b})'' ≠ B'' for all b ∈ B then
            return B
return ⊥
```

The problem of computing the family of all minimal generating sets was investigated by several authors, see, e.g., Nehmé *et al.* [166]. The TITANIC algorithm [206], which customizes ideas from the APRIORI algorithm [2], offers a level-wise solution. This approach limits the amount of (sequential) accesses to the data, which is especially important for datasets so large that they cannot be kept completely within main memory.

TITANIC combines two ideas that are both of interest. The first is a method of generating an arbitrary order ideal of sets in a level-wise manner, the level being defined by the size of the sets it contains.

For a set of size $k+1$ to belong to the $(k+1)$st level of the order ideal, a necessary condition is that all its k-element subsets belong to the kth level $\mathcal{K}$ of the ideal (see Proposition 5). Sets satisfying this condition are called **candidate sets**. Algorithm 8 generates all candidate sets.

Algorithm 8 NEXT CANDIDATE SET$(C, \mathcal{K}, k)$

Input: An integer $k > 0$, a lectically ordered list $\mathcal{K}$ of k-subsets of the linearly ordered set M, and a $(k+1)$-subset C that has all its k-subsets in $\mathcal{K}$.

Output: The lectically next $(k+1)$-subset after C that has all its k-subsets in $\mathcal{K}$ if it exists; $\perp$, otherwise.

$B := C \setminus \{\max C\}$
$A := C \setminus \{\max B\}$
NEXT_IN$(A, \mathcal{K})$
repeat
  $D := B \setminus \{\max B\}$
  while $A \cap B = D$ **do**
    if $(A \cup B) \setminus \{x\} \in \mathcal{K}$ for all $x \in D$ **then**
      return $A \cup B$
    NEXT_IN$(A, \mathcal{K})$
  if $A = B$ **then**
    $A := D \cup \max\{x \in M \mid D \cup \{x\} \in \mathcal{K}\}$
  else
    $A := B$
until not NEXT_IN$(B, \mathcal{K})$
return $\perp$

Here, NEXT_IN$(X, \mathcal{K})$ returns **true** and replaces X by the lectically first set in the list $\mathcal{K}$ after X if such a set exists, and it returns **false** otherwise.
To get the first candidate set, invoke the algorithm with C equal to the union of the first two consecutive entries of $\mathcal{K}$ that have the same $(k-1)$-prefix.

Algorithm 8 relies on the fact (already used by APRIORI) that, with respect to some arbitrary linear order of the base set, each candidate set $C := \{c_0 < \ldots < c_{k-1} < c_k\}$ can be obtained as a union of $A := \{c_0 < \ldots < c_{k-2} < c_k\}$ and $B := \{c_0 < \ldots < c_{k-2} < c_{k-1}\}$. Algorithm 8 receives a $(k+1)$-set C as a parameter and computes such sets A and B from it, so that $C = A \cup B$. The first $k-1$ elements of B, shared by A, are denoted by D in the pseudocode.

To get the next candidate set, the algorithm first tries to advance A (in the list $\mathcal{K}$), but only if the intersection of A and B remains equal to D. If successful, it checks whether the union of the new A and B is a candidate set and repeats if necessary. When A reaches B, the algorithm resets A to the lectically first superset of D in $\mathcal{K}$, advances B, and continues. If, on the other hand, D is not a subset of the new A, the algorithm sets $A := B$ before advancing B.

To obtain the $(k+1)$st level of the order ideal, it is sufficient to test for each candidate set if it actually belongs to the order ideal.

As an example, we determine all minimal generating sets of (absolute) support greater than 33 (relative support $> 1\%$) for the formal context in Figure 2.9. For each set C, we compute two parameters: its support, support(C), and

$$\text{predsupp}(C) := \min\{\text{support}(C \setminus \{c\}) \mid c \in C\}.$$

The predsupp parameter gives an easy test for generating sets: predsupp$(C) \neq$ support(C) is the crucial condition, as the following proposition shows.

Proposition 6 *A set with $k+1>0$ elements is a minimal generating set iff all its k-subsets are minimal generating sets of strictly larger support.*

The empty set is the only generating set for level zero, and its support in this example is 3316. All singleton sets are candidate sets for level one. We get, in lectic order,

C	$\{h\}$	$\{g\}$	$\{f\}$	$\{e\}$	$\{d\}$	$\{c\}$	$\{b\}$	$\{a\}$
support(C)	826	386	250	348	123	533	373	2043
predsupp(C)	3316	3316	3316	3316	3316	3316	3316	3316

Observe that all candidate sets have sufficient support and an even higher value of predsupp. Thus, all singleton sets survive the test, and we may proceed to level two. There, it turns out that all two-element subsets of M are candidate sets in this example. We list only those of sufficient support:

C	$\{g,h\}$	$\{f,h\}$	$\{e,h\}$	$\{e,f\}$	$\{d,h\}$	$\{c,h\}$	$\{c,g\}$	$\{c,f\}$
support(C)	77	69	90	36	44	183	35	63
predsupp(C)	386	250	348	250	123	533	386	250
C	$\{c,e\}$	$\{b,h\}$	$\{a,h\}$	$\{a,g\}$	$\{a,f\}$	$\{a,e\}$	$\{a,d\}$	$\{a,c\}$
support(C)	129	60	364	178	92	211	36	212
predsupp(C)	348	373	826	386	250	348	123	533

Since the support in all these cases is less than the value of the predsupp parameter, all these sets are minimal generating sets. We compute the next level from them:

C	$\{c,e,h\}$	$\{a,g,h\}$	$\{a,e,h\}$	$\{a,c,h\}$	$\{a,c,e\}$
support(C)	54	40	39	60	57
predsupp(C)	90	77	90	183	129

Eventually, for level four, there is a single candidate set:

C	$\{a, c, e, h\}$
support(C)	18
predsupp(C)	39

.

However, its support of 18 does not exceed the given threshold 33.

So 29 minimal generating sets of support $\geq 1\%$ were found. It is an immediate consequence of the definition that the support of a set always equals the support of its closure. Thus, all these generating sets generate "frequent" intents. It is however a peculiarity of the present example that these closures are all different. In general, the same closure may be generated by many different minimal generating sets.

The TITANIC algorithm uses the support function for computing closures, thereby avoiding invocations of the closure operator. This is based on the following observation:

Proposition 7 *Let W be a set (of "weights") and let $s : \mathfrak{P}(M) \to W$ be a function satisfying for all $X, Y \subseteq M$*

(*i*) $s(X) = s(X'')$,

(*ii*) $X \subseteq Y$ *and* $s(X) = s(Y)$ *together imply* $X'' = Y''$.

Then, for all $X \subseteq M$, it holds that

$$X'' = X \cup \{m \in M \setminus X \mid s(X) = s(X \cup \{m\})\} .$$

The support function satisfies the conditions of this proposition and may thus be used for computing closures. For example, consider the attribute set $X := \{a, e, g, h\}$ in Figure 2.9. The data necessary for an application of Proposition 7 is the following:

X	$\{a, e, g, h\}$	$\{a, b, e, g, h\}$	$\{a, c, e, g, h\}$	$\{a, d, e, g, h\}$	$\{a, e, f, g, h\}$
support(X)	4	0	4	0	2

As the result, we get

$$\{a, e, g, h\}'' = \{a, c, e, g, h\}.$$

The pseudocode of Algorithm 9 shows the high-level structure of TITANIC. It uses subroutines CLOSURE, WEIGH, and CANDIDATE-SETS. The implementation of CLOSURE can be based on Proposition 7, although additional bookkeeping is needed in order to have the required information about sets of the form $X \cup \{m\}$ and be able to use it efficiently. In particular, the computation of closures

Algorithm 9 TITANIC ((G, M, I), `minsupp`, kmax)

Input: A formal context (G, M, I), a number `minsupp` $\in [0, 1]$, and a positive integer kmax.
Output: The set of all concept intents of relative support $\geq$ `minsupp` that have a generating set of size $\leq$ kmax.

support($\emptyset$) := 1
$\mathcal{K}_0 := \{\emptyset\}$
$\mathcal{C} := \{\{m\} \mid m \in M\}$ in lectic order
for all $X \in \mathcal{C}$ **do**
  predsupp(X) := 1

for $k := 1$ to kmax **do**
  WEIGH($\mathcal{C}$)
  $\mathcal{K}_k := \{X \in \mathcal{C} \mid \text{predsupp}(X) > \text{support}(X) \geq \texttt{minsupp}\}$
  if $\mathcal{K}_k = \emptyset$ **then**
    exit for
  $\mathcal{C}$:= CANDIDATE-SETS($\mathcal{K}_k$)

return $\bigcup_{i=0}^{k}\{\text{CLOSURE}(X) \mid X \in \mathcal{K}_i\}$

Here, support(X) denotes the relative support of a set X, while predsupp(X) is the minimum relative support of all proper subsets of X. With `minsupp` := 0, TITANIC produces all closed sets with a generating set of size $\leq$ kmax. Invocation with kmax $\geq |M|$ yields all "frequent" concept intents, i.e., those of support $\geq$ `minsupp`.

should be done level by level rather than all at once in the end of the algorithm (see [206] for more details).

WEIGH($\mathcal{C}$) determines the relative supports of all $X \in \mathcal{C}$ with *a single pass* of the context: it iterates over $g \in G$ and, for each $X \in \mathcal{C}$, checks whether $X \subseteq \{g\}'$ incrementing the support counter of X if so.[3] This, together with the fact that only $\max\{|X| \mid X \subseteq M \text{ is candidate set}\}$ passes are needed, makes TITANIC suitable for data sets so large that they cannot be kept in the main memory of the computer.

CANDIDATE-SETS($\mathcal{K}_k$) uses Algorithm 8 to produce the list of $(k+1)$-sets that have all their k-element subsets in $\mathcal{K}_k$. The algorithm should be modified to

[3]Note that the same approach can be used to compute the closures of all $X \in \mathcal{C}$: just like WEIGH maintains the support counter, it could maintain a set C_X for the closure of X and intersect C_X with $\{g\}'$ each time X is found to be a subset of $\{g\}'$.

compute, with every candidate set C, the parameter

$$\text{predsupp}(C) := \min\{\text{support}(C \setminus \{c\}) \mid c \in C\}.$$

The intermediate results computed by the TITANIC algorithm, such as the support and predsupp-values and the minimal generating sets themselves, may be of interest too. For example, TITANIC can produce a list of all minimal generating sets that are minimal with respect to not being frequent. But this can be rephrased: such sets are just the minimal non-frequent sets. They will turn out useful later.

2.3.4 Only one of each kind

We begin this subsection with a motivating example. Suppose we want a list of all graphs on four vertices $\{a, b, c, d\}$. There are six possible edges:

$$\{a,b\}, \{a,c\}, \{a,d\}, \{b,c\}, \{b,d\}, \{c,d\},$$

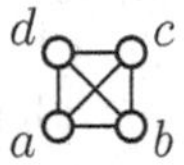

d c
a b

and every choice of edges gives a graph. We may actually apply the NEXT CLOSURE algorithm to generate them all, because the set of all subsets is a closure system. The first results in lectic order would be

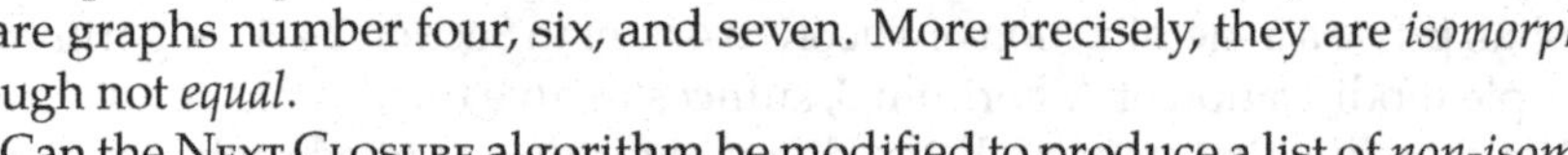

and the total number of graphs is 64.

But this is perhaps not what one wants. Many graphs occur several times in the output. Graphs number two, three, and five are essentially the same, and so are graphs number four, six, and seven. More precisely, they are *isomorphic*, though not *equal*.

Can the NEXT CLOSURE algorithm be modified to produce a list of *non-isomorphic* graphs? Yes, it can, and the modification is surprisingly simple. Moreover, the modified algorithm for *finding closed sets under symmetry* will be useful later.

An **automorphism** of a closure system on a set M is a permutation of M that maps closed sets to closed sets and non-closed sets to non-closed sets. **Context automorphisms**, that is, automorphisms of a formal context (G, M, I), are defined as pairs (α, β) of permutations

$$\begin{aligned} \alpha &: G \to G, \\ \beta &: M \to M, \end{aligned}$$

which are both **incidence preserving** and **incidence reflecting**, meaning that they satisfy

$$g\ I\ m \iff \alpha(g)\ I\ \beta(m).$$

Whenever (α, β) is such a context automorphism and (A, B) is some formal concept of (G, M, I),

$$(\alpha(A), \beta(B)),$$

which is short for

$$(\{\alpha(g) \mid g \in A\}, \{\beta(m) \mid m \in B\}),$$

is a formal concept as well. In this manner, α induces an automorphism of the closure system of extents, and similarly β one of intents.

The totality of all automorphisms of a given closure system forms the **automorphism group** of that system. Often, only a subgroup Γ of the automorphism group is considered. The example below will show why this can be meaningful. Two closed sets are considered **Γ-equivalent** if there is an automorphism in Γ mapping one to the other. The family of all closed sets that are Γ-equivalent to a given one is called its **Γ-orbit**. In a more formal setting, the orbit of A under Γ is denoted by A^Γ and is formally defined as

$$A^\Gamma := \{\alpha(A) \mid \alpha \in \Gamma\}.$$

What we are aiming at is an algorithm that generates exactly one closed set from each orbit. There is a natural choice: the lectic order that we use for the NEXT CLOSURE algorithm orders the sets in each orbit. We shall produce the lectically maximal closed set of each Γ-orbit, the **Γ-orbit-maximal** set.

Theorem 4 *Let $A \mapsto A''$ be a closure operator on M and let Γ be a subgroup of the automorphism group on M. The smallest Γ-orbit-maximal closed set larger than a given set $A \subset M$ is*

$$A \oplus m,$$

m being the largest element of M such that $A <_m A \oplus m$ and $A \oplus m$ is Γ-orbit-maximal.

We shall prove a generalization of Theorem 4 later in Section 2.4.1. Algorithm 10, a simple modification of Algorithm 3, suffices to give the desired result.

We show the first steps of the modified algorithm applied to the graphs on four elements for the case when Γ is the set of all automorphisms on the set of edges induced by graph automorphisms (i.e., by permutations of the vertex set preserving the edge–vertex connectivity). Note that the closure operator is trivial: the closure of any set is the set itself, nothing is added. The lectically smallest graph is the graph with no edges.

$\{a,b\}$	$\{a,c\}$	$\{a,d\}$	$\{b,c\}$	$\{b,d\}$	$\{c,d\}$	
						d∘ ∘c a∘ ∘b

Algorithm 10 NEXT ORBIT-MAXIMAL CLOSURE$(A, M, '', \Gamma)$

Input: A closure operator $X \mapsto X''$ on a finite linearly ordered set M, a subset $A \subseteq M$, and a subgroup Γ of automorphisms.
Output: The lectically next closed Γ-orbit maximal set after A if it exists; $\perp$, otherwise.

for all $m \in M$ in reverse order **do**
  if $m \in A$ **then**
    $A := A \setminus \{m\}$
  else
    $B := (A \cup \{m\})''$
    if $B \setminus A$ contains no element $< m$ **and** B is Γ-orbit-maximal **then**
      return B
return $\perp$

The next solution is obtained by adding one edge (since the closure operator is trivial). The edge must be Γ-orbit-maximal, and so $\{a, b\}$ must be chosen:

$\{a,b\}$	$\{a,c\}$	$\{a,d\}$	$\{b,c\}$	$\{b,d\}$	$\{c,d\}$	
						d c a b
×						

The next possible choice is $B := \{\{a,b\},\{c,d\}\}$, and this is indeed Γ-orbit-maximal.

$\{a,b\}$	$\{a,c\}$	$\{a,d\}$	$\{b,c\}$	$\{b,d\}$	$\{c,d\}$	
						d c a b
×						
×					×	

The largest element that can be added here is $\{b, d\}$, and we have to test

$$\{\{a,b\},\{c,d\}\} \oplus \{b,d\} = \{\{a,b\},\{b,d\}\}$$

for being Γ-orbit-maximal. The result is negative, because the graph automorphism that interchanges a with b and c with d maps $\{\{a,b\},\{b,d\}\}$ to $\{\{a,b\},$

$\{a, c\}\}$, and this is lectically larger:

$$\{\{a, b\}, \{b, d\}\} < \{\{a, b\}, \{a, c\}\}.$$

Indeed, this is a Γ-orbit-maximal set and our next solution.

$\{a, b\}$	$\{a, c\}$	$\{a, d\}$	$\{b, c\}$	$\{b, d\}$	$\{c, d\}$	
						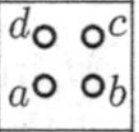 d c a b
×						
×					×	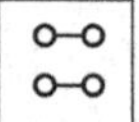
×	×					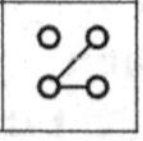

We add an element as large as possible, which is $\{c, d\}$, and obtain . Again this is not Γ-orbit-maximal, because (obtained by interchanging b and c) is lectically larger and is included as a Γ-orbit-maximal solution. The full output of the algorithm is shown in Figure 2.11.

2.4 More about the algorithms

2.4.1 Generalizations and proofs

The Next Closure algorithm actually works not only for closure systems, but for arbitrary families $\mathcal{F}$ of sets. The only requirement is a set operator $\mathcal{L}$ that produces for any given subset $A \subseteq M$ of the linearly ordered base set M the lectically smallest member of $\mathcal{F}$ containing A if it exists. Such an operator $\mathcal{L}$ is, of course, uniquely determined by $\mathcal{F}$, and, in particular, it always exists for any given $\mathcal{F}$. It may, however, be difficult to compute, unless certain conditions are fulfilled. In the case that $\mathcal{F}$ is a closure system, $\mathcal{L}$ is the corresponding closure operator.

We shall now present a rather general version of the algorithm, including a proof. The precise formulations are not difficult, but a little technical. We therefore sketch the main idea in advance. The technicalities occur mainly because we work with sets in the lectic order. If we simplify phrases like

"the lectically smallest set that is larger than A with respect to the lectic order"

to

" the next set after A ",

$\{a,b\}$	$\{a,c\}$	$\{a,d\}$	$\{b,c\}$	$\{b,d\}$	$\{c,d\}$	
						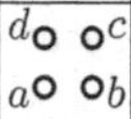 d c a b
×						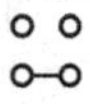
×					×	
×	×					
×	×			×		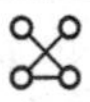
×	×			×	×	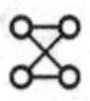
×	×		×			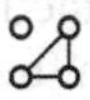
×	×	×				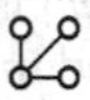
×	×	×	×			
×	×	×	×	×		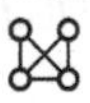
×	×	×	×	×	×	

Figure 2.11: Graphs on four vertices up to isomorphism, as generated by the NEXT CLOSURE algorithm with symmetries

then the basic idea is this: the task to

find for any given set $A \subseteq M$ the next set in $\mathcal{F}$ after A

is easy to solve provided you can

find for any given set $A \subseteq M$ the lectically smallest set in $\mathcal{F}$ containing A.

As before, we assume that the base set $M := \{m_1 < m_2 < \ldots < m_n\}$ is finite and linearly ordered, so that we get an induced lectic order $\leq$ on the power set $\mathfrak{P}(M)$. We will use $\perp$ to denote a distinguished element outside M. It is convenient to include $\perp$ in the lectic order as the largest element, i.e., to assume $A < \perp$ for all $A \subseteq M$.

Let $\mathcal{F} \subseteq \mathfrak{P}(M)$ be a family of subsets of M and let $\mathcal{L} : \mathfrak{P}(M) \to \mathfrak{P}(M) \cup \{\bot\}$ be the operator that maps each subset X of M to the lectically smallest set in $\mathcal{F}$ containing X if such a set exists and to $\bot$ otherwise. Define

$$A \oplus m_i := \mathcal{L}((A \cap \{m_1, \ldots, m_{i-1}\}) \cup \{m_i\}).$$

Proposition 8

1. *If $A <_{m_i} B$, then $(A \cap \{m_1, \ldots, m_{i-1}\}) \cup \{m_i\} \subseteq B$.*
2. *If $A <_{m_i} B$ and $B \in \mathcal{F}$, then $A \oplus m_i \leq B$.*
3. *If $m_i \notin A$, then $A < A \oplus m_i$.*
4. *If $A <_{m_i} A \oplus m_i$, $A <_{m_j} A \oplus m_j$, and $i < j$, then $A \oplus m_j < A \oplus m_i$.*

Proof *1.* is immediate from the definition of the $<_m$ relation. *2.* follows, observing *1.*, from the definition of $A \oplus m_i$. *3.* is trivially true when $A \oplus m_i = \bot$, but also if $A \oplus m_i$ is a subset of M, because all elements of A that are less than m_i are contained in $A \oplus m_i$. *4.* is a direct consequence of Proposition 4. □

Theorem 5 *The lectically smallest element in $\mathcal{F}$ that is larger than A, if such an element exists, is $A \oplus m_i$, where m_i is the largest element of M satisfying $A <_{m_i} A \oplus m_i$.*

Proof Suppose that B is the lectically smallest set in $\mathcal{F}$ larger than A. We have $A < B$ and thus $A <_{m_i} B$ for some $m_i \in M$. From Proposition 8, we know that B contains $(A \cap \{1, \ldots, m_{i-1}\}) \cup \{m_i\}$. But since $A \oplus m_i$ is the lectically smallest element of $\mathcal{F}$ containing $(A \cap \{1, \ldots, m_{i-1}\}) \cup \{m_i\}$, we get $A < A \oplus m_i \leq B$, which forces $B = A \oplus m_i$. So we have proved that the lectically smallest element in $\mathcal{F}$ greater than A, if such a set exists, is of the form $A \oplus m_i$ for some $m_i \in M$ satisfying $A <_{m_i} A \oplus m_i$. Of course, if $A <_{m_i} A \oplus m_i$, then $A \oplus m_i$ is lectically larger than A and is in $\mathcal{F}$ anyway. It may happen that there are two such elements, say, $m_j < m_i$, with $A <_{m_i} A \oplus m_i$ and $A <_{m_j} A \oplus m_j$. As stated in Proposition 8, $A \oplus m_i$ is lectically smaller. This completes the proof of the theorem. □

An interesting case is when the family $\mathcal{F}$ of sets is part of a closure system in a particular way. The next theorem shows that we may then use the closure operator instead of the operator $\mathcal{L}$ occurring in Theorem 5.

Theorem 6 *Let $X \mapsto X''$ be a closure operator on $M := \{m_1 < \ldots < m_n\}$ and let, for $A \subseteq M$ and $m_i \in M$,*

$$A \oplus m_i := ((A \cap \{m_1, \ldots, m_{i-1}\}) \cup \{m_i\})''.$$

Now let $\mathcal{F}$ be a family of closed sets with the property

$$A \in \mathcal{F} \text{ and } 1 \leq i \leq n \quad \Rightarrow \quad (A \cap \{m_1, \ldots, m_{i-1}\})'' \in \mathcal{F}.$$

Then we obtain for every subset $A \subseteq M$ the lectically next set in $\mathcal{F}$ (if it exists) as

$$A \oplus m_i,$$

where m_i is the largest element of M for which $A <_{m_i} A \oplus m_i$ and $A \oplus m_i \in \mathcal{F}$.

Proof Suppose that A is not the largest element of $\mathcal{F}$, and let B be the next set in $\mathcal{F}$ after A. Then $A <_{m_i} B$ for some i. Since B is a closed set containing $(A \cap \{m_1, \ldots, m_{i-1}\}) \cup \{m_i\}$, it also contains the closure $A \oplus m_i$ of this set. If $B \neq A \oplus m_i$, there must be some smallest $m_j \in B \setminus (A \oplus m_i)$, and $A <_{m_i} B$ forces $m_j > m_i$. Since m_j was chosen minimal, we get $B \cap \{m_1, \ldots, m_{j-1}\} \subseteq A \oplus m_i$, and, furthermore,

$$(B \cap \{m_1, \ldots, m_{j-1}\})'' = A \oplus m_i.$$

But then $(B \cap \{m_1, \ldots, m_{j-1}\})'' \in \mathcal{F}$ and

$$A <_{m_i} (B \cap \{m_1, \ldots, m_{j-1}\})'' <_{m_j} B,$$

contradicting the assumption that B is the first element of $\mathcal{F}$ after A. Thus we know that the next $\mathcal{F}$-set after A must be of the form $A \oplus m_i$, and, according to Proposition 8, we must choose m_i as large as possible. □

Two instances of Theorem 6 have already been presented without a proof:

- Theorem 3, where it is assumed that $\mathcal{F}$ is an order ideal of closed sets. The condition from Theorem 6 is obviously fulfilled, because the closure of any subset of $A \in \mathcal{F}$ is contained in A and therefore is in $\mathcal{F}$.
- Theorem 4, where $\mathcal{F}$ is the family of Γ-orbit-maximal closed sets. To show that this family satisfies the condition of Theorem 6, we prove Proposition 9. This proposition states that when $(A \cap \{m_1, \ldots, m_i\})''$ is not Γ-orbit-maximal, A cannot be one either. By contraposition, we get that if A is Γ-orbit-maximal, then so is $(A \cap \{m_1, \ldots, m_i\})''$, as required in Theorem 6.

Proposition 9 *If A is a closed set and $B := (A \cap \{m_1, \ldots, m_{i-1}\})''$, then*

$$B < \alpha(B) \Rightarrow A < \alpha(A)$$

holds for every automorphism α of the closure system.

Proof $B < \alpha(B)$ implies $B <_{m_j} \alpha(B)$ for some j. For $i \leq j$, this would mean

$$B \cap \{m_1, \ldots, m_j\} \subseteq \alpha(B),$$

which, because of

$$B = (B \cap \{m_1, \ldots, m_{i-1}\})'',$$

would imply $B \subseteq \alpha(B)$. Since $|B| = |\alpha(B)|$, we obtain $B = \alpha(B)$, contradictory to $B < \alpha(B)$. Hence, $j < i$ must hold. From the definition of B, we get

$$A \cap \{m_1, \ldots, m_{i-1}\} = B \cap \{m_1, \ldots, m_{i-1}\},$$

which, since $j < i$, implies

$$A \cap \{m_1, \ldots, m_{j-1}\} = B \cap \{m_1, \ldots, m_{j-1}\}.$$

Moreover, since $m_j \notin B$, we have $m_j \notin A$. Combining all this, we obtain

$$\begin{aligned} A \cap \{m_1, \ldots, m_{j-1}\} &= B \cap \{m_1, \ldots, m_{j-1}\} \\ &= \alpha(B) \cap \{m_1, \ldots, m_{j-1}\} \\ &\subseteq \alpha(A), \end{aligned}$$

which, together with $m_j \in \alpha(A)$, proves $A < \alpha(A)$. □

2.4.2 The order relation and the lattice diagram

The closed sets of any closure operator form, as was mentioned above, a complete lattice. Several implementations of the Next Closure algorithm offer lattice diagrams, automatically or semi-automatically drawn from data. For that, the order relation of the lattice is needed, more precisely, the *neighborhood relation*, which was introduced in Definition 6 (p. 24) for concept lattices.

There are several options for computing the neighborhood relation. One possibility is to determine for each closed set $C \subseteq M$ a set $N \subseteq M \setminus C$ such that each $n \in N$ corresponds to an upper neighbor $(C \cup \{n\})''$ of C and each upper neighbor is obtained from exactly one such $n \in N$. Such a set is very easy to determine; Algorithm 11 does just that.

For closure systems of reasonable size, it may be advisable to store a list of all closed sets and perform binary search to determine the addresses of each upper neighbor in the list. The lectic order in which Next Closure generates closed sets makes it straightforward. Algorithm 12 can find the index of any upper neighbor C of a closed set F_m in the list produced by Next Closure. Because binary search is used, the loop requires at most $\log_2 n = O(|M|)$ iterations, and the complexity of the algorithm is thus $O(|M|^2)$.

Algorithm 11 UPPER NEIGHBORS($C, M, ''$)

Input: A closure operator $X \mapsto X''$ on a finite set M and a set $C = C'' \subseteq M$.
Output: A minimum-size set $N \subseteq M$ such that $\{(C \cup \{n\})'' \mid n \in N\}$ is the family of upper neighbors of C.

$N := M \setminus C$
for all $m \in M \setminus C$ **do**
    if $(C \cup \{m\})'' \cap N \neq \{m\}$ **then**
        $N := N \setminus \{m\}$
return N

Algorithm 12 FIND INDEX($C, m, \mathcal{F}$)

Input: The list $\mathcal{F} = [F_1, \ldots, F_n]$ of closed sets of a closure operator on a finite set M in the lectic order, an index m of a closed set in $\mathcal{F}$, and a closed set $C \in \mathcal{F}$, where $F_m \subset C$.
Output: The index k of C in $\mathcal{F}$.

$i := m + 1$
$j := n$
repeat
    $k := \lfloor (i + j)/2 \rfloor$
    if $F_k < C$ **then**
        $i := k + 1$
    else
        $j := k - 1$
until $C = F_k$
return k

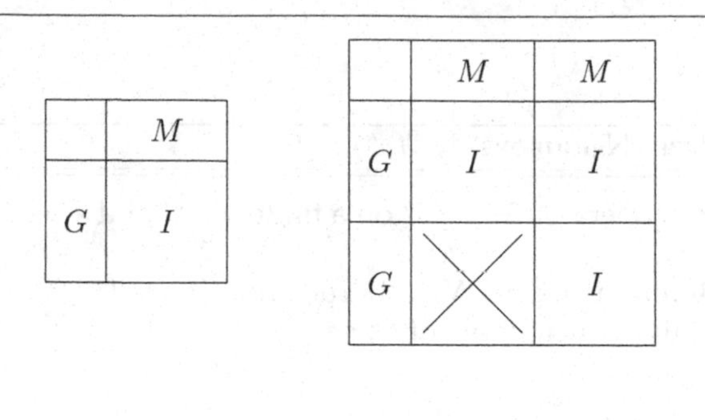

Figure 2.12: A formal context for the order relation

The *order relation* is not so often needed, but is also easy to compute for concept lattices. It can be obtained by means of a simple context construction depicted in Figure 2.12 and more formally described in the next proposition. The context construction used there is the **context product** with the **ordinal scale**

	1	2
1		
2	×	

,

see [110].

Proposition 10 *Let (G, M, I) be a formal context. The formal concepts of the formal context*

$$(G \times \{1,2\}, M \times \{1,2\}, \nabla),$$

where

$$(g,i) \nabla (m,j) :\iff g\, I\, m \text{ or } (i = 2 \text{ and } j = 1),$$

are precisely the pairs

$$(A \times \{1\} \cup C \times \{2\}, B \times \{1\} \cup D \times \{2\}),$$

for which (A, B) and (C, D) are formal concepts of (G, M, I) satisfying

$$(A, B) \leq (C, D).$$

A small example is presented in Figures 2.13 and 2.14. As an example of how to read the diagrams, consider in Figure 2.14 the rightmost element in the middle layer. In the left diagram, this element is labeled (b, c). It reflects the fact that $b \leq c$ holds in the lattice in Figure 2.13 and therefore (b, c) is an element of the order relation. In the right diagram, this element represents a formal concept with the extent $\{(b,1),(b,2),(c,2)\}$ and the intent $\{(b,1),(c,1),(c,2)\}$. This represents the pair

$$((\{b\},\{b,c\}),(\{b,c\},\{c\}))$$

of formal concepts of the small context in the middle of Figure 2.13, and these are indeed the lattice elements labeled b and c in the small lattice.

2.4.3 Computing closures in the reverse order

For any closed set, we know how to determine the lectically next closed set. What about the previous closed set? Theorem 2 tells us that the set following A is of the form $A \oplus m$, where m is an element satisfying certain properties. So, the first step in generating the lectically previous set could be to identify m given $A \oplus m$. Proposition 11 allows us to do this.

For convenience, we denote $X|_m = \{n \in X \mid n < m\}$.

Proposition 11 *Let A be a closed set, B be the lectically next closed set, and m be the largest element such that $A <_m A \oplus m$. Then m is also the largest element such that $m \in B$ and $m \notin (B|_m)''$.*

Proof $B = A \oplus m$ by Theorem 2 and $m \in B$ by the definition of $\oplus$. Note also that $B|_m \subseteq A$. If $m \in (B|_m)''$, then $(B|_m \cup \{m\})'' = (B|_m)''$. This leads to $B \subseteq A$ – for $(B|_m)'' \subseteq A$ and $B = (B|_m \cup \{m\})''$ – which is a contradiction.

For all $n \in B$ such that $n > m$, we have $B|_m \cup \{m\} \subseteq B|_n$ and, consequently, $n \in (B|_n)'' = B$. Therefore, m is indeed the largest element in B with the desired property. □

This suggests how m could be found: iterate over all elements of B in the reverse order until an element is found that is not contained in the closure of the corresponding prefix of B.

The elements before m are exactly the same in A and B, but there may be other elements in A to the right of m.

Proposition 12 *Let $m \in M$, A be a closed set, and $B = A \oplus m$ be the lectically next closed set. Then*

1. $A|_m = B|_m$,
2. $m \notin A$,
3. *an element $n > m$ is in A if and only if $m \notin A \oplus n$ and $(A \oplus n)|_m = A|_m$.*

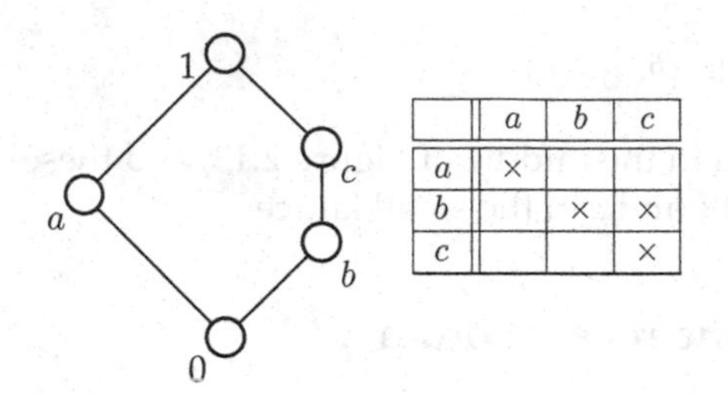

	a	b	c
a	×		
b		×	×
c			×

	$(a,1)$	$(b,1)$	$(c,1)$	$(a,2)$	$(b,2)$	$(c,2)$
$(a,1)$	×			×		
$(b,1)$		×	×		×	×
$(c,1)$			×			×
$(a,2)$	×	×	×	×		
$(b,2)$	×	×	×		×	×
$(c,2)$	×	×	×			×

Figure 2.13: A small lattice, its standard context, and the formal context for its order relation, according to Figure 2.12. See also Figure 2.14 below

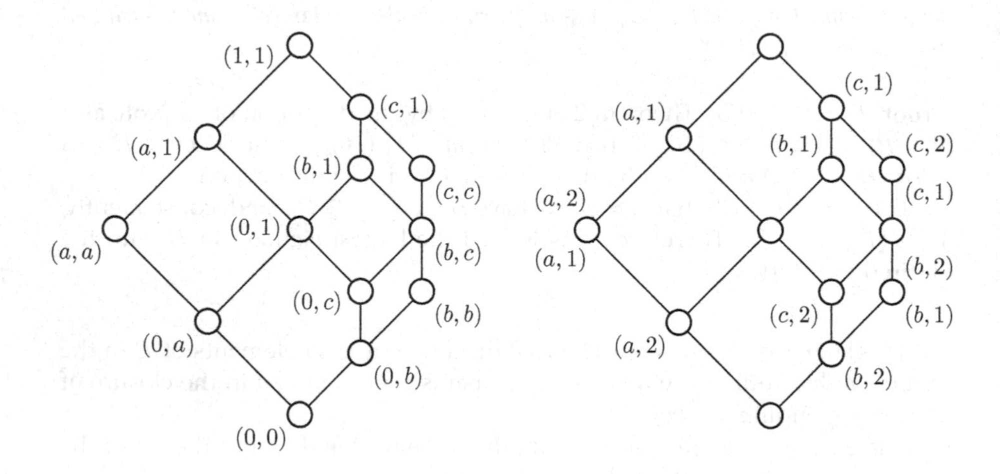

Figure 2.14: The order relation of the lattice in Figure 2.13 as a lattice. The two diagrams differ only in their labels. In the left diagram, each element is labeled as an ordered pair of elements of the small lattice from Figure 2.13. The other diagram is labeled as the concept lattice of the larger formal context in Figure 2.13

Proof The first two items directly follow from the definition of $\oplus$ and Theorem 2. The third item says that none of the attributes greater than m should generate new attributes that are less than or equal to m, which is obvious, and that all attributes that do not do this are in A, which might need an explanation. Suppose that there are elements violating this condition and let n be the first of them, i.e., the first element not in A such that $n > m$, $m \notin A \oplus n$, and $(A \oplus n)|_m = A|_m$. First, we show that $(A \oplus n)|_n = A|_n$. If this is not the case, then $(A \oplus n)$ contains an element k such that $m < k < n$ and $k \notin A$. From our choice of n, it follows that $m \in A \oplus k$ or $(A \oplus k)|_m \neq A|_m$. Together with $k \in A \oplus n$, this makes impossible, respectively, $m \notin A \oplus n$ or $(A \oplus n)|_m = A|_m$ and, thus, leads to a contradiction.

Now, it can be easily seen that $A <_n A \oplus n$, but according to Theorem 2, m must be the largest element with this property, which contradicts $n > m$. □

Proposition 12 describes all the attributes in A both to the left of m and to the right of m (and also says that m is not in A). This suggests Algorithm 13 for computing the lectically previous closed set for a given closed set B.

Algorithm 13 Previous Closure($B, M, ''$)

Input: A closure operator $X \mapsto X''$ on a finite linearly ordered set M and a closed subset $B \subseteq M$.
Output: The lectically previous closed set before B if it exists; $\bot$, otherwise.

```
for all m ∈ B in reverse order do
    B := B \ {m}
    if m ∉ B'' then
        for all n ∈ M such that n > m do
            if n ∉ B then
                C := (B ∪ {n})''
                if C \ B contains no element ≤ m then
                    B := C
        return B
return ⊥
```

2.4.4 Other algorithms

Next Closure is a general algorithm in the sense that it can be used for enumerating closed sets no matter how the closure operator is specified. For the case when this operator is specified by a formal context, many other algorithms have been proposed. Here, we briefly mention a few referring to [115, 119, 144] for more detailed comparative analysis.

One of the advantages of NEXT CLOSURE is that, since it does not have to store the closures it computes, it requires very little memory. The downside of using this approach for generating intents or extents of a formal context is that the algorithm has to do quite a bit of repetitive work. For example, every concept extent besides G is the intersection of a larger concept extent and an attribute extent. Thus, if an algorithm could store previously generated concepts, it would be possible to produce a new extent by applying a single intersection operation (assuming that attribute extents are pre-computed). For NEXT CLOSURE, it means more efficient computation of

$$B := (A \cup \{m\})''.$$

Normally, computing $(A \cup \{m\})''$ would first require computing $(A \cup \{m\})'$ by intersecting the extents of all attributes from $A \cup \{m\}$. However, if A', the extent of A, is already known, the same can be achieved by simply intersecting A' and $\{m\}'$. For a large A, this can make a difference from the efficiency point of view.

There are several ways to implement this idea. It is possible to store all concepts in a hash table or a trie with intents as keys. Or one could introduce a stack into NEXT CLOSURE, which would contain at most $|M|$ (or $|G|$ for the NEXT EXTENT version) concepts at every single moment. This is clearly more economical in terms of memory than the hash table solution. A similar approach is implemented in the CLOSE BY ONE algorithm [141], where a tree rather than a stack is used for storing concepts, which helps efficiently constructing the neighborhood relation. If the neighborhood relation is not required, a stack (corresponding to one branch of the tree) or a recursive implementation as in [144] is more appropriate. This optimization does not affect the asymptotic complexity of the algorithm, which remains the same as that of NEXT CLOSURE. A further optimization of CLOSE BY ONE (which can, in a sense, be extended to NEXT CLOSURE) was proposed in [174].

In some cases, it becomes necessary to update an already constructed lattice by introducing a new object or attribute into the context. Incremental algorithms can do this without rebuilding everything from scratch. A very simple but efficient incremental algorithm was developed by Norris [168, 167]. Given a lattice and a new object intent, it computes the intersection of every intent already in the lattice and the new object intent. Of course, some of these intersections may be identical. To avoid multiple generation of the same intent, the algorithm uses exactly the same test as in NEXT CLOSURE and CLOSE BY ONE. Having produced D as the intersection of the new object intent $\{g\}'$ and B, the intent of an existing concept (A, B), it checks whether D' equals to $A \cup \{g\}$, and, if not, D gets rejected. (For adding a new attribute, the algorithm works symmetrically.) When used to build the concept set of a formal context from scratch, the algorithm performs essentially the same operations as Close by One, but in a different order. The disadvantage of this order is that it makes it necessary to have access to all generated concepts for the algorithm to proceed. An obvious advantage is the ability to update the concept set incrementally.

Another incremental algorithm was proposed in [115]. Its main difference is how duplicate concepts are detected: instead of the test described above, a hash function is used to partition concepts into relatively small buckets and then simply search for a duplicate of a newly generated concept in the corresponding bucket. The asymptotic complexity of this algorithm depends on the hash function. If, for example, the function used is the cardinality of the concept intent as in [115], the algorithm becomes quadratic in the number of concepts in the worst case.

As far as we know, the algorithm developed by Nourine and Raynaud [169] currently features the best worst-case computational complexity among all lattice construction algorithms: $O((|G|+|M|)|G|)$ per concept. This is achieved by combining the basic incremental strategy with a smart data structure, a trie (also known as prefix tree), for fast concept lookup. Despite the optimistic theoretical complexity estimate, experimental tests suggest that it might not be that efficient in practice [144], but it is possible that the algorithm is still waiting for a good implementation.

Bordat's algorithm [55] constructs the lattice level-wise starting with the top concept and generating lower neighbors for already computed concepts. It uses a tree to search for duplicate concepts. The version of the algorithm described in [144] involves a different, kind of cache-based, strategy which avoids tree maintenance and makes it possible to check whether a concept has already been generated in time $O(|M|)$. The complexity of this modified version is $O(|G||M|^2)$ per concept. Because of the way the algorithm builds the lattice, constructing the neighborhood relation comes naturally and does not take much time (and does not affect the asymptotic complexity).

An incremental variation of Bordat's algorithm was proposed by Ferré [85]. From experiments, it seems that Ferré's algorithm is particularly suited for very sparse contexts (i.e., such where $|I|$ is significantly less than $|G|\times|M|$), whereas Bordat's algorithm is more appropriate for contexts of average density. For dense contexts, the algorithm of Norris and Close by One are quite good.

AddIntent [162] differs from other incremental algorithms in its intensive usage of the neighborhood relation, which is built in the process. Although maintaining the neighborhood relation involves a certain computational overhead, the advantages of having access to the information it represents sometimes allow AddIntent to be faster than most other algorithms. When constructing the neighborhood relation is part of the task, AddIntent seems to be a natural choice.

Among other work on computing concept lattices, it is worth mentioning the "divide and conquer" approach developed by Valtchev *et al.* and aimed at lattice construction based on the apposition of binary relation fragments [212]. This is a generalization of the incremental approach: instead of adding a new object intent to an already constructed lattice, the task is now to merge two concept lattices over the same object or attribute set [211].

One of the most widely used techniques in data analysis is association rule mining, which involves search for frequent closed item sets [1] discussed in Section 2.3.2. Recall that these are exactly concept intents such that corresponding extents have size equal to or above a given threshold. Many algorithms have been developed for finding such item sets within the association rule mining community [116]. These algorithms are usually based on rather different intuitions than the algorithms mentioned above. A notable exception is TITANIC [206], which was developed for constructing iceberg lattices (see Section 2.3.3), but was inspired by APRIORI, a well-known algorithm for finding frequent item sets [2]. Obviously, these algorithms can be used to compute the set of all concept intents: it is enough to set the threshold at zero. On the other hand, many concept lattice construction algorithms can be easily adapted to computing frequent closed item sets for a given threshold. A systematic theoretical and experimental comparison of the two families of algorithms would be in order.

2.5 Further reading

Closure systems were introduced by E. H. Moore [165]. In his honour they are also called **Moore families**. The number of Moore families on an n-element set (see Section 2.1.4) can be computed for $n \leq 6$ using Algorithm 1, and the number of non-isomorphic Moore families (A193674 in the On-Line Encyclopedia of Integer Sequences, `oeis.org`) can be computed using Algorithm 10. For $n = 7$ the number of closure systems is too large to be found by an enumeration algorithm, and a structural analysis [65] was necessary to obtain the result of [64]. The asymptotic size is $\mathcal{O}(2^{\binom{n}{\frac{n}{2}}})$ [4, 59].

The NEXT CLOSURE algorithm and some of its generalizations are due to the first author and his coauthors, see [92], [94], [139], [105], and [127]. References to some of the many other algorithms are given in Section 2.4.4 above.

Hypotheses, as defined in Exercise 2.7, are related to a Formal Concept Analysis-based approach to supervised machine learning; see [103] for references.

2.6 Exercises

2.1 Compute all formal concepts of the formal context in Figure 2.15.

	car	*hobby*	*guests*	*sewing*	*entertainment*	*walking*
170		×	×		×	×
195		×	×			
290						
410					×	×
431			×		×	×
456		×				
459			×		×	×
466			×	×		×
467			×		×	×
481		×	×	×	×	×

Figure 2.15: A formal context describing how respondents of a questionnaire tend to spend their free time

2.2 Draw the line diagram of the concept lattice from Exercise 2.1.

2.3 Suppose that nodes of the diagram in Figure 2.16 correspond to employees in some organization and the diagram shows who reports to whom: those from below report to those from above, e.g., a reports to b and e, b reports to d, etc. Define the closure operator $\phi(X)$ as follows: it maps $X \subseteq M = \{a, \ldots, n\}$ to the set of all employees in M who either belong to X or report – directly or indirectly – to some employees from X (in other words, to the set of employees collectively controlled by the members of X). For example,

$$\phi(\{b, f, k, l\}) = \{a, b, c, d, e, f, g, h, i, k, l\}$$

Prove that ϕ is indeed a closure operator. Find the upper neighbors of $\phi(\{b, f, k, l\})$ in the closure system defined by ϕ, i.e., closed sets immediately above $\phi(\{b, f, k, l\})$ in the line diagram of the closed sets. You do not have to build the line diagram; you may want to use the algorithm Upper Neighbors instead.

2.4 Interpret the graph below as a fragment of an online social network with edges connecting "friends". Let $\mathcal{F}$ be the collection consisting of *maximal* groups of people such that every two different members of a group are either friends or have at least one common friend.

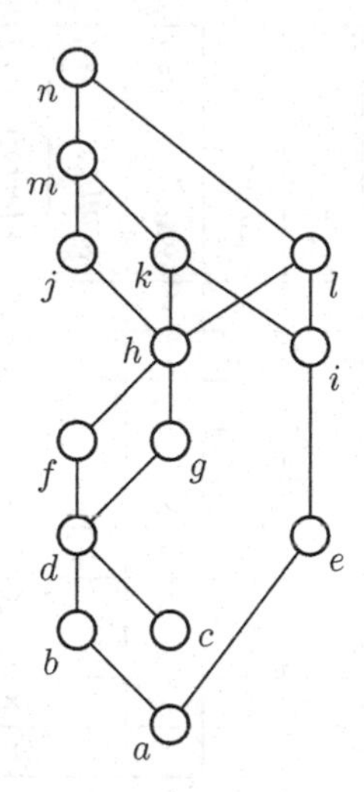

Figure 2.16: With reference to Exercise 2.3

List the members of $\mathcal{F}$. Is $\mathcal{F}$ a closure system? If not, add as few sets to $\mathcal{F}$ as possible to make it a closure system.

2.5 Consider the following naïve algorithm computing all the concepts of $\mathbb{K} = (G, M, I)$:

for all $A \subseteq G$ **do**
  if $A = A''$ **then**
    output (A, A')

What is its time complexity? Suppose that $|G| = 1000$ and $|M| = 10$ and consider the "transposed" context $\mathbb{K}^{-1} = (M, G, I^{-1})$. Which context, $\mathbb{K}$ or $\mathbb{K}^{-1}$, has more concepts? On which of the two contexts the naïve algorithm will spend more time?

2.6 Let $X \mapsto X''$ be a closure operator on M and let P and Q be sets such that $P \subseteq Q \subseteq M$. The task is to find all closed sets C satisfying $P \subseteq C$ and $Q \nsubseteq C$.

What linear order should be chosen on M so that NEXT CLOSURE produces all such sets consecutively, and what set should the algorithm be started with?

2.7 Let $\mathbb{K}_+ := (G_+, M, I_+)$ and $\mathbb{K}_- := (G_-, M, I_-)$ be two contexts with the same finite attribute set M. The objects of $\mathbb{K}_+$ are positive examples of some property p (e.g., they are all poisonous mushrooms), and the objects of $\mathbb{K}_-$ are negative examples of the same property (say, they are non-poisonous, edible mushrooms). An intent of $\mathbb{K}_+$ that is not contained in any object intent of $\mathbb{K}_-$ is called a (positive) **hypothesis**. Such hypotheses can be used for classification of previously unseen objects: if a description of a new object contains a positive hypothesis, it may be suggested that the object has the property p.

Design an algorithm that produces all hypotheses. Modify the algorithm to produce all inclusion-minimal hypotheses.

Chapter 3

The canonical basis

Have another look at the concept lattice shown in Figure 1.11 (p. 14). The six attributes describe how two unit squares can be placed with respect to each other. Each of the ten objects is a pair of unit squares representing a possible placement. These ten pairs are representatives for an infinite set of possible positions that such pairs of squares may have. It is not stated, but perhaps expected by the reader, that these ten examples cover all possible combinations of the given attributes.

Such a situation occurs often: attributes are given, but objects are not known or are too many to enumerate them explicitly. We then have to study the feasible attribute combinations, the *attribute logic* of the respective situation. Whenever possible we restrict ourselves to a certain fragment of this logic, which can be expressed by (material) *implications*.

We will show that the implications of a formal context describe precisely the system of its concept intents. We will also explain why implications are particularly easy to handle. Proofs and further details can be found in [110].

3.1 Implications

Let M be some set. We call the elements of M *attributes*, so as if we consider a formal context (G, M, I). However, we do not require such a context to be given or specified. An **implication between attributes** in M is a pair of subsets of M, denoted by $A \to B$. The set A is the **premise** of the implication $A \to B$, and B is its **conclusion**. We also say that $A \to B$ is an implication **over** M.

A subset $T \subseteq M$ **respects** an implication $A \to B$ if $A \not\subseteq T$ or $B \subseteq T$. We then also say that T is a **model** of the implication $A \to B$ and denote this by $T \models A \to B$. T **respects a set** $\mathcal{L}$ of implications (T is a **model** of $\mathcal{L}$, $T \models \mathcal{L}$) if T respects every single implication in $\mathcal{L}$. The implication $A \to B$ **holds** in a set $\{T_1, T_2, \ldots\}$ of subsets of M if each of these subsets respects $A \to B$. $\mathrm{Imp}\{T_1, T_2, \ldots\}$ denotes the set of all implications that hold in $\{T_1, T_2, \ldots\}$.

3.1.1 Implications of a formal context

We say that $A \to B$ **holds in a context** (G, M, I) if every object intent respects $A \to B$, that is, if each object that has all the attributes from A has all the attributes from B as well. We then also say that $A \to B$ is a *(valid) implication of* (G, M, I).

Proposition 13 *An implication $A \to B$ holds in (G, M, I) if and only if $B \subseteq A''$, which is equivalent to $A' \subseteq B'$. It then automatically holds in the set of all concept intents as well.*

An implication $A \to B$ holds in (G, M, I) if and only if each of the implications

$$A \to m, \qquad m \in B,$$

holds ($A \to m$ is short for $A \to \{m\}$). We can read this off from a concept lattice diagram in the following manner: $A \to m$ holds if the infimum of the attribute concepts corresponding to the attributes in A is less general than or equal to the attribute concept of m; formally, if

$$\bigwedge\{\mu a \mid a \in A\} \leq \mu m.$$

Example 5 Consider the context of triangles in Figure 3.1, which was already discussed in Section 1.4. It is easy to see that the implication $\{a, b\} \to \{c\}$ holds in this context: the only object that has both attributes a and b also has attribute c or, slightly more formally,

$$\{T_4\} = \{a, b\}' \subseteq \{c\}' = \{T_3, T_4, T_6\}.$$

This implication can be interpreted as follows: *every equilateral isosceles triangle is acute-angled*. Note that the implication $\{a\} \to \{c\}$ also holds, which is not surprising, since $\{a\} \to \{b\}$ is a valid implication, i.e., *every equilateral triangle is* trivially *isosceles*. On the other hand, the implication $\{b\} \to \{c\}$ does not hold, because

$$\{T_1, T_2, T_4, T_6\} = \{b\}' \nsubseteq \{c\}' = \{T_3, T_4, T_6\},$$

or, in words, *not every isosceles triangle is necessarily acute-angled*. The right-angled isosceles triangle T_2 is one counterexample.

Example 6 Figure 3.1 also shows a line diagram of the concept lattice of the triangles. To verify that the implication $\{a, b\} \to \{c\}$ holds, we need to find the infimum of the attribute concepts of a and b, which is, in fact, the attribute concept of a, and observe that it is below the attribute concept of c. On the other hand, the attribute concept of b is not below the attribute concept of c and, thus, the implication $\{b\} \to \{c\}$ does not hold.

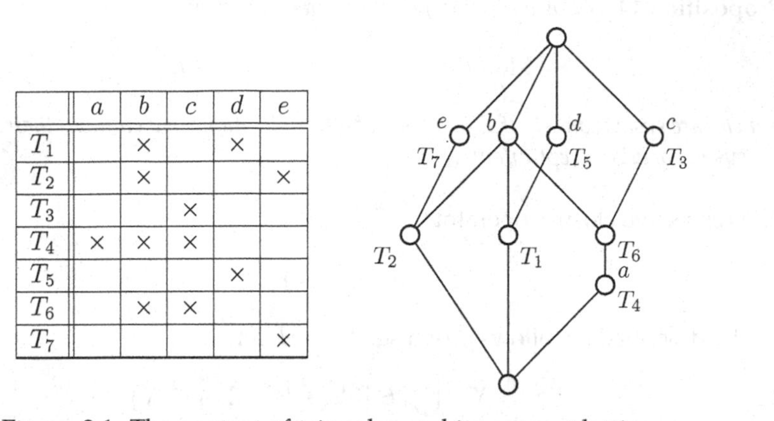

	a	b	c	d	e
T_1		×		×	
T_2		×			×
T_3			×		
T_4	×	×	×		
T_5				×	
T_6		×	×		
T_7					×

Figure 3.1: The context of triangles and its concept lattice

Example 7 More generally, $A \to B$ holds in (G, M, I) if

$$\bigwedge\{\mu a \mid a \in A\} \leq \bigwedge\{\mu b \mid b \in B\}.$$

As an example, consider the implication $\{a\} \to \{b, c\}$. The infimum of the attribute concepts of b and c is the node labelled by T_6, and this node is above the node labelled by a. Hence, the implication is valid.

There is a special case that sometimes causes confusion and for which we introduce extra notation. Consider again the example of triangles shown in Figure 3.1. The implication

$$\{c, e\} \to \{a, b, c, d, e\}$$

holds in this example for the trivial reason that there is no object intent containing the premise $\{c, e\}$ and, consequently, no object intent refuting this implication. Implications stating that certain attribute combinations imply *all* other attributes occur when a certain attribute combination is *contradictory*, i.e., when these attributes do not hold together for any object. It is sometimes more suggestive to use the symbol $\perp$ for the set M of all attributes, but only if there is no object g with $g' = M$. Thus we define

$$\perp := M \qquad \text{provided} \qquad M' = \emptyset,$$

and express the above implication as $\{c, e\} \to \perp$.

Proposition 14 *If $\mathcal{L}$ is a set of implications over M, then*

$$\text{Mod}\,\mathcal{L} := \{T \subseteq M \mid T \text{ respects } \mathcal{L}\}$$

is a closure system on M. If $\mathcal{L}$ is the set of all implications of a context, then Mod $\mathcal{L}$ *is the system of all concept intents.*

The respective closure operator

$$X \mapsto \mathcal{L}(X)$$

can be described as follows. For a set $X \subseteq M$, let

$$X^{\mathcal{L}} := X \cup \bigcup\{B \mid A \to B \in \mathcal{L}, A \subseteq X\}.$$

Form the sets $X^{\mathcal{L}}, X^{\mathcal{L}\mathcal{L}}, X^{\mathcal{L}\mathcal{L}\mathcal{L}}, \ldots$ until[1] a set $\mathcal{L}(X) := X^{\mathcal{L}\ldots\mathcal{L}}$ is obtained with $\mathcal{L}(X)^{\mathcal{L}} = \mathcal{L}(X)$. Later on we shall discuss how to do this computation efficiently.

We can also give an equivalent non-iterative definition of $\mathcal{L}(X)$:

$$\mathcal{L}(X) = \bigcap\{Y \mid X \subseteq Y \subseteq M \text{ and } Y \in \text{Mod}\,\mathcal{L}.\}$$

It is not difficult to construct, for any given set $\mathcal{L}$ of implications over M, a formal context such that Mod $\mathcal{L}$ is the set of concept intents of this formal context. In fact, $(\text{Mod}\,\mathcal{L}, M, \ni)$ will do.

3.1.2 When does an implication follow from other ones?

Definition 10 An implication $A \to B$ **follows (semantically)** from a set $\mathcal{L}$ of implications over M (notation: $\mathcal{L} \models A \to B$) if each subset of M respecting $\mathcal{L}$ also respects $A \to B$. A family of implications is called **closed** if every implication following from $\mathcal{L}$ is already contained in $\mathcal{L}$. Closed sets of implications are also called **implication theories**. A set $\mathcal{L}$ of implications of (G, M, I) is called **complete** with respect to a closure operator if every implication that holds in the associated closure system follows from $\mathcal{L}$. A set $\mathcal{L}$ is *complete* with respect to (G, M, I) if it is complete with respect to the associated closure operator on M or, equivalently, if every implication that holds in (G, M, I) follows from $\mathcal{L}$.

In other words, an implication $A \to B$ follows semantically from $\mathcal{L}$ if it holds in every model of $\mathcal{L}$.

[1] If both $\mathcal{L}$ and M are infinite, this may require infinitely many iterations and the usual transfinite constructions.

Example 8 Consider again the context of triangles in Figure 3.1. Below is a complete set $\mathcal{L}_\triangle$ of implications for this context:

$$\{a\} \to \{b,c\} \qquad \{c,e\} \to \{a,b,d\}$$
$$\{d,e\} \to \{a,b,c\} \qquad \{c,d\} \to \{a,b,e\}$$

Although it is complete (this might be a bit tricky to prove), it is not closed, since it does not contain, for example, the implication $\{a,b\} \to \{c\}$, which, as we saw earlier, holds in the context. However, it is trivial to see that every subset of M respecting $\{a\} \to \{b,c\}$ must respect $\{a,b\} \to \{c\}$; hence, this implication follows from the above set.

Another implication that follows from $\mathcal{L}_\triangle$ is $\{a,e\} \to \{d\}$. Indeed, suppose that A is a model of $\mathcal{L}_\triangle$. If $\{a,e\} \not\subseteq A$, then A is obviously a model of $\{a,e\} \to \{d\}$. If $\{a,e\} \subseteq A$, then $c \in A$ because of the implication $\{a\} \to \{b,c\}$. But then the implication $\{c,e\} \to \{a,b,d\}$ guarantees that $d \in A$ and, thus, A is a model of $\{a,e\} \to \{d\}$. One can check that this implication holds in the context of triangles, simply because there are no objects that have both attributes a and e, i.e., there are no triangles that are at the same time equilateral and right-angled.

The semantic definition of implication inference has a syntactic counterpart. We can give a sound and complete system of inference rules (known as **Armstrong rules**[2] [15]) and an efficient algorithm for inference testing.

Proposition 15 *A set $\mathcal{L}$ of implications over M is closed if and only if the following conditions*[3] *are satisfied for all $W,X,Y,Z \subseteq M$ and, for an arbitrary index set T, for all $Y_t, t \in T$:*

1. $X \to X \in \mathcal{L}$,
2. *If* $X \to Y \in \mathcal{L}$, *then* $X \cup Z \to Y \in \mathcal{L}$,
3. *If* $X \to Y \in \mathcal{L}$ *and* $Y \cup Z \to W \in \mathcal{L}$, *then* $X \cup Z \to W \in \mathcal{L}$.
4. *If* $X \to Y_t \in \mathcal{L}$ *for all* $t \in T$, *then* $X \to \bigcup_{t\in T} Y_t \in \mathcal{L}$.

Sketch of the **Proof** : It is straightforward to show that every closed set of implications satisfies the conditions of the proposition. For the other direction, let $\mathcal{L}$ be a family of implications satisfying the conditions, and let $\mathcal{C}$ be the family of all sets that respect all implications in $\mathcal{L}$. The implications that hold in $\mathcal{C}$ form a closed set $\mathcal{L}_0$ containing $\mathcal{L}$. Suppose $X \to Y_0$ is in $\mathcal{L}_0$, and let

$$\{X \to Y_t \mid t \in T\}$$

[2]The Armstrong rules have been developed for functional dependencies, but these are closely related to implications, even though they have different semantics.
[3]The fourth condition is needed only for the infinite case.

be the set of all implications in $\mathcal{L}$ that have premise X, assuming $0 \notin T$. By the first condition, this set contains $X \to X$, and by the fourth we conclude that

$$X \to \bigcup_{t \in T} Y_t =: Y$$

is in $\mathcal{L}$ as well. If $Y_0 \subseteq Y$, then it follows from the conditions that $X \to Y_0$ is in $\mathcal{L}$. If Y_0 is not a subset of Y, then Y cannot respect $\mathcal{L}$, since it does not respect $X \to Y_0$, which follows from $\mathcal{L}$. So, there must be some $Z \to W \in \mathcal{L}$ not respected by Y. From $Z \subseteq Y$ and $X \to Y \in \mathcal{L}$, we infer, using conditions 1–3, that $X \to Z \in \mathcal{L}$. From the third condition, we obtain that $X \to W \in \mathcal{L}$, but, given that $W \not\subseteq Y$, this is a contradiction to the definition of Y. Thus, $\mathcal{L}_0 = \mathcal{L}$, and $\mathcal{L}$ must be closed. □

Readers with a background in computational logic may prefer a different notation for these Armstrong rules:

$$\frac{}{X \to X}, \quad \frac{X \to Y}{X \cup Z \to Y}, \quad \frac{X \to Y, \quad Y \cup Z \to W}{X \cup Z \to W}, \quad \frac{X \to Y_t, \quad t \in T}{X \to \bigcup_{t \in T} Y_t}.$$

The proposition means, in particular, that the set of all implications of a formal context must be closed with respect to these rules. An implication follows from other implications if and only if it can be derived from them by successive applications of these rules. Thus, semantic and syntactic inference are the same.

However, these rules do not always suggest the best proof strategy. Instead, we may note the following:

Proposition 16 *An implication $X \to Y$ follows from a list $\mathcal{L}$ of implications if and only if $Y \subseteq \mathcal{L}(X)$.*

Proof $\mathcal{L}(X)$ is a model of $\mathcal{L}$ containing X. If $Y \not\subseteq \mathcal{L}(X)$, then $\mathcal{L}(X)$ is not a model of $X \to Y$. In this case, $X \to Y$ does not follow from $\mathcal{L}$. Conversely, any model of $\mathcal{L}$ that contains X must also contain $\mathcal{L}(X)$, since $\mathcal{L}(X)$ is, by definition, the intersection of models of $\mathcal{L}$ containing X. This proves the proposition. □

3.1.3 Computing the closure $\mathcal{L}(X)$ of a set X

We present a simple algorithm, Closure, that computes the closure $\mathcal{L}(X)$ for any given set X, as Algorithm 14. The inner **for all** loop of the algorithm iterates through all implications $A \to B$ in $\mathcal{L}$ and adds attributes from B to X whenever A is already contained there. If this happens for an implication $A \to B$, it is removed from $\mathcal{L}$, since it cannot be used to enlarge $\mathcal{L}(X)$ any further. If X gets bigger as the result of the **for all** loop, the loop has to be repeated, which is the reason for the outer **repeat** loop. The algorithm terminates only when the closure of X remains unchanged after a complete iteration of the **repeat** loop. At each but the last iteration, at least one implication is removed from $\mathcal{L}$ and at least one attribute is added to X. Therefore, the maximal number of iterations of the outer loop is $\min(|\mathcal{L}|, |M|)$, while the maximal number of iterations of the inner loop is, obviously, $|\mathcal{L}|$. Thus, the algorithm CLOSURE is essentially quadratic in the number of implications.[4]

Algorithm 14 CLOSURE($X, \mathcal{L}$)

Input: An attribute set X and a set $\mathcal{L}$ of implications.
Output: The closure of X w.r.t. implications in $\mathcal{L}$.

```
repeat
    stable := true
    for all A → B ∈ L do
        if A ⊆ X then
            X := X ∪ B
            stable := false
            L := L \ {A → B}
until stable
return X
```

Such algorithms are used in the theory of relational databases for the study of functional dependencies (which can be interpreted as implications, see [110], Proposition 28). There are better algorithms than Algorithm 14, e.g., LINCLOSURE (Algorithm 15, from [35], see also [159][5]), which has complexity linear in the size of its input.

3.1.4 Linear complexity of implication inference

Summarizing these considerations we learn that implication inference is easy: to check if an implication $X \to Y$ follows from a list $\mathcal{L}$ of implications, it suffices to check if $Y \subseteq \mathcal{L}(X)$ (by Proposition 16), and this can be done in time linear in the size of the input.

In other words, implications are easy to use, much easier than many other logical constructs. This may be a reason why implications are popular and

[4]To be a bit more precise, the worst-case scenario is when exactly one implication is applied at each iteration of the **while** loop, which means that there are in total $\frac{|\mathcal{L}|(|\mathcal{L}|+1)}{2}$ iterations of the **for all** loop, each requiring $O(|M|)$ time.

[5]The version of LINCLOSURE given in [159] assumes that the premises of all implications are not empty. We present a slightly modified version, free from this assumption.

Algorithm 15 LINCLOSURE($X, \mathcal{L}$)

Input: An attribute set X and a set $\mathcal{L}$ of implications.
Output: The closure of X w.r.t. implications in $\mathcal{L}$.

```
for all A → B ∈ ℒ do
    count[A → B] := |A|
    if |A| = 0 then
        X := X ∪ B
    for all a ∈ A do
        add A → B to list[a]
update := X

while update ≠ ∅ do
    choose m ∈ update
    update := update \ {m}
    for all A → B ∈ list[m] do
        count[A → B] := count[A → B] − 1
        if count[A → B] = 0 then
            add := B \ X
            X := X ∪ add
            update := update ∪ add

return X
```

intuitive and, perhaps, part of an explanation why our simple theory of formal concepts is so useful.

3.2 Pseudo-closed sets

The number of implications that hold in a given situation can be very large. For example, if there is only one closed set, M, then *every* implication holds. If M has n elements, then these are some 2^{2n} implications. But this is ridiculous, because all these implications can be inferred from a single one, namely from $\emptyset \to M$.

We see from this trivial example that the set of *all* implications of a given formal context may be highly redundant. It is a natural question to ask for a small *implication basis*, from which everything else follows. More precisely, we ask, for any given formal context (G, M, I), for a list $\mathcal{L}$ of implications that is

- **sound**, i.e., each implication in $\mathcal{L}$ holds in (G, M, I),

- **complete,** i.e., each implication that holds in (G, M, I) follows from $\mathcal{L}$, and
- **nonredundant,** i.e., no implication in $\mathcal{L}$ follows from other implications in $\mathcal{L}$.

Nonredundancy is an internal property of implication sets, while the notions of soundness and completeness can also be defined with respect to an arbitrary closure operator $(\cdot)''$: for $\mathcal{L}$ to be sound, we require that $B \subseteq A''$ for every implication $A \to B \in \mathcal{L}$, and, if every implication for which this is the case follows from $\mathcal{L}$, then $\mathcal{L}$ is said to be complete.

It is easy to see that (for finite M) such sets $\mathcal{L}$ exist. We may start with some sound and complete set of implications, for example, with the set of all implications that hold in (G, M, I). We then can successively remove redundant implications from this set until we obtain a sound, complete, and nonredundant set.

But this can be very time-consuming. We therefore look for a better way to construct an implicational basis. Duquenne and Guigues [120] have shown that there is a natural choice, the *canonical basis* or *stem base*.

3.2.1 A recursive definition

The following recursive definition is rather irritating at first glance. We define a pseudo-closed set to be a set which is not closed, but contains the closure of every pseudo-closed proper subset. More precisely,

Definition 11 Let $X \mapsto X''$ be a closure operator on the set M. We call a finite subset $P \subseteq M$ **pseudo-closed** if (and only if)

- $P \neq P''$ and,
- if $Q \subsetneq P$ is a pseudo-closed proper subset of P, then $Q'' \subseteq P$.

$\Diamond$

This *is* a valid definition.[6] It is not circular, because the pseudo-closed set Q, as a proper subset of P, must have fewer elements than P. The empty set is pseudo-closed iff it is not closed. An n-element set is pseudo-closed if it is not closed and contains the closure of every pseudo-closed subset *with fewer than n elements*.

Reformulating this definition for formal contexts, we obtain

Definition 12 Let (G, M, I) be a formal context. A finite subset $P \subseteq M$ is a **pseudo-intent** of (G, M, I) iff

[6]Compare with the following: a natural number is *prime* iff it is greater than 1 and not divisible by any smaller prime number.

- P is not a concept intent and,
- if $Q \subsetneq P$ is a pseudo-intent and a proper subset of P, then $Q'' \subseteq P$.

Pseudo-extents are defined dually. $\Diamond$

Example 9 Consider the context of triangles from Section 1.4.2. The empty attribute set, $\emptyset$, is not a pseudo-intent, because it is closed. For the same reason, $\{b\}$, $\{c\}$, $\{d\}$, and $\{e\}$ are not pseudo-intents, either. The set $\{a\}$, however, is a pseudo-intent, since it is not closed ($\{a\}'' = \{a, b, c\}$), while its only proper subset, $\emptyset$, is closed and, thus, not pseudo-closed. In this case, the second condition in the definition of the pseudo-intent trivially holds. This condition does not hold for $\{a, b\}$, which properly contains a pseudo-intent $\{a\}$, but does not contain its closure. Therefore, $\{a, b\}$ is not a pseudo-intent.

The object set $\{T_5\}$ gives an example of a pseudo-extent with closure $\{T_1, T_5\}$. Since the set $\{T_2, T_4, T_5, T_7\}$ contains the pseudo-extent $\{T_5\}$, but not its closure, it is not a pseudo-extent.

3.2.2 The canonical basis

Theorem 7 (Guigues and Duquenne) *Let M be finite and let $X \mapsto X''$ be some closure operator on M. The set of implications*

$$\{P \to P'' \mid P \text{ is pseudo-closed }\}$$

is sound, complete (with respect to $(\cdot)''$), and nonredundant.

The **Proof** is simple enough to give it here. Obviously, all implications of the form $X \to X''$ hold in the given closure system; therefore, the set is sound.

Suppose that it is not complete. Then there is some implication $X \to Y$ that holds, but does not follow from the set $\mathcal{L}$ of implications given in the theorem. In particular, $Y \subseteq X''$ (because the implication holds), but $Y \not\subseteq \mathcal{L}(X)$ (because the implication does not follow from $\mathcal{L}$). Then $\mathcal{L}(X) \neq X''$, and thus $\mathcal{L}(X)$ is not closed. Now let $Q \subset \mathcal{L}(X)$ be any pseudo-closed proper subset of $\mathcal{L}(X)$. Since the implication $Q \to Q''$ belongs to $\mathcal{L}$, the set Q'' must be a subset of $\mathcal{L}(X)$.

Therefore, $\mathcal{L}(X)$ is not closed but contains the closure of every pseudo-closed proper subset. By definition then, $\mathcal{L}(X)$ must be pseudo-closed. But this gives a contradiction, because, in this case, the implication $\mathcal{L}(X) \to \mathcal{L}(X)''$ would be in $\mathcal{L}$, which would imply $\mathcal{L}(X)'' \subseteq \mathcal{L}(X)$ and, consequently, $\mathcal{L}(X) = \mathcal{L}(X)''$.

To see that the list $\mathcal{L}$ is nonredundant, remove one of the implications, say, $P \to P''$, to obtain a smaller list $\mathcal{L}^-$. With respect to $\mathcal{L}^-$, the set P is closed (because P is a model of all implications $Q \to Q''$ in $\mathcal{L}^-$), and thus $\mathcal{L}^-(P) = P$. By Proposition 16, $P \to P''$ does not follow from $\mathcal{L}^-$. $\square$

The implication set in the theorem deserves a name. It is sometimes called the *Duquenne–Guigues basis*. Duquenne and Guigues called it the **canonical basis**, Ganter and Wille the *stem base* of the closure operator (or of a formal context, if the closure operator is given that way). In practice, one uses a slightly different version of the canonical basis, namely,

$$\{P \to P'' \setminus P \mid P \text{ is pseudo-closed }\}.$$

As before, we will make use of the symbol $\perp$ and write $P \to \perp$ when $P' = \emptyset$. With these shorthand notations, the canonical basis of the triangles context from Figure 1.14 (p. 23) is presented in Figure 3.2. It will be explained later how it was computed, see Figure 3.5 below.

$$\begin{array}{rcl} \{\textit{obtuse-angled, right-angled}\} & \to & \perp \\ \{\textit{acute-angled, right-angled}\} & \to & \perp \\ \{\textit{acute-angled, obtuse-angled}\} & \to & \perp \\ \{\textit{equilateral}\} & \to & \{\textit{isosceles, acute-angled}\} \end{array}$$

Figure 3.2: The canonical basis for the triangles

The canonical basis is not the only implication basis, but it plays a special rôle. For example, no implication basis can consist of fewer implications, as the next proposition shows:

Proposition 17 *Every sound and complete set of implications contains at least one implication $A \to B$ with $A'' = P''$ per pseudo-closed set P.*

3.2.3 The size of the canonical basis

It is easy to give an example of a closure system with many pseudo-closed sets. For a given base set M with $|M| = 2n > 2$, take as closed sets all subsets with at most n elements plus M itself. The pseudo-closed sets then are precisely the subsets with $n+1$ elements. There are $\binom{2n}{n+1}$ of them, a number that is exponential in n. Thus, the number of pseudo-closed sets can be exponential in comparison to the size of the base set.

But any formal context describing this closure system must have at least $\binom{2n}{n}$ objects and thus itself be of a size exponential in n. In other words, the number of pseudo-closed sets in this example is large in comparison with the base set, but small in comparison with the formal context. Are there (infinite series of) formal contexts where the number of pseudo-closed sets is exponential in the size of the contexts? A positive answer to this question – but formulated

	m_0	$m_1, \ldots, m_n$	$m_{n+1}, \ldots, m_{2n}$
g_1			
$\vdots$		$\neq$	$\neq$
g_n			
g_{n+1}	$\times$		
$\vdots$	$\vdots$		
$\vdots$	$\vdots$	$\neq$	
$\vdots$	$\vdots$		
g_{3n}	$\times$		

Figure 3.3: A schematic representation of a formal context with an exponential number of pseudo-intents

in rather different terms – was given in [131]. Here we present a solution from [143], where a series of contexts is described such that a context with $3n$ objects and $2n+1$ attributes has 2^n pseudo-intents. Such a context is schematically represented in Figure 3.3, which slightly abuses notation by using the "$\neq$" symbol in the following sense: an object g_i has an attribute m_j if and only if

1. $i \leq n, j \neq 0, j \neq i$ and $j \neq i+n$;

2. or $i > n$ and $j \neq i - n$.

It can be easily checked that the pseudo-intents of this context are exactly the sets $\{m_{i_1}, \ldots, m_{i_n}\}$, where $i_j \in \{j, j+n\}$. Obviously, there are 2^n such sets.

On the other hand, one may ask if a small number of pseudo-closed sets can always be realized with a small number of objects. More precisely, is there an infinite series of object-reduced formal contexts (G, M, I), i.e., contexts with only irreducible objects, for which the number of pseudo-closed sets is polynomial in the size of M, but the size of G is not?

Consider a context with the set of attributes $M_1 \cup \cdots \cup M_n$, where $|M_i| = n$ and $M_i \cap M_j = \emptyset$ for all $i, j \leq n, i \neq j$. Let us define the object set of this context so that the object intents $\{g\}'$ will be all possible attribute combinations with $|\{g\}' \cap M_i| = n-1$ for all $i \leq n$. Then, there are n^n objects and, since their intents are all different and all of the same size, $n(n-1)$, the context is object-reduced. This context is, in fact, the semi-product of n contexts of the form $(M, M, \neq)$ with $|M| = n$, see Definition 22 (p. 227).

Now, every attribute subset that includes neither of M_i is closed, but, for each $i \leq n$, the closure of M_i equals $M_1 \cup \cdots \cup M_n$. Thus, the pseudo-intents of this context are exactly M_i for all $i \leq n$. Therefore, we have a context with n^2 attributes, n^n objects, and only n pseudo-intents.

Another example of a context with an exponential (in terms of $|M|$) number of objects, but a polynomial number of pseudo-intents is given by the context whose concept lattice is dual to the lattice of equivalence relations. The objects of this context are all partitions of a base set S into two non-empty subsets. Since such a partition is uniquely defined by each of the two subsets, the context has $2^{|S|-1} - 1$ objects. The $|S|^2$ attributes are all pairs of elements from S. An object has an attribute if both elements of the corresponding pair belong to the same subset in the corresponding partition. None of the object intents can be obtained as the intersection of other object intents; therefore, all the objects are irreducible. The concept intents of this context are all equivalence relations on S, and the canonical basis consists of the implications summarizing the familiar properties of equivalence relations. Namely, for $s, t, u \in S$:

1. $\emptyset \to \{(s, s)\}$;
2. $\{(s, t)\} \to \{(t, s)\}$;
3. $\{(s, t), (t, u)\} \to \{(s, u)\}$.

Hence, there are fewer than $|S| + |S|^2 + |S|^3$ pseudo-intents, while the size of the object set is exponential in the number of attributes.

The above results show that the formal context is neither an always more compact nor an always less compact representation of the closure operator than the canonical basis.

In general, determining the number of pseudo-closed sets is a #$\mathcal{P}$-hard problem [143]. Another interesting question is how difficult it is to decide if a given set is pseudo-closed (see Figure 3.4). In [145, 146], it was shown that this problem is in class co$\mathcal{NP}$ and, after a few related results in [187, 196, 197, 74], it was finally proved that it is co$\mathcal{NP}$-hard (and, thus, co$\mathcal{NP}$-complete) [23]. This has some unpleasant consequences for computing pseudo-intents. For example, there cannot be a polynomial-delay (see Section 2.2.4) algorithm for enumerating pseudo-intents in lectic order or in reverse lectic order unless $\mathcal{P} = \mathcal{NP}$ [74, 23, 24].

3.2.4 Making canonical basis implications shorter

We have stated above that the canonical basis is nonredundant. This means that no implication in the canonical basis is dispensable. It is however possible that an implication in the basis can be shortened, for example, because it has a redundant premise. Such are occasionally counter-intuitive, in particular, for mathematicians, who are trained to base their proofs on minimal resources.

The following problem is co-$\mathcal{NP}$-complete:

INSTANCE: A formal context (G, M, I) and a set $P \subseteq M$.

QUESTION: Decide if P is pseudo-closed.

Figure 3.4: The problem of recognizing pseudo-closed sets

Similarly, the conclusion of an implication may be redundant. It may suffice to prove only part of it and then use other implications to obtain the rest.

It seems to be straightforward what to do: if an attribute is redundant in a premise or a conclusion, just omit it. But things get more complicated if two or more attributes are redundant. If each of m and n is redundant, it does not mean that they may be omitted *simultaneously*.

Example 10 Consider the set of implications in Example 8 (p. 91). This is the canonical basis of the context of triangles. The conclusion of the implication $\{d, e\} \rightarrow \{a, b, c\}$ is redundant, because both a and c are redundant. Indeed, we can replace $\{d, e\} \rightarrow \{a, b, c\}$ with either $\{d, e\} \rightarrow \{b, c\}$ or $\{d, e\} \rightarrow \{a, b\}$. In the former case, we can use $\{c, d\} \rightarrow \{a, b, e\}$ to infer a from $\{d, e\}$, and, in the latter case, we can use $\{a\} \rightarrow \{b, c\}$ to infer c. Yet, it is not possible to remove both the redundant attributes a and c and replace the existing implication with $\{d, e\} \rightarrow \{b\}$, since the remaining implications do not allow us to infer a or c from $\{b, d, e\}$.

In general, there is no guarantee that a set A contains a *unique* minimal generating set $S \subseteq A$ satisfying $A'' = S''$. There may be many such sets, and they may be difficult to find. For example, polynomial time is not sufficient to find such a set with minimum possible cardinality unless $\mathcal{P} = \mathcal{NP}$. This can be derived from the following observations.

The problem of whether a graph (V, E) contains a set V_1 of vertices of size at most k such that every edge from E is incident to some $v \in V_1$ is called the vertex cover problem and is known to be $\mathcal{NP}$-complete. Given a graph (V, E), we build the context $(E, V, \not\ni)$, taking the edges as objects and the vertices as attributes; an edge e is related to a vertex v in the context if e is not incident to v in the graph. It turns out that there is a vertex cover of size k in the graph (V, E) if and only if V has a generating set of size k in the context $(E, V, \not\ni)$. Let us prove this.

Suppose that V_1 is a vertex cover of (V, E). Every $e \in E$ is incident to some $v \in V_1$. In other words, there is no e such that $e \not\ni v$ for all $v \in V_1$. Therefore, in our context, $V_1' = \emptyset$ and $V_1'' = V$, i.e., V_1 is a generating set of V.

Conversely, V_1 can be a generating set of V only if there is $e \in E$ with $V_1 \subseteq \{e\}' = V$ or there is no $e \in E$ with $V_1 \subseteq \{e\}'$. The first case is impossible, since it would mean that there is an edge that is not incident to any vertex of the graph. The second case happens only when every edge is incident to at least some $v \in V_1$, that is, when V_1 is a vertex cover.

Thus, V_1 is a vertex cover if and only if it is a generating set of V. This gives a polynomial reduction from the vertex cover problem to the problem of finding a minimum-cardinality generating set, which means that the latter is $\mathcal{NP}$–hard.

3.3 Finding pseudo-closed sets

As before, consider a closure operator $X \mapsto X''$ on a finite set M. We start with a harmless definition:

Definition 13 A set $Q \subseteq M$ is **preclosed** if it contains the closure of every preclosed set that is properly contained in Q.

Formally, Q is preclosed iff we have $Q_0'' \subseteq Q$ for every preclosed set $Q_0 \subseteq Q$ with $Q_0 \neq Q$. $\Diamond$

This is a simple (but convenient) renaming, as the next proposition shows.

Proposition 18 *A set is preclosed iff it is either closed or pseudo-closed.*

Observe that, if Q contains the closure of every preclosed subset, then Q must be closed.

3.3.1 Preclosed sets form a closure system

The first crucial step towards finding pseudo-closed sets is this:

Proposition 19 *The intersection of preclosed sets is preclosed.*

Proof Let Q_t, $t \in T$, be preclosed and let $Q := \bigcap_{t\in T} Q_t$. We have to show that Q is preclosed. Actually, we show more: either $Q = Q_t$ for some $t \in T$, or Q is closed. Assume $Q \neq Q_t$ for all t and let Q_0 $(0 \notin T)$ be some preclosed subset of Q. Then Q_0 is properly contained in each Q_t, $t \in T$, because Q is. Since each Q_t is preclosed, the closure of Q_0 must also be contained in Q. Thus, Q contains the closure of every its preclosed subset, which makes Q itself preclosed. But then Q must contain its own closure. Therefore, Q is closed. $\square$

In other words, preclosed sets form a closure system. We have described an algorithm to compute all closed sets for a given closure operator. We can use this algorithm to compute all preclosed sets if we can access the corresponding closure operator. This is easy. We prepare the result with a proposition that is an immediate consequence of Definition 13 and Proposition 18.

Proposition 20 *Q is preclosed iff Q satisfies the following condition:*

If $P \subseteq Q$, $P \neq Q$, is pseudo-closed, then $P'' \subseteq Q$.

Let us call the operator that takes a set to its smallest preclosed superset the *preclosure operator*. The preclosure operator is a closure operator due to Proposition 19. Proposition 20 shows how to find the preclosure of an arbitrary set $Q \subseteq M$: as long as the condition in the proposition is violated, we (are forced to) extend the set Q, until we finally reach a fixed point.

Let $\mathcal{L}$ be the canonical basis.[7] Define, for $X \subseteq M$,

$$X^{\mathcal{L}^\bullet} := X \cup \bigcup \{P'' \mid P \to P'' \in \mathcal{L}, P \subseteq X, P \neq X\},$$

iterate by forming

$$X^{\mathcal{L}^\bullet \mathcal{L}^\bullet}, X^{\mathcal{L}^\bullet \mathcal{L}^\bullet \mathcal{L}^\bullet}, \ldots$$

until a set

$$\mathcal{L}^\bullet(X) := X^{\mathcal{L}^\bullet \mathcal{L}^\bullet \ldots \mathcal{L}^\bullet}$$

is obtained that satisfies

$$\mathcal{L}^\bullet(X) = \mathcal{L}^\bullet(X)^{\mathcal{L}^\bullet}.$$

Proposition 21 ([92]) *$\mathcal{L}^\bullet(X)$ is the smallest preclosed set containing X.*

Note that in order to find the preclosure, we use only pseudo-closed sets that are properly contained in the preclosure and, therefore, are lectically smaller. Thus, the same result is obtained if $\mathcal{L}$ is a subset of the canonical basis containing the implications $P \to P''$ for which P is lectically smaller than $\mathcal{L}^\bullet(X)$.

Example 11 Consider the following set $\mathcal{L}$ of implications:

$$\begin{array}{ll} \{a\} \to \{b\} & \{a,b,d\} \to \{e\} \\ \{b,c\} \to \{d\} & \{a,b,c,d,e\} \to \{f\} \end{array}$$

It is easy to see that the premises of these implications are pseudo-closed with respect to the closure operator defined by $\mathcal{L}$. How do we compute the preclosure of $\{a,c\}$? Observing that $\{a\} \subsetneq \{a,c\}$, we use the implication $\{a\} \to \{b\}$ to obtain $\{a,b,c\}$. Similarly, we get $\{a,b,c,d\}$ by applying $\{b,c\} \to \{d\}$, which is then extended to $\{a,b,c,d,e\}$, the preclosure of $\{a,c\}$, due to $\{a,b,d\} \to \{e\}$. Note that we cannot use the implication $\{a,b,c,d,e\} \to \{f\}$, since its premise is not a *proper* subset of $\{a,b,c,d,e\}$.

[7] The reader may wonder why we use the canonical basis to construct the canonical basis. As we shall see soon, this works due to the recursive definition of the canonical basis.

3.3.2 An algorithm for computing the canonical basis

Now it is easy to give an algorithm to compute all pseudo-closed sets for a given closure operator. We use the NEXT CLOSURE algorithm applied to the closure system of preclosed sets. For short, we shall refer to this as the **next preclosure** after a given set A. This produces all preclosed sets in the lectic order. We record only those that are not closed. This yields a list of all pseudo-closed sets.

Since preclosed sets are generated in lectic order, we have, at each step, the full information about all lectically smaller pseudo-closed sets. We have seen that this suffices to compute the preclosure operator. Algorithm 16 uses a dynamic list $\mathcal{L}$. Whenever a pseudo-closed set P is found, the corresponding implication $P \to P''$ is included in the list. At the point when the algorithm has to compute $\mathcal{L}^{\bullet}(A)$, the premises of all implications in $\mathcal{L}$ are lectically smaller than A, and, in particular, there is no implication $B \to C \in \mathcal{L}$ with $A \subseteq B$. This means that $\mathcal{L}^{\bullet}(A) = \mathcal{L}(A)$ and we can use Algorithm 14 or Algorithm 15 (or, for that matter, any algorithm that computes the closure of an attribute set with respect to a set of implications) without modification to compute $\mathcal{L}^{\bullet}(A)$. In the pseudocode of Algorithm 16, we pass the closure operator defined by $\mathcal{L}$ as the third argument of the call to NEXT CLOSURE.

Algorithm 16 CANONICAL BASIS($M, ''$)

Input: A closure operator $X \mapsto X''$ on a finite set M, for example, given by a formal context (G, M, I).
Output: The canonical basis for the closure operator.

$A := \emptyset$
$\mathcal{L} := \emptyset$
while $A \neq M$ **do**
  if $A \neq A''$ **then**
    $\mathcal{L} := \mathcal{L} \cup \{A \to A''\}$
  $A :=$ NEXT CLOSURE$(A, M, \mathcal{L})$
return $\mathcal{L}$

3.3.3 An example and optimizations

We compute the canonical basis for the context of triangles given on page 23. The steps are shown in Figure 3.5. The first column contains all preclosed sets in lectic order. The pseudo-closed sets are precisely those that are not closed (see the middle column). Each pseudo-closed set gives rise to an entry in the canonical basis (last column, short form). This canonical basis is given again, in a slightly modified form, in Figure 3.2 (p. 97).

preclosed set	closed?	canonical basis implication
$\emptyset$	yes	
$\{e\}$	yes	
$\{d\}$	yes	
$\{d,e\}$	no	$\{d,e\} \to \{a,b,c\}$
$\{c\}$	yes	
$\{c,e\}$	no	$\{c,e\} \to \{a,b,d\}$
$\{c,d\}$	no	$\{c,d\} \to \{a,b,e\}$
$\{b\}$	yes	
$\{b,e\}$	yes	
$\{b,d\}$	yes	
$\{b,c\}$	yes	
$\{a\}$	no	$\{a\} \to \{b,c\}$
$\{a,b,c\}$	yes	
$\{a,b,c,d,e\}$	yes	

Figure 3.5: Steps in the CANONICAL BASIS algorithm

Since the closure operator is given in terms of a formal context, we may speak of *pseudo-intents* instead of pseudo-closed sets. We see that the algorithm generates all preclosed sets to find the canonical basis. In other words, to compute all pseudo-intents, we also compute all intents, possibly exponentially many. This looks like a rather inefficient method. Is it possible to be faster?

First, consider what happens after the implication $\{c,d\} \to \{a,b,e\}$ is produced in the example in Figure 3.5. The algorithm is trying to find the next preclosed set after $\{c,d\}$. Although this is not shown in the table, the first step in this search is to find the $\mathcal{L}^\bullet$-closure of the lectically next set after $\{c,d\}$, which is $\{c,d,e\}$, and check if this closure contains an element smaller than the added element e. But we know that it will contain such an element, because $\{a,b,c,d,e\}$, the closure of $\{c,d\}$ computed at the previous step, contains $a < d < e$. So, computing the $\mathcal{L}^\bullet$-closure of $\{c,d,e\}$ is unnecessary. It will be quick in this particular example, but, in general, it may introduce a serious overhead.

More generally, suppose that we have computed the preclosure $\mathcal{L}^\bullet(A)$ of a set A and added the implication $\mathcal{L}^\bullet(A) \to \mathcal{L}^\bullet(A)''$. What should be done next to compute the lectically next pseudo-closed set after $\mathcal{L}^\bullet(A)$? Algorithm 16 will simply pass $\mathcal{L}^\bullet(A)$ as the input to NEXT CLOSURE, but we can be a bit more selective. Let i be the maximal element of A and j be the minimal element of

$A'' \setminus \mathcal{L}^{\bullet}(A)$. Consider the following two cases:

$j < i$: For any $m > i$, the preclosure of $\mathcal{L}^{\bullet}(A) \cup \{m\}$ will be rejected by NEXT CLOSURE, since it will contain j. Hence, it is possible to skip all $m > i$ and continue as if $\mathcal{L}^{\bullet}(A)$ has been rejected. In the example, $A = \mathcal{L}^{\bullet}(A) = \{c, d\}$, $\mathcal{L}^{\bullet}(A)'' = \{a, b, c, d, e\}$, i is d, and j is a. Since $a < d$, we do not have to consider the preclosure of $\{c, d, e\}$, but can jump directly to $\{b\}$ instead.

$i < j$: This is, in a sense, the "normal" case, when everything new that gets generated by the preclosure and closure operators is "to the right" of A. However, even in this case, a lot of unnecessary work gets done sometimes. Consider what happens after $\{a\} \to \{b, c\}$ is produced. Here, $A = \mathcal{L}^{\bullet}(A) = \{a\}$, i is a, and j is b. The next set that is the input to NEXT CLOSURE is $\{a, e\}$. Of course, its preclosure contains b and c, both of which are less than e; so, this preclosure is rejected. The same then happens with $\{a, d\}$ and $\{a, c\}$. A reasonable shortcut is to skip all these obviously fruitless attempts and continue directly with $\{a, b, c\}$, the closure of A. We can do so because of the following fact:

Proposition 22 *If P is pseudo-closed and no element of $P'' \setminus P$ is smaller than any element of P, then P is the lectically largest pseudo-closed set with closure P''.*

Proof If $P = \emptyset$, it is the only pseudo-closed set with closure P''. Otherwise, if Q is a lectically larger pseudo-closed set with $Q'' = P''$, it must contain an element $q \notin P$ that is smaller than some element in P; but then q would also be contained in $Q'' = P''$ and in $P'' \setminus P$. □

Thus, when $i < j$, we know that the lectically next preclosed set after $\mathcal{L}^{\bullet}(A)$ is A'' and can proceed accordingly. The case when $A = \emptyset$ should be treated in the same way.

Algorithm 17 takes into account these considerations. Because of the somewhat more intricate interaction between the variables, we do not divide the algorithm into subprograms, but it is not too difficult to do this if needed.

Note that, although this algorithm skips many candidate sets as compared to the original version, it still generates all closed sets in addition to implications. In the worst case, the algorithm may require exponential time, while producing not a single implication, because all attribute sets are closed and the canonical basis is empty (see the context family $\mathbb{N}_n^c$ in Section 2.2.4).

3.3.4 Computing the canonical basis from given implications

Other complete but not necessarily small sets of implications can be found in the literature. A popular and frequently discovered complete set is the family of **proper implications**

$$\mathcal{L}_p := \{A \to A^{\bullet} \mid A^{\bullet} \neq \emptyset\},$$

Algorithm 17 CANONICAL BASIS(M, $''$), an optimized version

Input: A closure operator $X \mapsto X''$ on a finite set M, for example, given by a formal context (G, M, I).
Output: The canonical basis for the closure operator.

$A := \emptyset''$
if $A = \emptyset$ **then**
  $\mathcal{L} := \emptyset$
else
  $\mathcal{L} := \{\emptyset \to A\}$
$i :=$ the largest element of M

while $A \neq M$ **do**

  for all $j \leq i \in M$ in reverse order **do**
    if $j \in A$ **then**
      $A := A \setminus \{j\}$
    else
      $B := \mathcal{L}(A \cup \{j\})$
      if $B \setminus A$ contains no element $< j$ **then**
        $A := B$
        $i := j$
        exit for

  if $A \neq A''$ **then**
    $\mathcal{L} := \mathcal{L} \cup \{A \to A''\}$

  if $A'' \setminus A$ contains no element $< i$ **then**
    $A := A''$
    $i :=$ the largest element of M
  else
    $A := \{m \in A \mid m \leq i\}$

return $\mathcal{L}$

where

$$A^\bullet := \mathcal{L}(A) \setminus (A \cup \bigcup_{B \subsetneq A} \mathcal{L}(B)),$$

and $X \mapsto \mathcal{L}(X)$ is the closure operator under investigation. Our earliest reference is [108] (but [110], Prop. 22 ff., may be easier to access).

The premises of implications in this set, that is, the sets A with $A^\bullet \neq \emptyset$, are called **proper premises**, so that

$$\mathcal{L}_p = \{A \to A^\bullet \mid A \text{ a proper premise}\}.$$

In essence, a proper premise is a set that, as a whole, implies attributes not implied by any of its proper subsets. It was shown in [108, 110] that for finite M this set of *implications with proper premises*[8] is sound, complete, and **iteration-free** (or, synonymously, **direct**), meaning that

$$\mathcal{L}(S) = S \cup \bigcup \{A^\bullet \mid A \text{ a proper premise, } A \subseteq S\}$$

holds for all $S \subseteq M$. One practical consequence is that Algorithm 14 CLOSURE needs only one iteration of the outer loop to compute the closure of set X under these implications; thus, it becomes linear in the number of implications (as opposed to being quadratic when applied to an arbitrary implication set).

Moreover, it can be shown that the set of implications with proper premises is optimal in the following sense: any sound, complete, and iteration-free set of implications must contain, for every proper premise E, implications $E \to F_1$, $\ldots, E \to F_n$ such that $E^\bullet \subseteq F_1 \cup \ldots \cup F_n$.

Taouil and Bastide [208] describe an algorithm for constructing $\mathcal{L}_p$. They focus on **unit implications** of the form

$$\{A \to m \mid A \text{ a proper premise}, m \in A^\bullet\},$$

which were also studied by Bertet and Nebut [42]. Bertet and Monjardet [41] unify several equivalent definitions and call the system the **canonical direct basis**.

Proposition 23 (below) shows a way to construct proper premises as hypergraph transversals. Recently, Ryssel, Borchmann, and Distel [190, 191] have obtained impressive results using proper premises, showing that, for some types of contexts, proper premises can be generated much faster than pseudo-intents.

Such complete sets of implications can be reduced to the canonical basis using algorithms from, e.g., [159] (Section 5.6.2) or [71] (Section 8). Below we present a simple Algorithm 18 for this task, which can be implemented to run in time quadratic in the size of its input.

This algorithm transforms an arbitrary set of implications into its canonical basis. The algorithm works in two steps. First, it transforms all implications

[8] This name seems to be more popular than *proper implications*.

Algorithm 18 MINIMAL COVER($\mathcal{L}$)

Input: A set $\mathcal{L}$ of implications over a set M.
Output: Transforms $\mathcal{L}$ into its canonical basis.

for all $A \to B \in \mathcal{L}$ **do**
  $\mathcal{L} := \mathcal{L} \setminus \{A \to B\}$
  $B := \mathcal{L}(A \cup B)$
  $\mathcal{L} := \mathcal{L} \cup \{A \to B\}$
for all $A \to B \in \mathcal{L}$ **do**
  $\mathcal{L} := \mathcal{L} \setminus \{A \to B\}$
  $A := \mathcal{L}(A)$
  if $A \neq B$ **then**
    $\mathcal{L} := \mathcal{L} \cup \{A \to B\}$

into their "full" form, that is, $A \to \mathcal{L}(A)$. A second run over $\mathcal{L}$ is then performed to "maximize" the premise A of each implication $A \to \mathcal{L}(A)$. This is achived by replacing A with $\mathcal{L}^-(A)$, where $\mathcal{L}^-$ is the same as $\mathcal{L}$ except that it does not contain the implication $A \to \mathcal{L}(A)$.

To see that the algorithm is correct, observe that it produces an equivalent set of implications. By Proposition 17, the resulting implication set must contain an implication $A \to B$ with $\mathcal{L}(A) = \mathcal{L}(P)$ for every pseudo-closed set P of $\mathcal{L}$. It remains to show that every premise is pseudo-closed. Suppose that the resulting $\mathcal{L}$ contains an implication $A \to \mathcal{L}(A)$ such that A is not pseudo-closed. Then, there is a pseudo-closed set $B \subsetneq A$ such that $\mathcal{L}(B) \not\subseteq A$. By Proposition 17, there is $C \subseteq B$ that is the premise of some implication $C \to \mathcal{L}(B) \in \mathcal{L}$. However, since $C \subseteq A$, the second loop of the algorithm ensures that $\mathcal{L}(B) \subseteq A$.[9]

This approach may be beneficial if an implication set can be generated fast, its size is not large (with respect to $2^{|M|}$), and the reduction procedure is efficient. However, the CANONICAL BASIS algorithm presented above has one property desirable for applications we are interested in: it generates implications according to the lectic order of their premises. Why this may be important should become clear from the next chapter, when we turn to attribute exploration.

[9] The algorithm also works if $\mathcal{L}$ is a list of implications that may contain multiple occurrences of the same implication and operations $\setminus$ and $\cup$ applied to $\mathcal{L}$ are interpreted as operations of removing and adding one occurrence of an implication.

3.4 Finding important implications only

The canonical basis is a compact representation of the implication theory. But we have learnt in Section 3.2.3 (p. 97) that this basis may be huge even for small data sets. Attribute exploration then easily reaches the limits of practicability and must be restricted. The natural solution is to reduce the number of attributes under consideration. An alternative may be not to consider all implications, but only a fragment. This becomes even more important when *association rules* are introduced as implications with a loosened notion of validity. Such rules may be of interest when an exploration is based on a large data set containing some errors, see Section 6.4.1. In this case, important implications could be overlooked in the exploration process, because there are erroneous counterexamples in the data. They may however be discovered as association rules and, if accepted as valid implications, be used to clean up the data.

3.4.1 Implications with small premise

In some applications, we may not be interested in the complete implication theory of a certain field, but only in finding out the main facts. In this case, implications with large premises may be hard to interpret and, overall, too complicated for our goals. Thus, we may want to limit the size of the premise by, say, a number k. Can the algorithm from the previous section be modified to compute only implications $A \to B$ with $|A| \leq k$ similarly to how the algorithm for computing all closed sets was adapted in Section 2.3.2 to computing only small closed sets?

The answer may seem obvious, for it is easy to generate all pseudo-intents of size $\leq k$. It was demonstrated in Section 2.3.2 that for generating all closed sets of size at most k, all we need is a small modification of the NEXT CLOSURE algorithm. The same trick works for pseudo-closed sets. For example, we can modify the optimized Algorithm 17 by replacing the line

if $B \setminus A$ contains no element $< j$ **then**

with

if $B \setminus A$ contains no element $< j$ and $|B| \leq k$ **then**

Unfortunately, generating pseudo-intents of size at most k is not sufficient for our goal. Let us collect some simple facts:

1. Not every valid implication with premise size $\leq k$ can be inferred from the canonical basis implications with premise size $\leq k$.

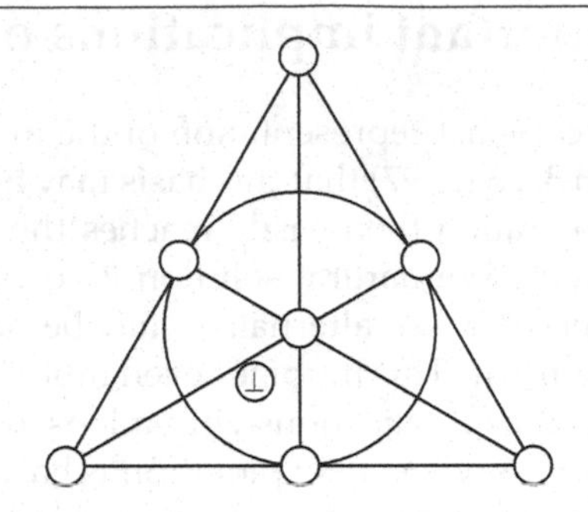

Figure 3.6: The projective plane of order 2 with an extra point $\bot$

A simple example is when the closure operator is given by the following set of two implications:

$$\emptyset \to \{a\}, \quad \{b\} \to \{c\}.$$

The pseudo-closed sets here are $\emptyset$ and $\{a, b\}$, and the canonical basis is

$$\emptyset \to \{a\}, \quad \{a, b\} \to \{c\}.$$

If we set $k = 1$, the only implication we obtain will be $\emptyset \to \{a\}$, although $\{b\} \to \{c\}$ also satisfies the threshold.

2. Unless we impose some order on attributes, there is no "canonical" nonredundant set of valid implications of premise size $\leq k$ from which all valid implications of premise size $\leq k$ can be inferred.

As an example, consider a finite geometry, the projective plane of order 2, say, plus some extra point $\bot$ contained in the plane, but in no line (see Figure 3.6). We get the following:

- $\bot$ implies everything else,
- each set of two points implies the third point on their line, and
- any three points not on a common line imply $\bot$.

Of the third type of implications, only one instance is needed, because all other ones can be inferred. But it is completely arbitrary which of the 28 non-collinear triples is chosen.

3. A nonredundant system of implications with premise size $\leq k$ may become redundant when an implication with a larger premise is added even if the new implication does not follow from this system (see [181] for an example).

Nevertheless, using implications with small premises in an interactive exploration is of interest, because these are more intuitive. The tasks then are

1. generating such implications efficiently,
2. keeping the list of accepted implications in a compact form in order to
3. decide if a newly generated implication can be inferred from accepted ones.

Generating from a given list $\mathcal{L}$ of implications the list

$$\mathcal{L}_k := \{A \to B \mid \mathcal{L} \models A \to B \text{ and } |A| \leq k\}$$

poses no fundamental problem, but this set is, of course, highly redundant. We shall therefore discuss some more focused approach below.

Whenever some list of implications (with premise size $\leq k$) is given, we may apply Algorithm 18 for transforming it into the canonical basis of the generated theory. The resulting implications may have larger premises, and we may use the method of Section 3.1.4 for deciding implication inference.

Reppe [181] has shown that the implications with proper premises of size at most k, that is, the implication set

$$\{A \to A^\bullet \mid A^\bullet \neq \emptyset, |A| \leq k\}$$

is equivalent to $\mathcal{L}_k$. This suggests restricting the search to proper premises, keeping in mind that this set of implications still may be highly redundant. From [110], we cite a characterization that is rather practical:

Proposition 23 ([108], [110] Prop. 23) *Let (G, M, I) be a formal context, $P \subseteq M$ and $m \in M \setminus P$. Then P is a proper premise with $m \in P^\bullet$ if and only if P is minimal with respect to the property that*

$$(M \setminus g') \cap P \neq \emptyset$$

holds for all $g \in G$ with $g \swarrow m$, where $g \swarrow m$ denotes that g' is maximal with respect to $(g, m) \notin I$.

To convince yourself that this proposition is true, observe P is a proper premise implying m if and only if P is a minimal subset of M that is included only in object intents containing m. This suggests the following method: to find all proper premises P with $m \in P^\bullet$, determine the family of minimal intent complements $(M \setminus g')$ of objects g that are not incident to m. The proper premises are precisely the minimal **hitting sets** of this family. Finding minimal hitting sets, also known as minimal **hypergraph transversals**, is a standard problem in computer science [79].

3.4.2 Implications with high support

In Section 2.3.2, we have introduced the support for sets of attributes. The **support** can be extended in a natural way to implications – and more generally, to any pair A, B of subsets of M – as the fraction of objects the intents of which contain $A \cup B$. Formally, we define the relative support of an implication by

$$\text{support}(A \to B) := \text{support}(A \cup B) = \frac{|(A \cup B)'|}{|G|} \qquad (G \neq \emptyset).$$

Note that, if $A \to B$ is a valid implication, $\text{support}(A \to B) = \text{support}(A)$. Usually a *minimal support* `minsupp` is specified and implications of support $\geq$ `minsupp` are considered. These are called **frequent implications** for short. To avoid misunderstanding, note that we are discussing *valid* implications here, i.e., implications with confidence equal to 1.

One may regard frequent implications particularly well founded in the data, and therefore give them priority in the interactive exploration process. But note that some implications of high importance have very small support. As a striking example, consider an implication of the form

$$\{m, n\} \quad \to \quad \bot,$$

which holds in (G, M, I) if the attributes m and n are mutually exclusive. The support then is zero, but every object intent is a model of this implication (since intents respect all valid implications).

We briefly discuss how to find frequent implications of a given formal context. Recall the iterative algorithm for implication inference as it was introduced on page 90. To derive an implication $A \to B$ from a list $\mathcal{L}$ of implications, one successively enriches A by using implications from $\mathcal{L}$, and each implication $C \to D$ used in this process must satisfy $C \cup D \subseteq A''$. In particular, we find that $\text{support}(C \to D) \geq \text{support}(A \to B)$. This is expressed in the next proposition.

Proposition 24 *If an implication $A \to B$ follows from a list $\mathcal{L}$ of implications that hold in (G, M, I), then it also follows from the sublist*[10]

$$\{C \to D \in \mathcal{L} \mid \text{support}(C \to D) \geq \text{support}(A \to B)\}.$$

Corollary 1 *If a frequent implication $A \to B$ follows from a list $\mathcal{L}$ of implications, then it also follows from the frequent implications in $\mathcal{L}$.*

It follows that *every valid implication can be inferred from canonical basis implications of the same or higher support.* In particular, we get the following:

[10]Note that implication inference is independent from any given formal context, while the sublist may vary with the data set.

Proposition 25 *All valid frequent implications follow from the set*

$$\{P \to P'' \mid P \text{ is a frequent pseudo-intent}\}.$$

This "high-support part" of the canonical basis can be computed with Algorithms 16 or 17, modified as described after Theorem 3 (p. 56) to produce frequent pseudo-intents only.

Note however that not all implications that follow from frequent implications are frequent themselves. We have mentioned on page 67 how to generate a list of all minimal non-frequent sets. Such a list solves the problem:

Proposition 26 *The valid frequent implications of (G, M, I) are precisely those implications that follow from the frequent implications in the canonical basis and the premises of which do not contain any non-frequent set.*

It would be useful if the frequent implications of a given formal context could be represented as the full set of implications of some other data set. This is not always possible, but the **iceberg lattices**, which were already mentioned in Section 2.3.2, offer something similar.

Definition 14 The **iceberg** of a concept lattice $\mathfrak{B}(G, M, I)$ (for a given minimal support) consists of all formal concepts with frequent intents, in their usual order. Icebergs are usually not lattices, but become lattices when $(\emptyset, M)$ is added as the smallest element. This structure is then called an **iceberg lattice**. $\diamond$

Each iceberg lattice is indeed a complete lattice. This follows from the fact that the concepts with frequent intents are closed under suprema. It also consists of formal concepts, but it is usually not a concept lattice because of its modified infimum operation. It is however *isomorphic* to the concept lattice of

$$(\mathcal{F}, M, \ni),$$

where $\mathcal{F}$ is the set of all frequent intents. Moreover, it has the same intents as this formal context. Thus, each implication valid in (G, M, I) must also hold in $(\mathcal{F}, M, \ni)$, but the converse is not true. Let $A \to B$ be an implication that holds in $(\mathcal{F}, M, \ni)$, but not in (G, M, I). Then there must be an intent of (G, M, I) refuting this implication, and this intent obviously cannot be frequent. On the other hand, if $A \to B$ is not frequent in (G, M, I), then A'' is not frequent either and, in the iceberg lattice, the smallest intent containing A is M. This proves the following:

Proposition 27 *Let (G, M, I) be a formal context,* `minsupp` $\in [0, 1]$, *and $\mathcal{F}$ be the family of frequent intents of (G, M, I). The implications of*

$$(\mathcal{F}, M, \ni)$$

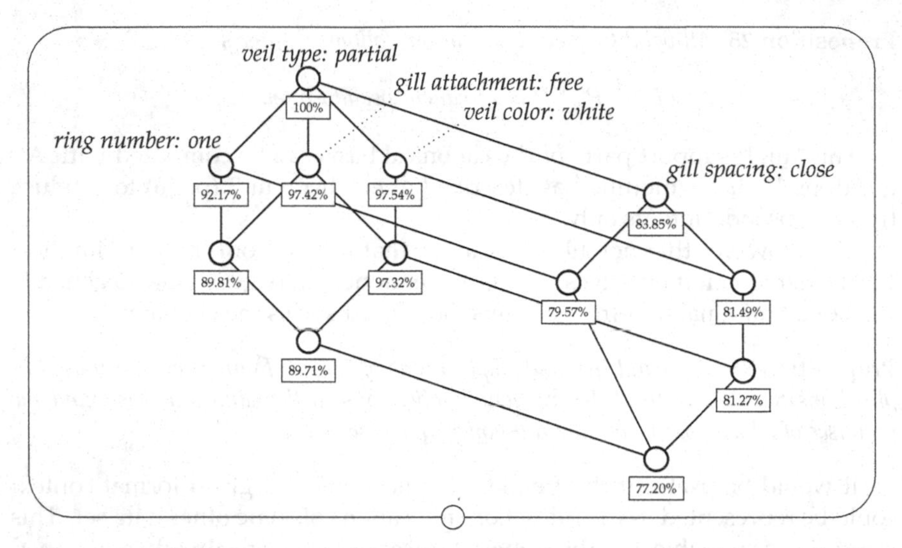

Figure 3.7: Iceberg lattice of the MUSHROOM data set with $\mathtt{minsupp} = 70\,\%$

are precisely the frequent implications of (G, M, I) *together with all the implications of the form*

$$A \to \bot,$$

where $A \subseteq M$ *is not frequent in* (G, M, I).

It is instructive, but quite imprecise, to say that *the implications of the iceberg lattice are the frequent implications.* Proposition 27 makes precise what this means.

We illustrate this by a small example (from [205]). The MUSHROOM data set from the *UC Irvine Machine Learning Repository* consists of 8,124 objects (mushrooms) and 22 (nominally valued) attributes [153]. We obtain a formal context by creating one (one-valued) attribute for each of the 125 possible values of the 22 original attributes. The resulting formal context has 8,124 objects and 125 attributes.

Figure 3.7 shows the iceberg lattice for a minimum support of 70 %. For each concept, its support is shown. One can clearly see that all mushrooms in the data set have the attribute 'veil type: partial'. Furthermore, the diagram tells us that the four next-frequent attributes are: 'veil color: white' (with 97.54 % support), 'gill attachment: free' (97.42 %), 'ring number: one' (92.17 %), and 'gill spacing: close' (83.85 %).

There is no other attribute having a support higher than 70 %. But even the combination of all these five attributes is frequent (with respect to our threshold of 70 %): 77.20 % of all mushrooms in our data set have one ring, a white partial veil, and free, close-standing gills. This concept with a quite complex description contains more objects than the concept described by the sixth-most attribute, which is not displayed in the diagram, since it has a support below our threshold of 70 %.

In the diagram, we can detect the implication

$$\{\text{ring number: one, veil color: white}\} \to \{\text{gill attachment: free}\}.$$

It is indicated by the fact that there is no concept having 'ring number: one' and 'veil color: white' (as well as 'veil type: partial') in its intent, but not 'gill attachment: free'. This implication has a support of 89.71 %.

The fact that the combination {veil type: partial, gill attachment: free, gill spacing: close} is not realized as a concept intent indicates another implication:

$$\{\text{gill attachment: free, gill spacing: close}\} \to \{\text{veil color: white}\}$$

This implication has 81.27 % support (the support of the most general concept having all three attributes in its intent).

3.4.3 Association rules

There is extensive literature on fault-tolerant analogues to implications, see Blockeel [48] for a recent point of view. Indeed, even small errors in the data given as a finite formal context (G, M, I) can drastically change the validity of implications. Luxenburger [158] investigated **partial implications**, i.e., implications with an extra parameter called[11] the **confidence**, which is the fraction of objects whose intents are models of $A \to B$ among those that support the premise. Formally:

$$\text{confidence}(A \to B) := \frac{|(A \cup B)'|}{|A'|} = \frac{\text{support}(A \cup B)}{\text{support}(A)} \qquad (A' \neq \emptyset).$$

For the trivial case of $A' = \emptyset$, we define the confidence to be 1.

Later, the support parameter was included and implications with given support and confidence were named **association rules** [1].

A partial implication thus is an implication together with its confidence; an association rule is a partial implication plus its support or, equivalently, an implication together with support and confidence values.

Implications are sometimes dubbed *exact association rules*, since they are precisely the association rules with confidence equal to 1. Association rules with a strictly lower confidence that nevertheless exceeds some given value `minconf` are also referred to as **approximate association rules**.

The "logic" of partial implications was studied by M. Luxenburger [158]. Later, Bastide *et al.* observed that association rule inference can seamlessly be

[11] Luxenburger used perhaps a more appropriate name, "relative conditional frequency".

combined with the support parameter [31]. This is necessary for *association rule mining*, where the task is to determine, for given minimum support and confidence thresholds, all association rules of a dataset. We are not going into detail here, although the theory is nice.

The two key observations are expressed by the following easy proposition:

Proposition 28

1. *For all* $A \subseteq B \subseteq C \subseteq M$,

$$\text{confidence}(A \to C) = \text{confidence}(A \to B) \cdot \text{confidence}(B \to C).$$

2. *In any finite formal context, the implications*

$$A \to B \qquad \textit{and} \qquad A'' \to (A \cup B)''$$

have the same support and the same confidence.

The two observations together imply that support and confidence of any association rule in a finite formal context can easily be computed, using the closure operator, from those of the associations rules of the form $A \to B$, where $A \subsetneq B$ are intents of neighboring concepts. This is often called the **Luxenburger basis** for association rules. In other words, the implications together with the association rules corresponding to edges of the lattice diagram (oriented downwards) suffice. Even this set is redundant, and a spanning tree of the neighborhood relation is enough, see [158] for details. Note that the confidence of an implication $A \to B$ with $A' \neq \emptyset$ and $A \subseteq B$ is simply the frequency ratio: $\text{confidence}(A \to B) = \frac{|B'|}{|A'|}$, and the support is just that of B.

To determine if an association rule $A \xrightarrow{c} B$ with confidence $c \geq \texttt{minconf}$ holds in a formal context (and to calculate its support and confidence), one can consider the rule $C \xrightarrow{c} D$ with $C := A''$ and $D := (A \cup B)''$, which has (by Proposition 28) the same support and the same confidence and is frequent only if D is. If $C = D$, then the confidence is equal to 1 and the support equals $\text{support}(C)$. If $C \neq D$, then there is a chain

$$C = C_1 \subsetneq C_2 \subsetneq \cdots \subsetneq C_n = D$$

of neighboring concept intents. The confidence of $C \to D$ is then given as

$$\prod_{i=1}^{n-1} \text{confidence}(C_i \to C_{i+1})$$

and the support is equal to that of the largest set, D.

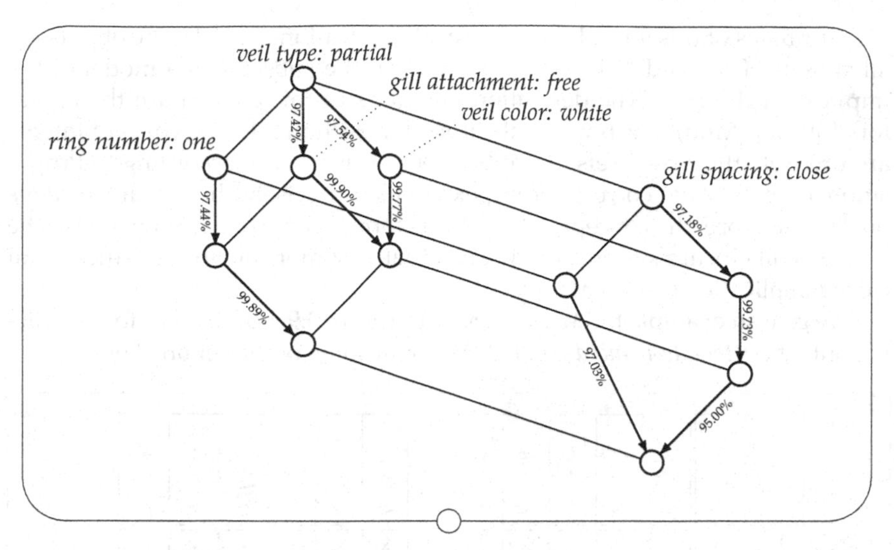

Figure 3.8: Visualization of the Luxenburger basis for `minsupp` $= 70\,\%$ and minconf $= 95\,\%$

The Luxenburger basis for approximate association rules can also be visualized directly in the line diagram of an iceberg concept lattice. Each approximate rule in the Luxenburger basis corresponds to exactly one edge in the line diagram. Figure 3.8 visualizes all rules in the Luxenburger basis of the Mushroom data set of Section 3.4.2 for minsupp = 70 % and minconf = 95 %. For instance, the rightmost arrow stands for the association rule

{veil color: white, gill spacing: close} $\to$ {gill attachment: free},

which holds with a confidence of 99.73 %. Its support is the support of the concept the arrow is pointing to: 81.27 %, as shown in Figure 3.7. Edges without label indicate that the confidence of the rule is below the minimum confidence threshold.

The derivation rules of Proposition 28 can also be applied in the diagram: the confidence of the association rule {gill spacing: close} $\to$ {gill attachment: free} is the product of the confidences of two edges in the diagram: $97.18\,\% \cdot 99.73\,\% \approx 96.92\,\%$.

3.4.4 Under symmetry

It was already mentioned in Section 2.1.4 (p. 43) that the closure systems on M themselves form a closure system. Consider the formal context

$$(\mathfrak{P}(M), \{A \to B \mid A, B \subseteq M\}, \models).$$

The attributes of this formal context are all implications over M, the objects are all subsets of M, and $T \models A \to B$ says that the subset T is a model of the implication $A \to B$. The derivation operators of this context are the operators Imp and Mod, which were introduced in Section 3.1. The concept intents are precisely the closed sets of implications over M, and, according to Proposition 14 (p. 89), the corresponding extents are precisely the closure systems on M. Therefore, the concept lattice of this formal context is isomorphic to the lattice of all closure systems on M and dually isomorphic to the lattice of all closed implication sets on M.

We give an example for $M := \{a, b, c\}$ in Figure 3.9, but display, for simplicity, only the *attribute-reduced* context. We omit some of the set brackets.

$\models$	$\emptyset \to a$	$\emptyset \to b$	$\emptyset \to c$	$a \to b$	$a \to c$	$b \to a$	$b \to c$	$c \to a$	$c \to b$	$a, b \to c$	$a, c \to b$	$b, c \to a$
$\emptyset$				×	×	×	×	×	×	×	×	×
$\{a\}$	×					×	×	×	×	×	×	×
$\{b\}$		×		×	×			×	×	×	×	×
$\{a, b\}$	×	×		×		×		×	×		×	×
$\{c\}$			×	×	×	×	×			×	×	×
$\{a, c\}$	×		×		×	×	×	×		×		×
$\{b, c\}$		×	×	×	×		×		×	×	×	
$\{a, b, c\}$	×	×	×	×	×	×	×	×	×	×	×	×

Figure 3.9: A formal context for the closure system of closure systems on the set $M := \{a, b, c\}$

Note that the extents of this formal context form a closure system, while each extent itself is a closure system. Even more brain-twisting seems the question which implications hold in this context. The attributes themselves are implications. So we get implications between sets of implications. This requires new notation, because expressions like

$$\{A_1 \to B_1, A_2 \to B_2, A_3 \to B_3, \ldots\} \to \{C_1 \to D_1, C_2 \to D_2, \ldots\}$$

may be difficult to read. Following the standard notation from mathematical logic, we therefore write

$$\frac{A_1 \to B_1, \quad A_2 \to B_2, \quad A_3 \to B_3, \quad \ldots}{C_1 \to D_1, \quad C_2 \to D_2, \quad \ldots}$$

in such a case, expressing that $A_1 \to B_1, A_2 \to B_2, A_3 \to B_3, \ldots$ together imply all of $C_1 \to D_1, C_2 \to D_2, \ldots$ But even with this notation, the canonical basis

$$\frac{c \to b}{a, c \to b} \qquad \frac{c \to b, \quad a, c \to b, \quad b, c \to a}{c \to a} \qquad \frac{c \to a}{b, c \to a}$$

$$\frac{c \to a, \quad a, c \to b, \quad b, c \to a}{c \to b} \qquad \frac{b \to c}{a, b \to c} \qquad \frac{b \to c, \quad a, b \to c, \quad b, c \to a}{b \to a}$$

$$\frac{b \to a}{b, c \to a} \qquad \frac{b \to a, \quad a, b \to c, \quad b, c \to a}{b \to c} \qquad \frac{a \to c}{a, b \to c}$$

$$\frac{a \to c, \quad a, b \to c, \quad a, c \to b}{a \to b} \qquad \frac{a \to b}{a, c \to b} \qquad \frac{a \to b, \quad a, b \to c, \quad a, c \to b}{a \to c}$$

$$\frac{\emptyset \to c}{a \to c, \quad b \to c, \quad a, b \to c} \qquad \frac{\emptyset \to b}{a \to b, \quad c \to b, \quad a, c \to b} \qquad \frac{\emptyset \to a}{b \to a, \quad c \to a, \quad b, c \to a}$$

$$\frac{\emptyset \to c, \quad a \to c, \quad b \to a, \quad b \to c, \quad c \to a, \quad a, b \to c, \quad b, c \to a}{\emptyset \to a}$$

$$\frac{\emptyset \to c, \quad a \to b, \quad a \to c, \quad b \to c, \quad c \to b, \quad a, b \to c, \quad a, c \to b}{\emptyset \to b}$$

$$\frac{\emptyset \to b, \quad a \to b, \quad b \to a, \quad c \to a, \quad c \to b, \quad a, c \to b, \quad b, c \to a}{\emptyset \to a}$$

$$\frac{\emptyset \to b, \quad a \to b, \quad a \to c, \quad b \to c, \quad c \to b, \quad a, b \to c, \quad a, c \to b}{\emptyset \to c}$$

$$\frac{\emptyset \to a, \quad a \to c, \quad b \to a, \quad b \to c, \quad c \to a, \quad a, b \to c, \quad b, c \to a}{\emptyset \to c}$$

$$\frac{\emptyset \to a, \quad a \to b, \quad b \to a, \quad c \to a, \quad c \to b, \quad a, c \to b, \quad b, c \to a}{\emptyset \to b}$$

Figure 3.10: The canonical basis of the formal context in Figure 3.9

of the context in Figure 3.9 is not very appealing, as can be seen in Figure 3.10. It becomes slightly better when the implications are simplified as described in Section 3.2.4, see Figure 3.11. A closer look at this list unveils that a much more radical simplification is possible. Each implication in the list is repeated several times, except that the letters a, b, c are permuted. There are actually only four different types of implications in the list, and these are given in Figure 3.12. This shortened form resembles the *Armstrong rules*, which were mentioned in Section 3.1.2 (p. 90), but is stated for single elements and not for sets.

The result in Figure 3.12 can be obtained directly, without computing the full canonical basis first and folding it afterwards. The reason is that the set of pseudo-intents and the canonical basis are preserved by all automorphisms of the closure system. And, of course, each permutation of $\{a, b, c\}$ induces an automorphism of the formal context in Figure 3.9 and hereby an automorphism of the closure system of its intents.

$$\frac{c \to b}{a, c \to b} \qquad \frac{c \to b, \quad b, c \to a}{c \to a} \qquad \frac{c \to a}{b, c \to a}$$

$$\frac{c \to a, \quad a, c \to b}{c \to b} \qquad \frac{b \to c}{a, b \to c} \qquad \frac{b \to c, \quad b, c \to a}{b \to a}$$

$$\frac{b \to a}{b, c \to a} \qquad \frac{b \to a, \quad a, b \to c}{b \to c} \qquad \frac{a \to c}{a, b \to c}$$

$$\frac{a \to c, \quad a, c \to b}{a \to b} \qquad \frac{a \to b}{a, c \to b} \qquad \frac{a \to b, \quad a, b \to c}{a \to c}$$

$$\frac{\emptyset \to c}{a \to c, \quad b \to c} \qquad \frac{\emptyset \to b}{a \to b, \quad c \to b} \qquad \frac{\emptyset \to a}{b \to a, \quad c \to a}$$

$$\frac{\emptyset \to c, \quad c \to a}{\emptyset \to a} \qquad \frac{\emptyset \to c, \quad c \to b}{\emptyset \to b} \qquad \frac{\emptyset \to b, \quad b \to a}{\emptyset \to a}$$

$$\frac{\emptyset \to b, \quad b \to c}{\emptyset \to c} \qquad \frac{\emptyset \to a, \quad a \to c}{\emptyset \to c} \qquad \frac{\emptyset \to a, \quad a \to b}{\emptyset \to b}$$

Figure 3.11: The simplified canonical basis of the formal context in Figure 3.9

$$\frac{x \to y}{x, z \to y} \qquad \frac{x \to y, \quad x, y \to z}{x \to z} \qquad \frac{\emptyset \to x}{y \to x, \quad z \to x} \qquad \frac{\emptyset \to x, \quad x \to y}{\emptyset \to y}$$

Figure 3.12: The simplified canonical basis of the formal context in Figure 3.9, folded "modulo symmetries". The list of Figure 3.11 is obtained when x, y, z are replaced in all possible (bijective) ways by a, b, c

The algorithm for finding the pseudo-intents is an instance of the Next Closure algorithm. In Section 2.3.4 (p. 67) a version of this algorithm "with symmetry" was introduced, and that can be applied to pseudo-closed sets as well. It computes, for a given group Γ of automorphisms, exactly one pseudo-closed set from each Γ-orbit. Using this as a premise, we get exactly one implication from each Γ-orbit in the canonical basis.

3.5 Further reading

Implications are very much related to propositional Horn clauses, which play an important role in artificial intelligence research [125]. We touch on the connection in Section 5.1.2.

Kautz *et al.* discuss representation of a Horn formula by the set of its characteristic models [131], which corresponds to representing a set of attribute implications by an object-reduced formal context with the same implications. Khardon studies the complexity of translating between Horn formulas and their characteristic models and relates this to the hypergraph transversal problem [133]. Distel and Sertkaya investigate similar problems in the context of formal context analysis and provide various complexity results regarding the enumeration of pseudo-closed sets [76].

The optimized version of the algorithm Canonical Basis of Section 3.3.3 is from [32], where it is experimentally compared with the original version in combination with several algorithms for computing the closure of an attribute set under implications. An incremental approach to the construction of the canonical basis is proposed in [171]: the algorithm described there updates the canonical basis and the set of formal concepts whenever a new attribute is added to the context. Valtchev and Duquenne discuss how the canonical bases of two subcontexts can be merged to form the canonical basis of the entire context [209]. Godin and Missaoui describe an algorithm for extracting implications from an already constructed concept lattice [114].

The canonical basis is minimal in the number of implications among all equivalent implication sets, and Algorithm 18 Minimal Cover efficiently transforms any set of implications into the corresponding canonical basis. But what if we want an equivalent implication set with the minimal total number of attributes in the premises and conclusions of its implications? In the theory of functional dependencies, such a set is called *optimal* [159]. This problem seems much harder: no polynomial-time algorithm exists unless $\mathcal{P} = \mathcal{NP}$ [122, Theorem 4.1].

Reasoning with implications of high confidence is *non-monotonic*. This was investigated by Borchmann [51].

3.6 Exercises

3.1 Recall what it means for an implication to hold or be valid in a formal context. Give an example of a formal context (G, M, I) where the implication

$$\emptyset \to \{d\},$$

$d \in M$, holds and of a formal context where it does not hold.

3.2 Describe a family of formal contexts (G, M, I) with $|M| = 1, 2, \ldots,$ where all implications over M hold and a family of contexts where none of them holds.

3.3 Recall what it means for an implication $A \to B$ to semantically follow from a set $\mathcal{L}$ of implications. Let $\mathcal{L}$ consist of the following five implications over $M = \{a, b, c, d, e\}$:

- $\{a\} \to \{c\}$;
- $\{c\} \to \{a\}$;
- $\{a, b, c\} \to \{d\}$.
- $\{d\} \to \{b\}$;
- $\{e\} \to \{a, b, c, d\}$;

Do the implications

$$\{b\} \to \{a, b, d, e\} \quad \text{and} \quad \{b, c, d\} \to \{a, b, c, d\}$$

follow from $\mathcal{L}$? Why or why not?

3.4 Now consider the following set of four implications over $M = \{a, b, c, d, e\}$:

- $\{a\} \to \{b, c\}$
- $\{c, e\} \to \{a, b, d\}$
- $\{d, e\} \to \{a, b, c\}$
- $\{c, d\} \to \{a, b, e\}$

Which of the implications below semantically follow from this set?

- $\{a\} \to \{a, b, c\}$
- $\{d\} \to \{a, e\}$
- $\{a, e\} \to \{d\}$

3.5 Recall Algorithm 15, which computes $\mathcal{L}(X)$, the closure of an attribute set X under the implications in $\mathcal{L}$. How long does it take this algorithm to compute $\mathcal{L}(\{1\})$ if

$$\mathcal{L} = \{\{i\} \to \{i+1\} \mid i \in \mathbb{N}, 0 < i < n\}$$

for some n assuming that the implications in $\mathcal{L}$ are processed in the *ascending* order of their one-element premises (first $\{1\} \to \{2\}$, then $\{2\} \to \{3\}$, etc.)? What if the implications are processed in the reverse, *descending*, order?

3.6 Compute the canonical basis of the context in Exercise 2.1 from Chapter 2 (p. 82).

3.7 Below is yet another formal context taken from an advertisement that describes a person's conditions positively affected by various ingredients in candies.

	mood	*immunity*	*eyesight*	*vigor*
raspberry	×			×
cherry				
strawberry		×		
blueberry	×		×	
cranberry		×		×
cowberry		×		
tomato and celery	×			
beetroot		×		
pumpkin		×		
carrot		×	×	
bell pepper			×	

(a) Compute all the concepts of this context.

(b) Draw a line diagram of its concept lattice.

(c) Construct its canonical basis.

3.8 What are the pseudo-closed sets of the closure system in Exercise 2.4 from Chapter 2 (p. 83)?

3.9 (Based on [131].) Let k be some natural number and consider the formal context

$$(\bigcup_{i=1}^{k} G_i, M \cup N, I),$$

where

$$G_i = \{x_i^j, y_i^j, z_i^j \mid 1 \le j \le k \text{ and } j \ne i\};$$
$$M = \{m_1, \ldots, m_k\}; \quad N = \{n_1, \ldots, n_k\};$$

$$\begin{array}{lcl} x_i^j I m_l & \iff & i \ne l; \\ x_i^j I n_l & \iff & i \ne l; \\ y_i^j I m_l & \iff & i \ne l \text{ and } j \ne l; \\ y_i^j I n_l & \iff & i \ne l; \\ z_i^j I m_l & \iff & i \ne l; \\ z_i^j I n_l & \iff & i \ne l \text{ and } j \ne l. \end{array}$$

How many objects and attributes are there in the context? What is the number of implications in its canonical basis? Describe all the pseudo-intents of this context.

3.10 Describe a family of contexts where the size of the canonical basis grows linearly with the size of the context (assuming that the size of a context (G, M, I) is $|G||M|$).

3.11 Give a complete proof of Proposition 25 (p. 113). That is, assuming that $\mathcal{L}$ is the set of all frequent implications in the canonical basis of $\mathbb{K}$ and $A \to B$ is a frequent implication valid in $\mathbb{K}$, prove that $A \to B$ follows from $\mathcal{L}$.

3.12 The *apposition* of two formal contexts (G, M_1, I_1) and (G, M_2, I_2), where $M_1 \cap M_2 = \emptyset$, is the formal context $(G, M_1 \cup M_2, I_1 \cup I_2)$.

Compute the iceberg lattice of the apposition of the formal contexts from Figure 1.22 (p. 36) and Exercise 1.7 from Chapter 1 (p. 38) for `minsupp` $= 0.2$. Use this iceberg lattice to compute association rules for the same `minsupp` and `minconf` $= 0.3$. For each association rule, indicate its support and confidence.

3.13 Call a set $A \subseteq M$ a *minimal premise* for $B \subseteq M$, if $A \to B$ holds, but $S \to B$ holds for no proper subset S of A. What is the relation to proper premises? Are all minimal premises proper? What about the converse?

Chapter 4

Attribute exploration

Have another look at the canonical basis in Figure 3.2 (p. 97), which was computed for the context of triangles. The implications in this set are obviously true, because *they hold for all triangles*, not only for the objects of the triangles context on page 23. This information is important, because it shows that the concept lattice given in Figure 1.16 (p. 27) will not change if more examples are added. The triangles in the formal context are representative of the entire theory. The canonical basis generates the implication theory of these five attributes for *all* triangles. Every implication that holds for all triangles follows from the canonical basis. For every implication that does not hold in general, there is already a counterexample among the seven triangles of the formal context.

Such completeness can be achieved through a process called attribute exploration, which, in its simplest version, can be summarized as follows. Given a subject domain and unified descriptions of its objects in terms of presence or absence of certain attributes, attribute exploration tries to build the implication theory of the entire domain and to collect a set of its objects that is representative in a very precise sense. Ideally, the implication theory is obtained in its natural minimal representation, the canonical basis, from which all valid implications can be inferred. The representative set of objects should respect all implications from the generated basis and provide a counterexample for every implication that cannot be inferred from the basis. This means, in particular, that the concept lattice of the domain is isomorphic to that of the representative objects.

4.1 The exploration algorithm

We design an algorithm that accomplishes the goals formulated above. First, we give an abstract version, which is then realized in the form of pseudocode as a basis for actual implementations. A short discussion follows on how to interpret the outcome.

4.1.1 Abstract exploration

Attribute exploration is an instance of a somewhat obvious general strategy: in order to explore a given domain of which some examples and some rules are known, one investigates *unsolved questions*. Answering such a question extends the knowledge either by a new rule or by another example refuting a conjectured rule.

In general terms, the ingredients of an exploration are as follows:

1. The **domain** to be explored. It is often also called the **universe** of the respective exploration.

   It is assumed that the domain is not directly accessible, but only through a *domain expert*, see below.

2. A **signature** that specifies the logical language used to explore the domain.

3. A family $\mathcal{F}$ of logical formulas (of the given signature), each of which may or may not hold in the domain under investigation.

4. An **exploration knowledge base** $(\mathcal{L}, \mathcal{E})$, which grows during exploration. Both parts may initially be empty.

   - $\mathcal{L}$ is a set of formulas of the given signature, all of which are known to hold in the domain under investigation.
   - $\mathcal{E}$ is some additional information about the domain, say, a collection of (correct) assertions about it ("examples").

5. The **inference engine**, an algorithm that tries to check if a proposed formula from $\mathcal{F}$ is a logical consequence of $\mathcal{L}$. The inference engine must be consistent in the sense that it never confirms false consequences, but it does not have to be complete: it may fail to verify that f follows from $\mathcal{L}$ even if this is the case. However, we assume that the inference engine is not completely trivial: it can at least deduce $\mathcal{L}$ from $\mathcal{L}$.

6. The **query engine**, an algorithm that finds, if it exists, a formula in $\mathcal{F}$ that cannot be deduced from $\mathcal{L}$ by the inference engine and that cannot be **refute**d by $\mathcal{E}$ (i.e., is logically consistent with $\mathcal{E}$).

7. A **domain expert**, which, when confronted with a formula $f \in \mathcal{F}$, either (reliably) confirms that f holds in the domain or provides evidence against f by extending $\mathcal{E}$ so that it refutes f as well.[1] The expert may be a person, a database equipped with an appropriate algorithm, etc.

[1] One can imagine an expert that answers "I don't know" to some questions, but we do not consider this possibility here.

We assume that the logic involved has a monotonic consequence relation and that the inference engine is also monotonic: if f can be inferred from $\mathcal{L}$, it can also be inferred from any superset of $\mathcal{L}$; and if f is refuted by $\mathcal{E}$, this cannot be changed by extending $\mathcal{E}$ with new evidence.

An exploration knowledge base $(\mathcal{L}, \mathcal{E})$ is called **$\mathcal{F}$-complete** with respect to the domain under investigation if the following three conditions are equivalent for all $f \in \mathcal{F}$:

1. f holds in the domain;

2. f is a logical consequence of $\mathcal{L}$;

3. f is consistent with $\mathcal{E}$.

The **exploration algorithm** repeatedly lets the query engine find undecided formulas $f \in \mathcal{F}$, each of which is either confirmed by the domain expert and added to $\mathcal{L}$ or disproved by additional evidence extending $\mathcal{E}$.

Theorem 8 *If $\mathcal{F}$ is finite, then the exploration algorithm will eventually lead to an $\mathcal{F}$-complete knowledge base.*

Proof Each formula $f \in \mathcal{F}$ is judged at most once by the domain expert, because the expert provides either a confirmation for f or evidence that refutes f, and this is stored in the knowledge base. Since the logic is monotonic, the algorithm will never need additional evidence against formulas that have been refuted. Neither it will need additional confirmation for formulas that have been confirmed, as these are added to $\mathcal{L}$ and thus the inference engine is able to deduce them from $\mathcal{L}$. So the algorithm terminates after finitely many queries. If f holds in the domain, but does not follow from $\mathcal{L}$, or if f does not hold in the domain, but is consistent with $\mathcal{E}$, then f can be found by the query engine. Therefore, the algorithm will not terminate as long as there is such f in $\mathcal{F}$. $\square$

All this is quite obvious as a general strategy. The crucial difficulties are to find (apart from a reliable domain expert) good inference and query engines. This goes surprisingly well in the case of basic attribute exploration, for which the above list is realized as follows:

1. The *domain* is a formal context (G, M, I) with finitely many attributes.

2. The *signature* is M, the logic is the propositional logic over M.

3. $\mathcal{F}$ is the set of all implications over M.

4. The *exploration base* $(\mathcal{L}, \mathcal{E})$ consists of

    - a list $\mathcal{L}$ of implications known to hold in (G, M, I);
    - a subcontext $\mathcal{E} := (E, M, J := I \cap E \times M), E \subseteq G$.

5. *Inference* is the propositional implication inference and thus is easy.

6. The *query engine* uses the NEXT CLOSURE algorithm to find the lectically smallest set $A \subseteq M$ for which

$$A = \mathcal{L}(A) \neq A^{JJ}.$$

If such a set exists, then $A \to A^{JJ}$ is posed as a question to the domain expert.

7. The *domain expert* acts as a mapping

$$p : \mathcal{F} \to \{(g, g') \mid g \in G\} \cup \{\top\}.$$

The remarkable advantage of basic attribute exploration is that questions are asked by the algorithm in such a way that eventually the canonical basis of the domain is produced as output, see Algorithm 19. Among other things, this implies that the number of implications the domain expert has to confirm is as small as possible.

4.1.2 A concrete algorithm

Algorithm 19 implements the basic version of attribute exploration. The algorithm starts with a context (E, M, J), which includes some examples from the domain (G, M, I), but it has access to the rest of the domain only through *interactive input* provided by the domain expert. The algorithm asks the expert whether an implication $A \to A^{JJ}$ has to be added or a new example refuting it is needed. In the pseudocode of Algorithm 19, the two lines demanding interactive input are marked with {*}. Algorithm 19 follows Algorithm 16 in how it computes the canonical basis, but it is perfectly possible to use the optimized procedure given by Algorithm 17.

Let us describe in words what the algorithm does. Since asking about implications $A \to A^{JJ}$ may be expensive, such interactive inquiries are avoided whenever possible. To this end, the algorithm uses the already acquired information, which is stored in the exploration base consisting of two lists: a list $\mathcal{L}$ of implications and a formal context $(E, M, J := I \cap E \times M)$ of examples with their attributes. Initially, both lists may be empty. Whenever a set $A \subseteq M$ is considered, these two lists yield partial information about the closure A^{II} of A, since this closure must contain $\mathcal{L}(A)$ and must be contained in A^{JJ}. Therefore, if $\mathcal{L}(A) = A^{JJ}$, then $A^{II} = A^{JJ}$ and there is no need to bother the expert.

Thus, the algorithm computes the lectically smallest set A closed under $\mathcal{L}$ (that is, $\mathcal{L}(A) = A$) for which $A \neq A^{JJ}$ and asks the domain expert the question

$$A \to A^{JJ}?$$

Algorithm 19 ATTRIBUTE EXPLORATION (basic version)

Input: A possibly empty subcontext $(E, M, J = I \cap E \times M)$ of a formal context (G, M, I), M finite.
Interactive input: $\{*\}$ Upon request, confirm that $C^{II} = D$ for $C, D \subseteq M$ or give an object $g \in G$ and its intent g^I such that $C \subseteq g^I$, but $D \nsubseteq g^I$, thus showing that $C^{II} \neq D$.
Output: The canonical basis $\mathcal{L}$ of (G, M, I) and a possibly enlarged subcontext $(E, M, J = I \cap E \times M)$ with the same implications.

```
L := ∅
A := ∅
while A ≠ M do
    while A ≠ A^JJ do
        if A^II = A^JJ then                                   {*}
            L := L ∪ {A → A^JJ}
            exit while
        else
            extend E by some object g ∈ A^I \ A^JJI           {*}
    A := NEXT CLOSURE(A, M, L)
return L, (E, M, J)
```

This is the same as asking whether $A^{II} = A^{JJ}$, and this is how the question is formulated in the pseudocode. The answer may be "yes", in which case the list $\mathcal{L}$ is extended by this implication. Or it may be an example that does not respect the implication, which is then included in the formal context (E, M, J) of examples.

After being presented with such input, the algorithm searches again for the "open question" of the form $A \to A^{JJ}$, where A is the lectically smallest set such that

$$A = \mathcal{L}(A) \neq A^{JJ}.$$

Meanwhile, due to the given input, either $\mathcal{L}(A)$ has become larger or A^{JJ} has become smaller, and this change is irreversible. Taking into account that the algorithm asks questions about implications in the lectic order of their premises and hence, once done with a premise, it never gets back to it, it is easy to see that no question will be asked twice.

Lemma 1 *At the beginning of each iteration of the outer **while** loop of Algorithm 19, the set A is preclosed in (G, M, I) and the set $\mathcal{L}$ consists of all implications of the form $P \to P^{II}$, where P is a pseudo-intent of (G, M, I) lectically smaller than A.*

Proof The statement is obviously true for the first iteration, when $A = \emptyset$ and $\mathcal{L} = \emptyset$. We show that the statement remains true after an iteration of the outer **while** loop provided that it was true before this iteration.

If A is closed in (G, M, I), we can exit the inner **while** loop only when, by extending E with zero or more new objects, we obtain $A^{II} = A = A^{JJ}$. Note that $\mathcal{L}$ does not change. We then assign A to the lectically next set closed under $\mathcal{L}$. There is no $\mathcal{L}$-closed set, and hence no (G, M, I)-pseudo-closed set, that is lectically between the sets corresponding to the old and the new values of A. Thus, the statement remains true for the new value of A.

If A is not closed, it is pseudo-closed. In this case, A never becomes equal to A^{JJ}, but when, at some point, we reach $A^{II} = A^{JJ}$, we add the implication $A \to A^{JJ} = A^{II}$ to $\mathcal{L}$, exit the inner **while** loop, and assign A to the lectically next $\mathcal{L}$-closed set, which is also the lectically next (G, M, I)-preclosed set. Obviously, the statement still holds for this new A. □

Corollary 2 *If Algorithm 19 asks a question about an implication $A \to A^{JJ}$ during attribute exploration of a formal context (G, M, I), then A is preclosed in (G, M, I).*

As argued above, no question is asked twice during execution of the algorithm. Since, for a finite M, the number of possible questions is finite, the algorithm eventually terminates with $A = M$. It follows from Lemma 1 that the resulting $\mathcal{L}$ is the canonical basis of $\mathbb{K}$. Therefore, we have

Theorem 9 *Let $\mathbb{K} = (G, M, I)$ be a formal context with finite M. Taking as input a subcontext (E, M, J) of $\mathbb{K}$ with $J = I \cap E \times M$, Algorithm 19 finds the canonical basis $\mathcal{L}$ of $\mathbb{K}$ after a finite number of questions.*

4.1.3 When the exploration terminates

So attribute exploration always terminates at some point, although it is often laborious, and it is a relief when the completion is confirmed:[2]

The strongest feature of the algorithm is not its efficiency, but its completeness. When the algorithm terminates, the implication theory of the respective formal context is completely known (provided the input was correct). No essential

[2]This is a screenshot taken with the Concept Explorer software, which implements the basic version of attribute exploration [229].

counterexample has been overlooked; each possible attribute combination is the intersection of attribute combinations of the examples; each implication that is not valid in general is already refuted by the examples.

We have advertised attribute exploration as a tool for classifying all feasible attribute combinations in a given domain. So do the acquired examples precisely reflect all attribute combinations that may occur?

Well, not exactly! Sometimes we are not even close to that, since the number of feasible solutions may be exponential in the number of examples, and the number of possible solution sets may even be doubly exponential. But fortunately the situation is very transparent, and the exploration result is sufficient for many purposes.

Attribute exploration uncovers the implication theory of the given domain. It collects sufficiently many implications to generate the theory and sufficiently many examples to refute all invalid implications. According to Proposition 14, the models of the implications form a closure system, the closure system of intents of the context of examples. The essential examples are the *irreducible* ones, as defined in Section 1.4.3 (p. 27). For determining the implications, it is necessary and sufficient to determine all meet-irreducible attribute combinations, as the following consideration shows.

Let A be a feasible attribute combination and let B_t, $t \in T$ for some index set T, be the family of all feasible attribute combinations properly containing A. We denote their intersection by $B := \bigcap_{t\in T} B_t$ and get

$$A \subseteq B = \bigcap_{t\in T} B_t.$$

A is meet-irreducible in the lattice of intents of the domain context if $B \neq A$, and this is true if and only if there is some element b that belongs to B but not to A. But then b must belong to each set B_t, independently of the index t. The implication

$$A \to \{b\}$$

is therefore *not* refuted by any feasible attribute combination other than A, because every refuting combination must contain A, but not b. This shows that each meet-irreducible attribute combination is indispensable for the implication theory. Conversely, if A is meet-reducible and $C \to D$ is an implication refuted by A, then $C \subseteq A$ and $d \notin A$ for some element d of D. But since A is meet-reducible, there must be some $B_t \supseteq A$ not containing d, and this B_t refutes $C \to D$ as well. Meet-reducible attribute combinations are thus not informative for the implication theory.

After all, which attribute combinations are possible? Here is the precise answer: the result of the exploration tells us that all possible attribute combinations of the domain are intersections of the ones given as examples. Which of these do actually occur in the given domain cannot be determined by implications alone. Indeed, choose *any* family of intents of the context of examples including all the irreducible ones. Then there are propositional logic formulas the models of which are precisely the intents in that family, but which entail precisely the implications valid in the context. In fact, for each reducible intent A, the formula

$$\bigwedge A \quad \to \quad \bigvee\{\bigwedge B \mid B \text{ is an upper neighbor of } A\}$$

excludes the attribute combination A, and all these formulas are independent of each other. Such formulas will play a role in Section 5.1.3 (p. 191).

The concept lattice of triangles in Figure 1.16 (p. 27) shows three concepts that are not labeled by objects, and indeed all three do not correspond to attribute combinations of any triangle. For the two upper ones, this is due to the fact that every triangle is one of *acute-*, *right-*, or *obtuse-angled*, that is,

$$\emptyset \to \textit{acute-angled} \vee \textit{right-angled} \vee \textit{obtuse-angled}.$$

The least element is excluded by, for example,

$$\{\textit{acute-angled}, \textit{right-angled}\} \to \bot.$$

The $\bot$ notation that we have used, e.g., in Figure 3.2 gives some extra information: $P \to \bot$ expresses that $P \to M$ holds, because $P' = \emptyset$. From the canonical basis as it is written in Figure 3.2, we can conclude that the least of the three unlabeled lattice elements in Figure 1.16 (p. 27) indeed does not occur.

To summarize: the possible attribute combinations are among the concept intents of the context of examples. Which of the reducible ones actually can occur must be decided one-by-one.

This has its expression in the concept lattice. Each join-irreducible concept (that is, each concept with a meet-irreducible intent) must be labeled with an object. Join-reducible concepts may be object-labeled too, but which ones are is completely independent of the implication theory and cannot be concluded from the exploration result (except, of course, that the examples in the exploration knowledge base cannot be excluded).

4.1.4 Exploration strategy and incompleteness

The number of implications to be confirmed equals the number of pseudo-closed sets and thus is pre-determined. The number of examples may vary, even for a fixed dataset. An example given later as input for the interactive algorithm may make redundant an example given earlier. Therefore, the number of examples depends on the answering strategy of the user. One can come up with examples where this takes a long time. For instance, consider the trivial closure operator with $A = A^{II}$ for all $A \subseteq M$. The least efficient answering strategy is to reply to each question

$$A \to A^{JJ}?$$

by giving the trivial counterexample A. This will result in an exponential number of questions for a rather poor result: there will be no pseudo-closed sets at all. The most efficient strategy in this case requires only $|M|$ examples (see Section 2.2.4 for an idea of what such strategy might be).

But experience shows that such cases are rather artificial. Usually, the algorithm is quite parsimonious in asking. Of course, every canonical basis implication and every irreducible example must be given as input, because otherwise the information would necessarily be incomplete. This suggests the lower bound of $k + |\mathcal{L}|$ on the number of questions asked during exploration, where $\mathcal{L}$ is the canonical basis and k is the number of objects in the reduced version of (G, M, I) with intents different from those initially recorded in the exploration knowledge base. The questions about implications from $\mathcal{L}$ are the only ones that are answered positively. Note that this lower bound is exact: it is always possible to provide counterexamples in such a way that only so many questions are asked. Indeed, whenever a question is asked about the validity of $A \to A^{JJ} \neq A^{II}$, it can always be answered with an irreducible counterexample from $G \setminus E$.

Reducible examples are somewhat superfluous, but hard to avoid. Have a look at Figure 4.10 (p. 149), which shows the counterexamples given for an attribute exploration on music. Roughly one third of the examples are reducible, and, in hindsight, the same result would have been obtained using only the remaining 20 examples. During an exploration, it is very difficult to predict if an example will eventually be reducible or not. If one has a choice, it may be advisable to prefer examples with many attributes, because they are less likely to be reducible.

Every single implication $P \to P''$ that the query engine proposes may be difficult to decide. Is it possible to somehow support the domain expert in the process? We see at least three possibilities:

1. Shortening the premise. This was discussed in Section 3.2.4 (p. 99).

2. Allowing to confirm only a part of the conclusion. If the domain expert confirms $P \to Q$, where Q is a subset of P'', then this answer can be treated as harmless background knowledge, see Section 4.3.2 below. The query engine, when invoked again, will then usually propose a new implication of the form $P \cup R \to P''$, where $Q \subseteq R$.

3. Offering possible counterexamples. Each counterexample to $P \to Q$ must be an object intent C such that $P \subseteq C$ and $Q \not\subseteq C$. In particular, C must

be closed under all validated implications. Finding all closed sets C with $P \subseteq C, Q \nsubseteq C$ for $P \subseteq Q$ was the task of Exercise 2.6 (p. 84).

Complete attribute exploration can be tiresome. Sometimes, it suffices to find only some important implications and to leave others open. It is, of course, a matter of interpretation what makes an implication *important*, but a natural choice is to concentrate on implications with a short premise, which were already mentioned in Section 3.4.1 (p. 109).

The question of modifying the attribute exploration process raises a plethora of theoretical questions, some of which will be treated in subsequent chapters. We conclude this section with a very abstract, but sometimes helpful, structural view on **incomplete explorations**.

In Sections 2.1.4 (p. 43) and 3.4.4 (p. 117), we have mentioned the lattice of all closure systems on a given attribute set M and how it can be described as a concept lattice. The extents of the concepts are the closure systems, while the intents are the corresponding implication theories.

The attribute exploration process collects implications and counterexamples. At any intermediate state of the process, we have a set of examples, a set of implications, and thereby two closure systems: the closure system generated by the given examples and the system of models of the confirmed implications. Since it is required that all examples respect the confirmed implications, the former closure system is contained in the latter. The two systems form an interval in the lattice of closure systems. When further examples or implications are added, the interval is restricted to a subinterval.

Therefore, the attribute exploration process results in a focussing sequence of intervals in the lattice of closure systems, which eventually converges to a single point. An incomplete exploration results in a nontrivial interval, the elements of which represent the still possible outcomes. This is visualized in Figure 4.1.

4.2 Examples

In Section 4.2.2, we complete our "running example" from the very first pages of the book: the classification of *pairs of squares* with respect to some simple attributes. We give this example primarily because it is small and clear enough to show all the steps of the algorithm correctly.

Most of the examples in this book come from mathematics, and for a good reason. Attribute exploration is not confined to mathematical concepts, but whether *an object has an attribute* can be controversial if not both are defined mathematically. A "cross" in a formal context with mathematical content represents a *fact* ("this triangle is isosceles"), while it may have *normative* significance in another context ("we agree that this research is important"). Such

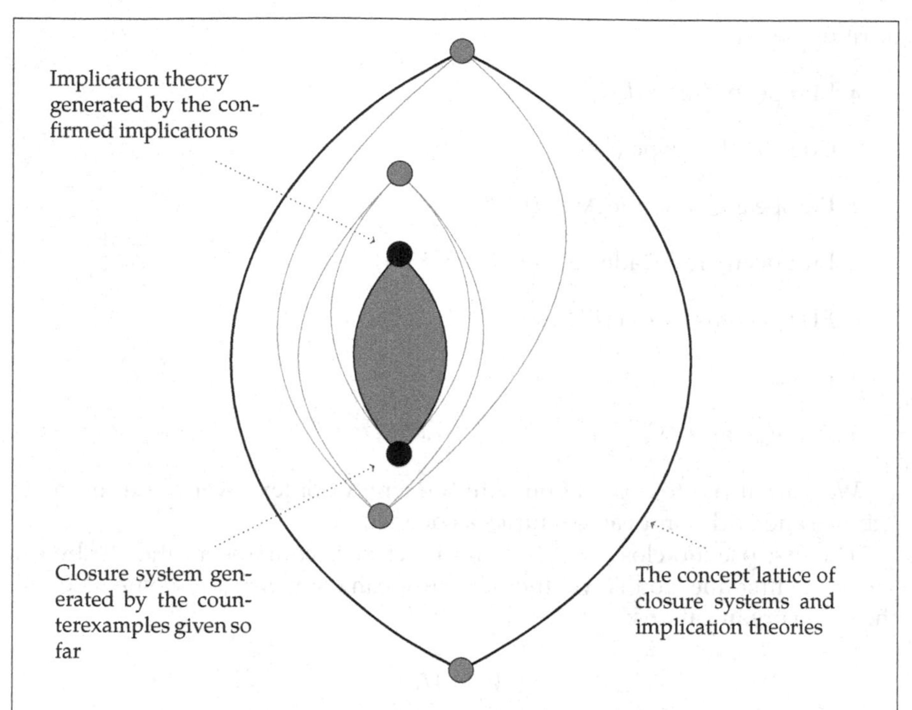

Figure 4.1: The attribute exploration procedure yields a converging sequence of intervals in the lattice of closure systems and implication theories. Ideally, the final result is a single element of this lattice, but an intermediate result may be an arbitrary interval

normative studies may be less convincing to the reader, because she may have a different opinion. Our experience is however that such investigations are considered particularly useful because they can help build consensus in a group of experts effectively. Therefore, as another example we present Rudolf Wille's prominent exploration on the perception of music, followed by a small example in information security.

As a warm-up, we start by exploring the membership of European countries in international organizations.

4.2.1 International organizations in Europe

As a first example, we are going to explore the structure of membership in various international organizations among the European countries. Here is the list of seven such organizations and associations that we will use as our attribute set M:

- European Union (*EU*)
- Council of Europe (*CoE*)
- European Economic Area (*EEA*)
- European Free Trade Association (*EFTA*)
- EU Customs Union (*EUCU*)
- *Eurozone*
- *Schengen* area

We start attribute exploration with the empty context over these seven attributes and add European countries as needed.

The first pseudo-closed set is $\emptyset$ and its closure contains all the attributes. Thus, the first question is whether all European countries are members of all the seven organizations:

$$\emptyset \to M.$$

This is not true, since, for example, Albania (currently the lexicographically first European country) is not a member of any of these organizations but the Council of Europe. So we add Albania:

	EU	*CoE*	*EEA*	*EFTA*	*EUCU*	*Eurozone*	*Schengen*
Albania		×					

Now, the closure of the empty set is equal to {*CoE*} and the next question is whether all the European countries are members of the Council of Europe:

$$\emptyset \to \{\textit{Council of Europe}\},$$

for which we name the Vatican City, which belongs to neither organization, as a counterexample:[3]

	EU	*CoE*	*EEA*	*EFTA*	*EUCU*	*Eurozone*	*Schengen*
Vatican							

[3]Vatican is not officially a member of the *Schengen area*, even though the open border with Italy makes it unnoticeable for many tourists visiting St. Peter's Basilica. Vatican is not a member of the *Eurozone* either, but uses the euro as its official currency.

Since our context contains no country from the *Schengen area* (represented by the largest attribute in the context according to the given order on attributes), the next question is whether such countries exist, which takes the form of the following implication:

$$\{\textit{Schengen area}\} \to M.$$

Here we reply with Switzerland, which is known for its reluctance to participate in international organizations, but which nevertheless joined the *Schengen area* in 2008:

	EU	*CoE*	*EEA*	*EFTA*	*EUCU*	*Eurozone*	*Schengen*
Switzerland		×		×			×

The set $\{\textit{Schengen area}\}$ is still pseudo-closed, but its closure has become smaller. Thus, the next implication we need to address is

$$\{\textit{Schengen area}\} \to \{\textit{Council of Europe}, \textit{EFTA}\}.$$

Again, the answer is "No", and we add Austria:

	EU	*CoE*	*EEA*	*EFTA*	*EUCU*	*Eurozone*	*Schengen*
Austria	×	×	×		×	×	×

The next question is whether the *Schengen area* includes only members of the Council of Europe:

$$\{\textit{Schengen area}\} \to \{\textit{Council of Europe}\},$$

which is indeed so. We accept the implication.

After that, we are asked whether being part of *Eurozone* implies membership in the European Union, Council of Europe, *EEA*, EU Customs Union, and the *Schengen area*:

$$\{\textit{Eurozone}\} \to \{\textit{EU}, \textit{Council of Europe}, \textit{EEA}, \textit{EUCU}, \textit{Schengen area}\}.$$

Cyprus is not yet in the *Schengen area*; therefore, it is a counterexample:

	EU	*CoE*	*EEA*	*EFTA*	*EUCU*	*Eurozone*	*Schengen*
Cyprus	×	×	×		×	×	

However, apart from that, the above implication is valid; so, we accept it with a smaller conclusion:

$$\{Eurozone\} \to \{EU, Council\ of\ Europe, EEA, EUCU\}.$$

In our context, which so far consists of five countries, each of them is either a member of both the *EUCU* and *Eurozone* or not a member of any of the two. Hence, the next implication we must consider is similar to the previous one, but is about the *EUCU* rather than the *Eurozone*:

$$\{EUCU\} \to \{EU, Council\ of\ Europe, EEA, Eurozone\}.$$

Here, unlike in the previous case, the answer is negative, because of, e.g., San Marino:

	EU	*CoE*	*EEA*	*EFTA*	*EUCU*	*Eurozone*	*Schengen*
San Marino		×			×		

Next, we accept the implication

$$\{EUCU\} \to \{Council\ of\ Europe\}$$

and proceed with

$$\{EFTA\} \to \{Council\ of\ Europe, Schengen\}.$$

Indeed, all the *EFTA* countries are members of the *Council of Europe*, and Liechtenstein was the last of them to join the *Schengen* area in 2011. However, it is part of neither the *EU* nor the *EUCU* nor the *Eurozone* and, therefore, serves a counterexample for the next implication:

$$\{EEA\} \to \{EU, Council\ of\ Europe, EUCU, Eurozone\}.$$

We add Liechtenstein to the exploration knowledge base:

	EU	*CoE*	*EEA*	*EFTA*	*EUCU*	*Eurozone*	*Schengen*
Liechtenstein		×	×	×			×

Since the *EEA* unites only countries from the *Council of Europe*, we accept the implication

$$\{EEA\} \to \{Council\ of\ Europe\}.$$

After this, we have to find a country that is a member of the *Council of Europe*, EU Customs Union, and *Schengen area*, but is outside the European Union or *EEA* or *Eurozone*. This would be a counterexample to the following implication:

$$\{\textit{Council of Europe}, \textit{EUCU}, \textit{Schengen area}\} \to \{\textit{EU}, \textit{EEA}, \textit{Eurozone}\}.$$

It has to be answered negatively due to the existence of Swedish kronas. We add Sweden:

	EU	*CoE*	*EEA*	*EFTA*	*EUCU*	*Eurozone*	*Schengen*
Sweden	×	×	×		×		×

and accept the relaxed implication

$$\{\textit{Council of Europe}, \textit{EUCU}, \textit{Schengen area}\} \to \{\textit{EU}, \textit{EEA}\}.$$

After this, we are asked to find a country that is a member of the Council of Europe, *EEA*, and *EUCU*, but is not in the European Union:

$$\{\textit{Council of Europe}, \textit{EEA}, \textit{EUCU}\} \to \{\textit{EU}\},$$

No such country can be found; so we accept the implication. We then agree that the "reverse" implication

$$\{\textit{EU}\} \to \{\textit{Council of Europe}, \textit{EEA}, \textit{EUCU}\}$$

also holds and discover that the European Union consists precisely of all the countries that are, at the same time, members of the Council of Europe, *EEA*, and *EUCU*.

Our current context does not contain any country that is a member of both the *EFTA* and *EUCU*. Since there is no such country outside our context, either, we accept the next implication:

$$\{\textit{EU}, \textit{Council of Europe}, \textit{EEA}, \textit{EFTA}, \textit{EUCU}, \textit{Schengen area}\} \to \{\textit{Eurozone}\}.$$

and we are done! The resulting concept lattice is shown in Figure 4.2.

4.2.2 Pairs of squares

We apply the technique to the data from Figure 1.11 (p. 14). The formal context for this diagram is given in Figure 4.3.

Do the given examples cover all possible cases? In order to find out, we apply attribute exploration. The formal context in Figure 4.3 is the *initial context of examples*, that is, the context (E, M, J) of Algorithm 19. The formal context (G, M, I) is infinite: its object set G is the set of *all* possible combinations of two squares (of equal size). This infinite context is the one we are interested in, and we would like to compute its canonical basis. Since we cannot access this infinite formal context directly, we start with the finite subcontext given in

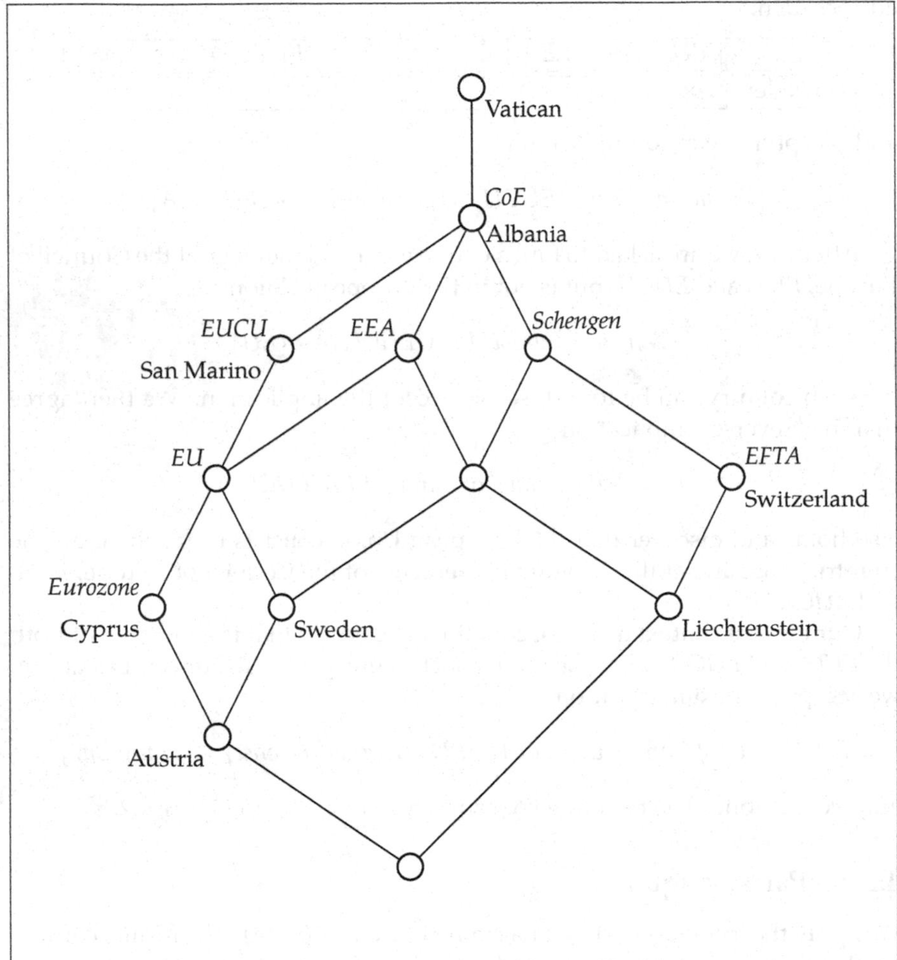

Figure 4.2: The result of the exploration in Section 4.2.1 as a concept lattice

	di	ov	pa	cv	cs	ce
	disjoint	*overlap*	*parallel*	*common vertex*	*common segment*	*common edge*
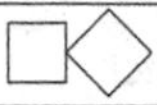						
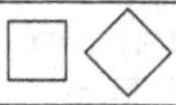	×					
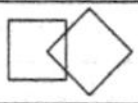		×				
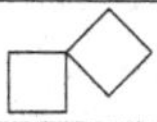				×		
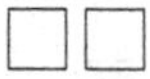	×		×			
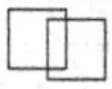		×	×			
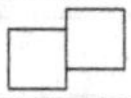			×		×	
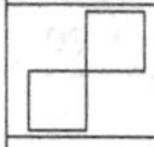			×	×		
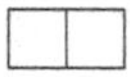			×	×	×	×
2		×	×	×	×	×

Figure 4.3: The formal context for Figure 1.11

Figure 4.3 and extend it when necessary. The steps of the exploration algorithm are given in Figure 4.4.

At start, our list $\mathcal{L}$ of implications is empty. The first preclosed set to consider is the empty set $\emptyset$. This set is closed in the context of examples and is therefore ignored. We compute the next preclosed set, using the (still empty) list $\mathcal{L}$. The result is {ce}. This set is not closed in the current context; its closure is {ce, pa, cv, cs}. In other words, the implication ce $\to$ pa, cv, cs holds for all examples considered so far. Obviously, this implication holds in general, because two squares that share a common edge must always be parallel, must share a vertex and a line segment. We therefore include this implication in $\mathcal{L}$.

The next preclosed set is {cs}, and again this set is not closed in the context of examples. Its closure is {cs, pa}. We observe that this is necessarily so for all possible examples (two squares with a common segment must be parallel to each other) and include the implication cs $\to$ pa in $\mathcal{L}$.

The next four preclosed sets are all closed. They are ignored. Then we discover that we should include the implication pa, cv, cs $\to$ ce in the list $\mathcal{L}$. This implication holds in general, because two squares of equal size that have a common vertex and a common line segment must share an edge. It is somewhat ir-

preclosed set A	closed? $A = A^{JJ}$?	general? $A^{JJ} = A^{II}$?	action	new object or implication
$\emptyset$	yes		next q	
$\{ce\}$	no	yes	new imp	$ce \to pa, cv, cs$
$\{cs\}$	no	yes	new imp	$cs \to pa$
$\{cv\}$	yes		next q	
$\{pa\}$	yes		next q	
$\{pa, cs\}$	yes		next q	
$\{pa, cv\}$	yes		next q	
$\{pa, cv, cs\}$	no	yes	new imp	$pa, cv, cs \to ce$
$\{pa, cv, cs, ce\}$	yes		next q	
$\{ov\}$	yes		next q	
$\{ov, cv\}$	no	no	new obj	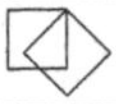
$\{ov, cv\}$	yes		next q	
$\{ov, pa\}$	yes		next q	
$\{ov, pa, cs\}$	no	no	new obj	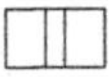
$\{ov, pa, cs\}$	yes		next q	
$\{ov, pa, cv\}$	no	yes	new imp	$ov, pa, cv \to cs, ce$
$\{ov, pa, cv, cs, ce\}$	yes		next q	
$\{di\}$	yes		next q	
$\{di, cv\}$	no	yes	new imp	$di, cv \to \bot$
$\{di, pa\}$	yes		next q	
$\{di, pa, cs\}$	no	yes	new imp	$di, pa, cs \to \bot$
$\{di, ov\}$	no	yes	new imp	$di, ov \to \bot$
M	yes		end.	

Figure 4.4: Steps in the exploration algorithm

ritating that the premise of this implication contains the assumption pa, which is unnecessary because it follows from cs. This is a typical feature of pseudo-closed sets: they tend to be "large".

After two more closed sets, we obtain the preclosed set $\{\text{ov}, \text{cv}\}$, the closure of which in the context of examples is $\{\text{ov}, \text{pa}, \text{cv}, \text{cs}, \text{ce}\}$. This suggests the implication ov, cv $\rightarrow$ pa, cs, ce, which does *not* hold in general. Here is a counterexample: 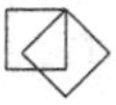 , with attributes ov, cv only. It seems that this possibility had been overlooked when the data was collected. We extend the formal context of examples by this new object and continue our work with this extended context. Now, the preclosed set $\{\text{ov}, \text{cv}\}$ is closed, and we may proceed.

The next situation that requires some action is when we meet the preclosed set $\{\text{ov}, \text{pa}, \text{cs}\}$, the closure of which in the (extended) context of examples contains cv, suggesting that a common segment is sufficient for overlapping parallel squares to share a vertex. Again, we find that this is not an instance of a generally valid implication, because we can give another new example: 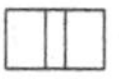 described by attributes ov, pa, cs only. In the extended context, the set $\{\text{ov}, \text{pa}, \text{cs}\}$ is closed.

The next preclosed set leads to a new entry in $\mathcal{L}$: if two squares (of equal size) overlap, are parallel, and have a common vertex, then they must be equal and, thus, have a common segment and a common edge.

Continuing with the algorithm, we find several other generally valid implications, which express that two squares cannot simultaneously be disjoint and have a vertex or a segment in common or otherwise overlap. Adding these to $\mathcal{L}$, the algorithm terminates.

The canonical basis is the set of the seven implications given in the last column of Figure 4.4. The context of examples now has 12 objects, because we have added the two examples in Figure 4.5. Three of these examples are, however, dispensable, because they are reducible and therefore can be omitted without effect on the implications. These are 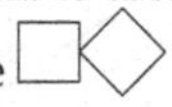 (which was reducible from the beginning), 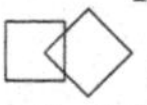 and 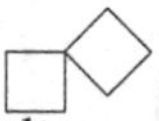 (which became reducible when the new examples were added). These dispensable objects are omitted in the concept lattice displayed in Figure 4.6. This lattice is our final result: its implication theory is the same as the general implication theory, i.e., the theory for *all* possible placements of two squares of equal size. In other words, this concept lattice has the same canonical basis as the infinite context (G, M, I).

Five of the formal concepts in Figure 4.6 are not object concepts. Three of them correspond to the reducible objects just mentioned. The other two do not correspond to an object intent, but only to a proper intersection of object intents. This comes from the fact that

$$\textit{parallel} \rightarrow \textit{disjoint} \vee \textit{comm. segment} \vee \textit{comm. vertex} \vee \textit{overlap}$$

	di	ov	pa	cv	cs	ce
	disjoint	*overlap*	*parallel*	*common vertex*	*common segment*	*common edge*
		×		×		
		×	×		×	

Figure 4.5: Additional objects for Figure 4.3

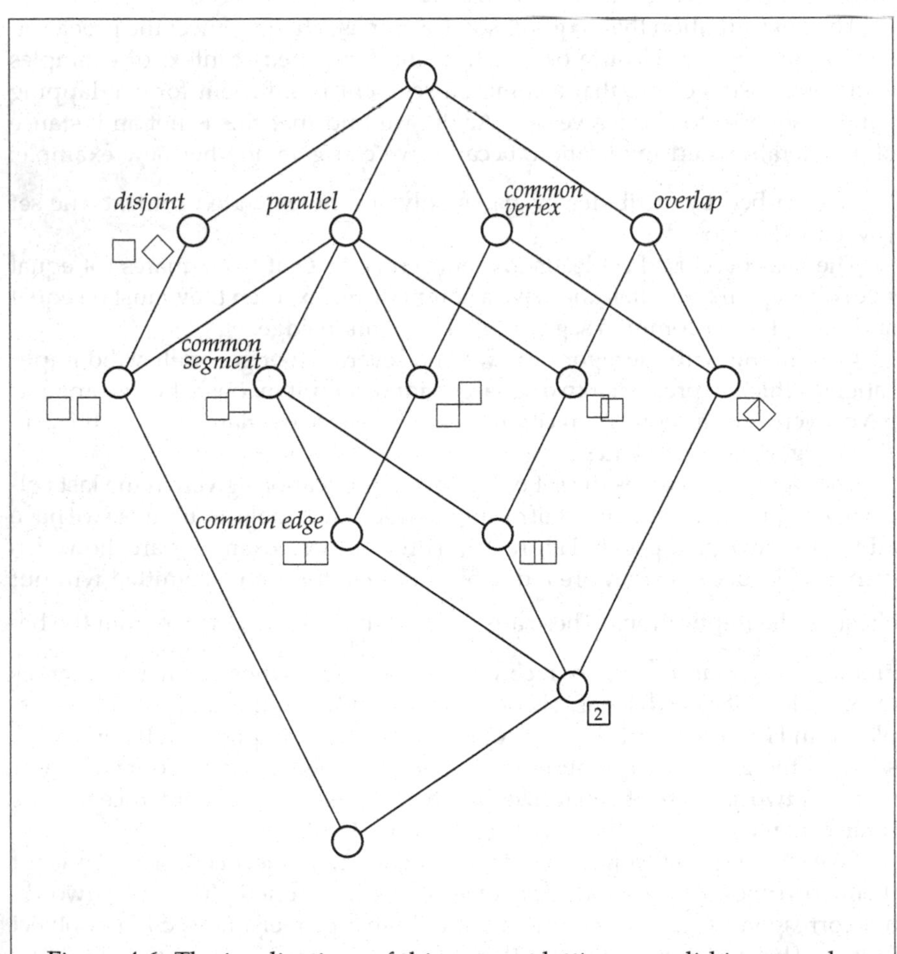

Figure 4.6: The implications of this concept lattice are valid in general

and

$$\textit{disjoint} \wedge \textit{overlap} \to \bot$$

hold as general rules, see Section 4.1.3 for a discussion.

A remark for later: although "pairs of squares" is already a toy example, we discuss its fragment in Section 6.1.1, where we consider only parallel unit squares and drop the attribute "common segment". The implications for this restricted case can easily be derived from the above result. The formal context in Figure 4.7 is a subcontext of the Figures 4.3 and 4.5 combined. We do not include the attribute "parallel", since all objects of the context have it.

	di	ov	cv	ce
	disjoint	*overlap*	*common vertex*	*common edge*
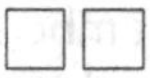	×			
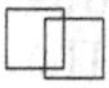		×		
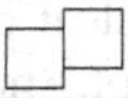				
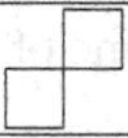			×	
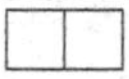			×	×
2		×	×	×
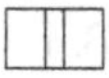		×		

Figure 4.7: Parallel squares only, and without the attribute cs

Figure 4.8 shows the canonical basis of the restricted example.

4.2.3 Components of musical experience

In his Ph.D thesis of 1976 the author Günther Batel presented an *"experimental psychological investigation"* into the *"components of musical experience."* He had selected 30 pieces of music, covering a wide range of styles from the 15th century Locham Songbook, including works by Bach, Mozart, Mussorgsky, and Mahler, to music by Ravi Shankar and Pink Floyd. He then asked probands to describe their musical feelings while hearing these pieces by descriptors from a list of 55 adjectives. Here are some of them:

balanced, brilliant, chaotic, damped, dedicated, dramatic, earnest, equable, erotic, ecstatic, excited, exotic, graceful, hammering, mesmerizing, pictorial, rounded,

common edge	$\to$	*common vertex*
disjoint, overlap	$\to$	$\bot$
disjoint, common vertex	$\to$	$\bot$
overlap, common vertex	$\to$	*common edge*

Figure 4.8: The canonical basis for parallel pairs of squares

stretched, transparent, well-structured, ... (35 more)[4]

Based on a statistical evaluation of the answers, the descriptors were hierarchically grouped using cluster analysis methods. In 1987, members of the Music Philosophical Seminar at Darmstadt University picked up this research when discussing how one can verbally describe musical pieces. They criticized Batel's methodology for its predetermined selection of musical pieces. Batel's findings are not phrased in the language of Formal Concept Analysis, but it is possible to derive from them implications in our sense. Some of these implications look rather accidental, like

$$slow \to earnest.$$

Of course, there are pieces of music that are slow, but not earnest. The implication survived simply because there was no such piece in Batel's list.

Rudolf Wille convinced the seminar to undertake an attribute exploration of this theme. They focused on eleven descriptors that had turned out most significant in Batel's investigation. These were

rounded, balanced, dramatic, transparent, well-structured, strong, lively, sprightly, rhythmizing, fast, and *playful.*[5]

The attribute exploration algorithm confronted the members of the Music Philosophical Seminar with questions like the following:

Is it true that every musical composition with the attributes

dramatic, lively, and *transparent*

[4]Batel used German words. Our translations may not render their meanings precisely. Moreover, word sense may change over time.

[5]These too may deviate from the meanings of the German words used in the exploration. For better readability, our translations are slightly different from Wille's. We write *balanced, rounded,* and *well-structured,* where Wille writes *well-balanced, well-rounded,* and *structured thoroughly* for German *abgerundet, ausgewogen,* and *durchstrukturiert.*

also has the attributes

sprightly, rhythmizing, and *fast*?

Each such question was thoroughly discussed by the panel of experts, and either confirmed or refuted by a concrete counterexample, i.e., by naming a piece of music that does not satisfy the suggested implication. The decision criterion was unanimity. For this specific question, the consensus in the seminar was "no", justified by the third movement of Beethoven's Moonlight Sonata as a counterexample, which was considered to be not *sprightly*, but to have the five other attributes (and additionally the attribute *strong*) [222].

Wille reports that it required some four hours of intensive work before the exploration was successfully finished. This is not very surprising, since a total of 50 questions had to be answered. Nineteen implications were confirmed and 31 counterexamples were given (of which 11 eventually turned out to be reducible), see Figures 4.9 and 4.10. Substantial attribute explorations tend to be laborious.

It would require too much space to present the entire order of events of this exploration. To give an impression, we have reconstructed some of the steps in Figure 4.11.

4.2.4 Lattice-based access control models

Multiuser computer systems must ensure that information they contain is accessible only to those who are authorized to use it. Therefore, choosing and implementing an access control model suitable for a particular computer system is an important part of its development.

Some of the well-known access control models make heavy use of lattices [192]. They are based on Denning's axioms formulating assumptions under which an information flow policy can be regarded as a lattice [73]. Such models have the advantage of enforcing consistency in the way access requests are handled, since all subjects and objects in the environment are bound by the same rules. Thus, they are easier to control and more reliable (albeit less flexible) than some other types of access control models where access rights are ad hoc decisions made by individual users.

In the access control setting, one speaks about active subjects (such as users or processes) accessing passive objects (such as files or resources). This separation is not absolute: an object such as a program can become a subject when trying to access another object such as a file.

The two best-known lattice-based access control models are those of Bell-LaPadula [36] and Biba [43]. In these models, security labels are assigned to entities (subjects and objects). The security labels are partially ordered and, in fact, form a lattice. There are good reasons for it to be a lattice rather than

1. *fast, playful* $\rightarrow$ *lively;*
2. *sprightly* $\rightarrow$ *lively, playful;*
3. *lively, rhythmizing, playful* $\rightarrow$ *sprightly;*
4. *strong, lively, fast, playful* $\rightarrow$ *transparent;*
5. *well-structured, strong, rhythmizing, playful* $\rightarrow$ *rounded, balanced, transparent;*
6. *well-structured, strong, rhythmizing, fast* $\rightarrow$ *transparent;*
7. *transparent, well-structured, rhythmizing, playful* $\rightarrow$ *rounded, balanced;*
8. *transparent, well-structured, lively, rhythmizing, fast* $\rightarrow$ *rounded, balanced;*
9. *transparent, well-structured, lively, sprightly, playful* $\rightarrow$ *rounded, balanced;*
10. *dramatic* $\rightarrow$ *strong;*
11. *dramatic, strong, playful* $\rightarrow$ *transparent, well-structured, lively, fast;*
12. *dramatic, well-structured, strong, rhythmizing* $\rightarrow$ *transparent;*
13. *balanced* $\rightarrow$ *rounded, transparent, well-structured;*
14. *rounded* $\rightarrow$ *balanced, transparent, well-structured;*
15. *rounded, balanced, transparent, well-structured, fast* $\rightarrow$ *lively;*
16. *rounded, balanced, transparent, well-structured, lively, playful* $\rightarrow$ *sprightly;*
17. *rounded, balanced, dramatic, transparent, well-structured, strong, rhythmizing* $\rightarrow$ *lively;*
18. *rounded, balanced, dramatic, transparent, well-structured, strong, lively, rhythmizing, fast* $\rightarrow$ $\perp$
19. *rounded, balanced, dramatic, transparent, well-structured, strong, lively, sprightly, fast, playful* $\rightarrow$ $\perp$

Figure 4.9: List of the confirmed implications from an attribute exploration in music

	rounded	*balanced*	*dramatic*	*transparent*	*well-structured*	*strong*	*lively*	*sprightly*	*rhythmizing*	*fast*	*playful*
• Beethoven: Romance for violin and orchestra F-major	×	×		×	×						×
• Bach: Contrapunctus I	×	×		×	×	×					
• Tchaikovsky: Piano concerto b flat minor, 1st movem.			×			×					
Mahler: 2nd Symphony, 2nd movement					×		×	×	×	×	×
Bartok: Concert for ochestra						×	×	×	×		×
• Beethoven: 9th symphony, 4th movement (presto)			×			×			×	×	
Bach: WTP 1, prelude c minor			×		×	×	×			×	
• Bach: 3rd Brandenburg Concerto, 3rd movement	×	×		×	×		×	×		×	×
• Ligeti: Continuum				×			×			×	×
Mahler: 9th symphony, 2nd movement (Ländler)					×	×	×		×		
Beethoven: Moonlight sonata, 3rd movement			×	×		×	×		×	×	
Hindemith: Chamber music No.1, finale				×		×	×	×	×	×	×
Bizet: Suite arlesienne	×	×		×	×	×			×		×
Mozart: Figaro, overture	×	×		×	×	×	×	×		×	×
Schubert: Wayfarer fantasy			×			×	×		×	×	
• Beethoven: Spring sonata, 1st movement	×	×		×	×		×	×			×
• Bach: WTP 1, fuge c minor	×	×		×	×		×	×	×		×
Shostakovich: 15th symphony, 1st movement					×	×	×	×			×
Wagner: Mastersinger, overture	×	×		×	×	×	×	×	×		×
Beethoven: String quartet op.131, final movement			×	×	×	×			×	×	
Johann Strauß: Spring voice waltz	×	×		×	×		×	×	×	×	×
Mozart: Il Seragio, "O, how I will triumph ..."			×	×	×	×	×			×	×
• Bach: Mathew's passion No.5 (choir)			×	×	×	×	×				
Brahms: Intermezzo op. 117, No.2	×	×		×	×	×	×		×	×	
Wagner: Ride of the valkyries			×	×	×	×	×		×		
• Mozart: Magic flute, "The hell revenge rages ..."			×	×	×	×	×			×	
Mendelssohn: 4th symphony, 4th movement	×	×		×	×	×	×	×	×	×	×
Brahms: 4th symphony, 4th movement	×	×	×	×	×	×					
Beethoven: Great fuge op. 133	×	×	×	×	×	×	×		×		
• Górecki: Symphony of Sorrowful Songs	×	×	×	×	×	×	×				
Verdi: Requiem, dies irae	×	×	×	×	×	×	×			×	

Figure 4.10: Formal context of the result of an attribute exploration in music. Reducible examples are marked with a small dot. From [222]

1.-8. The first eight questions asked by the exploration algorithm were all answered negatively. As counterexamples, the first eight lines of Figure 4.10 were given.

9. Question: Does *fast, playful* imply *well-structured, lively, sprightly*?

   This was refuted. Ligeti's "Continuum" served as a counterexample. It was agreed to be *transparent, lively, fast,* and *playful,* but neither *well-structured* nor *sprightly*.

10. Question: Does *fast, playful* imply *lively*?

    This was confirmed.

11. Question: Does *rhythmizing, playful* imply *lively, sprightly*?

    This was refuted, a counterexample being Bizet's "Suite arlesienne", which was agreed to be *rhythmizing* and *playful* (and, in addition, *rounded, balanced, transparent, well-structured,* and *strong*), but neither *lively* nor *sprightly*.

12. Question: Does *sprightly* imply *lively* and *playful*?

    This was confirmed.

13. Question: Does *lively, rhythmizing* imply *sprightly, playful*?

    This was refuted. The counterexample agreed upon was the 2nd movement of Mahler's 9th symphony, which is *well-structured, strong, lively,* and *rhythmizing,* but neither *sprightly* nor *playful*.

14. Question: Does *lively, rhythmizing,* and *playful* imply *sprightly*?

    This was confirmed.

15.–50. ...

Figure 4.11: Steps of an attribute exploration in music

just a partial order: the supremum and infimum operations play a role in, for instance, some versions of the Biba model. Information is allowed to flow only in one direction, e.g., from entities with security labels that are lower in the lattice to entities with security labels that are higher.

In many cases, a security label is a set of categories (such as project names, department names, etc.), augmented with a *security level* (*"unclassified"*, *"secret"*, *"top secret"*, etc.). The lattice of all possible combinations of security levels and categories grows linearly in the number of levels and exponentially in the number of categories. The size of the security lattice puts constraints on practical implementation, which is part of the reason why lattice-based models are not so widely used as they could be. Large models are hard to visualize and, therefore, hard to maintain: the assignment of categories to subjects and objects is likely to be error-prone, as, without a complete picture of the model, it is difficult to check the effect this or that assignment may have.

Usually, however, not all possible combinations of security levels and categories make sense as security labels. For instance, Smith describes a lattice from military practice based on four security levels and eight categories, which potentially give rise to 1024 labels, whereas the actual lattice contains only 21 elements: there are no labels consisting of more than three categories (apart from the label associated with the top element), and categories are used only in combination with two top-most security levels [199]. Hence, development of an access control model would involve identification of domain-specific dependencies between the security levels and categories and, by extension, their realistic combinations. Attribute exploration can help organize this process in a semiautomatic fashion and, in some sense, ensure the validity of its results.

Next, we briefly introduce the Bell-LaPadula model and show how to use attribute exploration to build a concrete instance of this model for a particular application.

Let H denote a totally ordered set of *security levels* (we use standard notation for this order: e.g., $<$ and $\leq$) and let C denote a set of *categories* such as names of departments within a company, project names, etc. Then, a *security label* is a pair (h, D) consisting of a security level $h \in H$ and a subset of categories $D \subseteq C$.

Security labels are partially ordered: $(h, D) \leq (g, E)$ if and only if $h \leq g$ and $D \subseteq E$. It is easy to see that this partial order is a lattice: it is the product of two lattices, $(H, \leq)$ and $(\mathfrak{P}(C), \subseteq)$. Such lattices are common in military security models.

The Bell-LaPadula model addresses the *confidentiality requirement*, which is formulated as "prevention of unauthorized disclosure of information" [117], by enforcing two rules. Assuming that $\lambda(e)$ is the security label of the subject or object e, the *simple-security property* is formulated as follows:

- Subject s can read object o only if $\lambda(o) \leq \lambda(s)$.

With security labels defined as above, a subject can read an object only if the subject is at least at the same security level as the object and the subject has access to all categories associated with the object. However, the simple-security property is not sufficient: some restrictions are necessary on how subjects can write or modify objects. If there are no restrictions, a secret subject can read a secret document and create its unclassified copy, thus providing unclassified subjects with access to secret information. To avoid this, the *$\star$-property* is introduced:

- Subject s can write object o only if $\lambda(s) \leq \lambda(o)$.

According to these two rules, information can flow only upwards: from less secure objects to more secure subjects and from less secure subjects to more secure objects.

It is straightforward to express this model in terms of FCA. The set H of security levels and the set C of categories constitute our attribute set. In fact, security level is a many-valued attribute. Many-valued attributes and contexts are explained in Chapter 5. For now, we just use the attribute "is at least at level h" for each level $h \in H$. For example, a "*secret*" user will get the attributes "*unclassified*" and "*secret*", but not "top secret". In the case of these three levels, the mapping of levels to attributes in the context is as follows (rows correspond to actual levels and columns correspond to attributes in the context):

	unclassified	*secret*	*top secret*
unclassified	×		
secret	×	×	
top secret	×	×	×

It is easy to see that the table above represents the context $(H, H, \geq)$. Here, it serves as a conceptual scale for the many-valued attribute "security level" (see Chapter 5 for details and definitions). The attribute *unclassified* is clearly redundant; so, in principle, we do not have to include the smallest security level as an attribute of the formal context.

The elements of G in our context are entities (subjects and objects) of the access control model and the incidence relation assigns categories and levels to them. Then, security labels correspond to concept intents, and the security label of an entity is the object intent of this entity. The resulting concept lattice is the reverse of the Bell-LaPadula lattice, since larger intents correspond to less general concepts; the set of concept intents with the usual subset order is precisely the Bell-LaPadula lattice.

The problem here is that a complete list of subjects and objects is usually unknown at the moment when the access control model is being developed (and, if a system is open to new users or new documents, a complete list never becomes available). To construct the lattice of security labels, the developer of the system must envision all types of potential subjects and objects and describe them in terms of security levels and compartments. Attribute exploration organizes the process of selecting these relevant combinations and ensures that all possible cases have been taken into account [173].

We consider an example from [154], where a specification of the Bell-LaPadula access control model for commercial applications is proposed. Two security levels are used: Audit Manager (*AM*) and System Low (*SL*). *AM* is the highest level of the two; it is reserved for system audit and management functions. Five categories are defined as follows:

- Development (*D*): production programs under development that are not yet in use;
- Production Code (*PC*): production processes and programs;
- Production Data (*PD*): data covered by integrity policy;

- System Development (*SD*): system programs in development but not yet in use;
- Software Tools (*T*): programs on production system not related to protected data.

We start with the context $(\emptyset, \{AM, D, PC, PD, SD, T\}, \emptyset)$. We chose not to include the *SL* attribute, since every user is at least at the system low level anyway. The first question asked by attribute exploration is whether all objects should have all attributes, to which the answer is, clearly, "no". The first counterexample we provide is of an object we call the ordinary user (**ou**), who can read production data and execute production code — all at the system low level; ordinary users do not write their own programs, but use existing programs and databases (hence, do not need software tools): $\{\mathbf{ou}\}' = \{PC, PD\}$.

Next, the algorithm asks whether all objects have at least all the same attributes as ordinary users, i.e., *PC* and *PD*. Since this is not true, e.g., for application developers, who should not have direct access to production data, we add an object **ad** to our context, where $\{\mathbf{ad}\}' = \{D, T\}$. Application developers work on their development system; if they need access to actual data (for testing purposes), they should use it only on the development system.

The next question is if everyone having access to software tools (*T*) must also have access to production programs in development (*D*), to which the counterexample is system programmers (**sp**): $\{\mathbf{sp}\}' = \{SD, T\}$. The algorithm then asks if everyone having access to system programs in development (*SD*) should have access to software tools (*T*), and it seems reasonable to agree with this, as software tools are needed for program development.

Proceeding as above, we accept six implications and provide two counterexamples before the attribute exploration stops (see Figure 4.12).

The resulting lattice is shown in Figure 4.13.

This lattice contains eight nodes, which is significantly less than 64 nodes theoretically possible for a lattice on two security levels and five categories. The information flow policy defined by this lattice can also be summarized by the six accepted implications:

$$
\begin{aligned}
\{SD\} &\to \{T\}\\
\{PD\} &\to \{PC\}\\
\{PC, T\} &\to \{D, PD, SD\}\\
\{D\} &\to \{T\}\\
\{D, SD, T\} &\to \{PC, PD\}\\
\{AM\} &\to \{D, PC, PD, SD, T\}
\end{aligned}
$$

The last implication suggests that there is only one security label involving the highest security level, *AM*, namely, the system top label; in other words,

preclosed set A	closed? $A = A^{JJ}$?	general? $A^{JJ} = A^{II}$?	action	new object or implication
$\emptyset$	no	no	new obj	Ordinary users: **ou**$' = \{PC, PD\}$
$\emptyset$	no	no	new obj	Application developers: **ad**$' = \{D, T\}$
$\emptyset$	yes		next q	
$\{T\}$	no	no	new obj	System programmers: **sp**$' = \{SD, T\}$
$\{T\}$	yes		next q	
$\{SD\}$	no	yes	new impl	$SD \rightarrow T$
$\{SD, T\}$	yes		next q	
$\{PD\}$	no	yes	new imp	$PD \rightarrow PC$
$\{PC\}$	no	no	new obj	Production code: **pc**$' = \{PC\}$
$\{PC\}$	yes		next q	
$\{PC, T\}$	no	no	new obj	System controllers: **sc**$' =$ **allcat**
$\{PC, T\}$	no	yes	new imp	$PC, T \rightarrow D, PD, SD$
$\{PC, PD\}$	yes		next q	
$\{D\}$	no	yes	new imp	$D \rightarrow T$
$\{D, T\}$	yes		next q	
$\{D, SD, T\}$	no	yes	new imp	$D, SD, T \rightarrow PC, PD$
allcat	yes		next q	
$\{AM\}$	no	yes	new imp	$AM \rightarrow$ **allcat**
M	yes		end.	

Figure 4.12: Steps in the exploration algorithm while building an access control model; **allcat** is an abbreviation for the set $\{D, PC, PD, SD, T\}$ of all categories

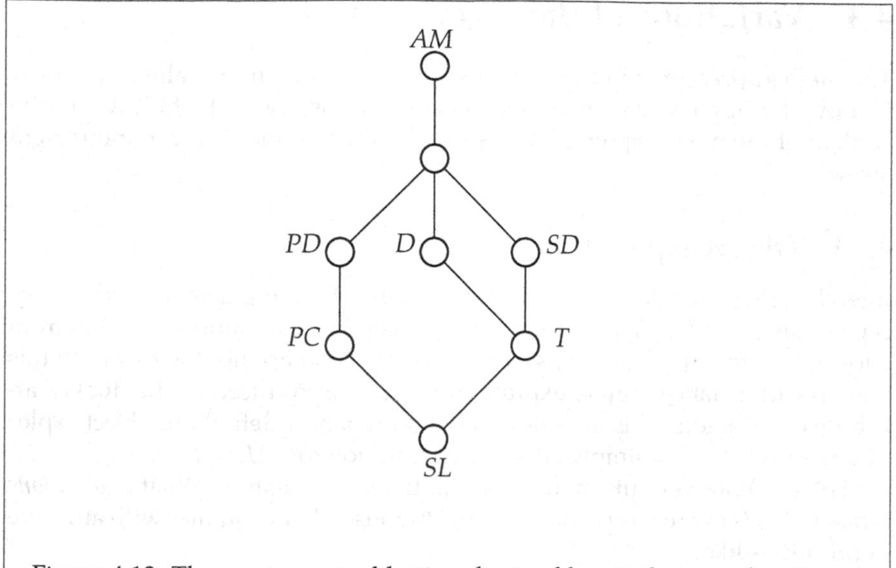

Figure 4.13: The access control lattice obtained by attribute exploration

subjects at the *AM* level have access to everything. Then, for example, implications $\{SD\} \to \{T\}$ and $\{D\} \to \{T\}$ mean that subjects dealing with system or production programs in development must also have access to software tools, since they need them to write programs.

The model we have built here is concerned only with the confidentiality requirement. Similarly, a separate model can be built for the *integrity requirement* understood as "prevention of unauthorized modification of information". This requirement is the concern of the Biba model [43]. It is, in a sense, in a conflict with the confidentiality requirement, since it essentially implies the information flow in the opposite direction: from more secure to less secure entities. In [172], it is suggested how the two requirements could be combined within one model using a specially-designed version of attribute exploration.

So far we used attribute exploration to build an access control model from a specified set of security levels and categories. A dual process called *object exploration*, which is discussed in Section 4.3.1, can help derive levels and categories suitable for a particular domain. Thus, one can start with a small set of subjects and objects and a small set of categories and apply object and attribute exploration interchangeably to enlarge both sets and build a realistic model.

4.3 Variations of the algorithm

The method presented in the previous two sections is simple, almost obvious. It allows for numerous variants and generalizations, some of which will be the content of the next chapters. We discuss the most immediate variations right away.

4.3.1 Object exploration

Interchanging the rôles of objects and attributes is no big deal in Formal Concept Analysis, as far as the formalism is concerned. One simply switches from a formal context (G, M, I) to its *dual* (M, G, I^{-1}) and applies the theory to this one. As an instance, **object exploration** can be introduced as the dual of attribute exploration. This does not even require a new definition: object exploration for (G, M, I) is simply attribute exploration for (M, G, I^{-1}).

There is however quite a difference in the interpretation. What *implications between objects* express is perhaps not so obvious. We are familiar with attribute implications like

furry, four-legged $\to$ *mammal,*

but what is an object implication such as

cat, cow, duck $\to$ *dog*

supposed to mean?

There is at least one natural and striking interpretation. Object implications are intuitive and useful when properly utilized. We did however not find many well-documented examples of conceptual object exploration in the literature. There are numerous related approaches which, in our understanding, could benefit from a more formal algorithmic framework, but our method seems not yet to be well prospected. We therefore limit our presentation here to a sketch of the method and its believed potential, but do not elaborate a complete example of an object exploration.

Imagine that you need to tidy up a messy situation, for example an unstructured collection of files on your computer. An obvious strategy is to group similar items together, putting them into folders and subfolders. Your computer's file system forces you to keep folders disjoint, a rather strong restriction.[6] In other situations there are possibilities of organizing your items more freely into a *classification*, an *ontology*, or, in the language of this book, into a *concept lattice*.

There are many possibilities of organizing your items into groups. It is your decision what you consider similar and what not. You decide which conceptual scheme to use for grouping the items. An object implication such as

[6]See [89] for a conceptual alternative.

$$\textit{cat, cow, duck} \rightarrow \textit{dog}$$

would stand for your decision that any group containing *cat, cow,* and *duck* must also contain *dog*. Why? Because you say so. Or, if you decide that this object implication does *not* hold, you specify a group containing *cat, cow, duck,* but not *dog*. Typically this is done by specifying an *attribute* distinguishing *cat, cow,* and *duck* from *dog*. Is there such an attribute? Of course, you can always make one up. But it is your decision which attributes you use for your classification and which not.

There are many applications that follow a similar strategy, but without the algorithmic support of object exploration. A well-tried idea is that of **semantic fields** (also called **lexical fields**), where words of related meaning are grouped.

MacQueen and Vanderveken [213] discuss several groups of English performative verbs, among them the group of *expressives*:

> *approve, compliment, praise, laud, extol, plaudit, applaud, acclaim, brag, boast, complain, disapprove, blame, reprove, deplore, protest, grieve, mourn, lament, rejoice, cheer, boo, condole, congratulate, thank, apologize, greet,* and *welcome.*

The structure they work out, called a **semantic tableau** (see Figure 4.14), may be interpreted in the language of Formal Concept Analysis as an ordered tree of object concepts. It satisfies the object implication

$$\textit{grieve, disapprove} \rightarrow \textit{protest},$$

but not

$$\textit{grieve, disapprove} \rightarrow \textit{greet}.$$

The latter implication is refuted by the group of expressives

> *complain, disapprove, blame, reprove, deplore, protest, grieve, mourn, lament,*

expressing discontent, to which *greet* does not belong. As said, their result was not obtained using object exploration, but is very similar in its structure.

Another tempting idea is to combine attribute and object exploration. We will do so in Section 6.3 as part of the *concept exploration* algorithm. It may result in an infinite exploration process, with both the set M of attributes and the set G of objects growing indefinitely. In Section 6.4.5, we shall briefly discuss C. Meschke's interesting approach. Meschke does not extend the sets G and M. Instead, he uses the information provided by the counterexamples for defining a family of *preconcepts*. He calls these *conceptual traces*.

Ganter has investigated which incidence relations are compatible with given sets of implications on G and M [99]. He found that they are just the so-called *dual bonds* between the respective closure systems.

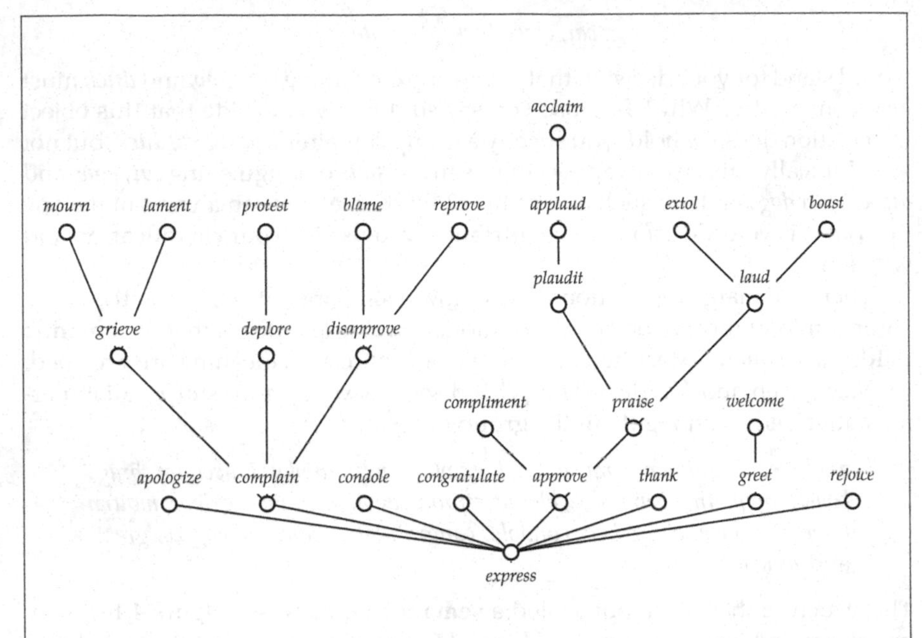

Figure 4.14: Semantic tableau for expressives. From Vanderveken and MacQueen [213]

4.3.2 Harmless background knowledge and exceptions

The user of an interactive attribute exploration starts with a (possibly empty) set of examples and then gives, on demand, further examples and implications until the theory is complete. It may happen that during the process the user gets aware of other examples or implications that hold in the respective situation. Does it require a major modification of the algorithm to allow the input of (correct) implications or examples without being asked for it?

No, this is harmless. Additional examples can be given at any time. They speed up the algorithm. Remember that the only function of the examples is to give an upper bound A^{JJ} for the closure of a set A; an additional example can only bring this upper bound closer to the true value $A^{II} \subseteq A^{JJ}$ of this closure.

Additional implications are also harmless, but this is not so obvious. If true implications are added to the list $\mathcal{L}$, they will increase the lower bound for closures. But if this is done, the output list of implications will no longer be the canonical basis of (G, M, I). Instead, we obtain a basis for the difference between the implication theory of (G, M, I) and the implication theory generated by the extra implications. The notion of a pseudo-closed set can be generalized to cover this situation. Our presentation follows that of Stumme [201]:

Definition 15 Let $\mathcal{C}_1$ and $\mathcal{C}_2$ be two closure operators on a finite set M satisfying

$$\mathcal{C}_1(A) \subseteq \mathcal{C}_2(A) \qquad \text{for all } A \subseteq M.$$

A set $P \subseteq M$ is called $(\mathcal{C}_1, \mathcal{C}_2)$**-pseudo-closed** iff

1. $P = \mathcal{C}_1(P) \neq \mathcal{C}_2(P)$, and
2. if $Q \subsetneq P$ is $(\mathcal{C}_1, \mathcal{C}_2)$-pseudo-closed, then $\mathcal{C}_2(Q) \subseteq P$.

◊

The set

$$\mathcal{L} := \{P \to \mathcal{C}_2(P) \mid P \text{ is } (\mathcal{C}_1, \mathcal{C}_2)\text{-pseudo-closed}\}$$

of implications is the canonical basis of $\mathcal{C}_2$ *relative to* $\mathcal{C}_1$ in the following sense: a set $C \subseteq M$ is $\mathcal{C}_2$-closed if and only if it is $\mathcal{C}_1$-closed and, at the same time, closed under all implications from $\mathcal{L}$, and $\mathcal{L}$ is minimal with respect to this property.

Instead of initializing $\mathcal{L} := \emptyset$ in the first line of the exploration algorithm, we may allow the algorithm to accept any set $\mathcal{L}$ of (valid) implications as a parameter. The result of the algorithm will then be the relative basis described above. It is sometimes called the basis relative to the implicational background knowledge $\mathcal{L}$. And even if implications are given at an intermediate stage of the algorithm, the result remains complete and correct (but not necessarily irredundant). The situation is sketched in Figure 4.1 (p. 135): attribute exploration restricted to the shaded interval follows the same algorithm as for the full lattice.

Example 12 One might ask if it is possible to get the $(\mathcal{C}_1, \mathcal{C}_2)$-pseudo-closed sets by just closing the $\mathcal{C}_2$-pseudo-intents in $\mathcal{C}_1$ and then removing the $\mathcal{C}_2$-closed results. The context in Figure 4.15 shows that in general this is not the case. Let $\mathcal{C}_2$ be the intent closure operator. We get the following pseudo-closed sets:

$$\{\{c, d\}, \{b\}, \{a, d\}, \{a, c\}\}.$$

If $\mathcal{C}_1$ is the closure operator corresponding to the background implication

$$c, d \to a,$$

we get as $(\mathcal{C}_1, \mathcal{C}_2)$-pseudo-closed sets the family $\{\{b\}, \{a, d\}, \{a, c\}\}$. The set $\{a, c, d\}$, which is the $\mathcal{C}_1$-closure of the $\mathcal{C}_2$-pseudo-intent $\{c, d\}$, is not $\mathcal{C}_2$-closed and not a $(\mathcal{C}_1, \mathcal{C}_2)$-pseudo-intent.

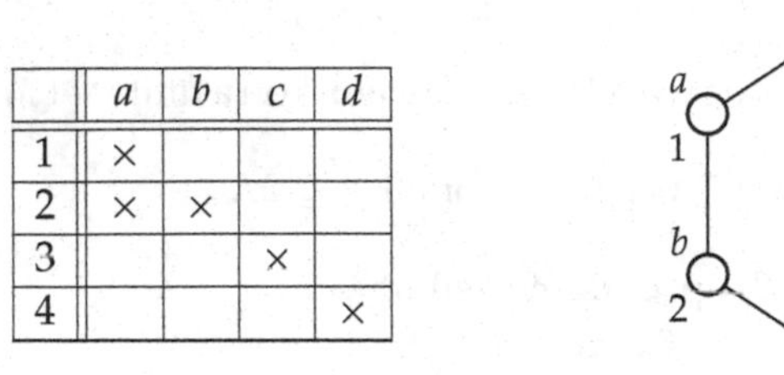

	a	b	c	d
1	×			
2	×	×		
3			×	
4				×

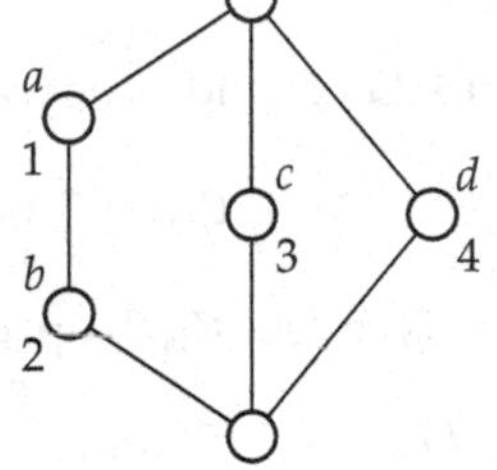

Figure 4.15: An example concerning relative pseudo-closed sets

Sometimes, during the exploration, the user may come across implications that hold for almost all objects of the universe and conflict only with some specific pathologic cases. In such a situation, it might be beneficial to remove these objects explicitly from the universe and list them separately in the result as exceptions, together with the list of accepted implications. Again, this is harmless.

As an example[7] for both background knowledge and exceptions, we discuss the exploration of the following attributes of (non-empty) undirected graphs:

$M :=$ {*connected (conn), disconnected (disc), bipartite (bip), complete (comp), complete bipartite (cbip), tree (tree), forest (for), planar (plan), Eulerian (eul), Hamiltonian (ham)*}

When we look at these attributes, we see that *connected* and *disconnected* are contradicting each other (i. e., no graph can have both attributes). Furthermore, we know that a *tree* is defined as a *connected forest* and that a *complete bipartite* graph is always *bipartite*. This justifies the following background implications:

$$\begin{aligned}
\{conn, disc\} &\rightarrow \bot \\
\{tree\} &\rightarrow \{conn, for\} \\
\{conn, for\} &\rightarrow \{tree\} \\
\{cbip\} &\rightarrow \{bip\}
\end{aligned}$$

During the exploration, the graphs G_1 to G_{20} in Figure 4.16 are given as counterexamples and the following implications are accepted:

[7]The example follows [201]. During the exploration, it becomes apparent that the author assumed a definition of the *Eulerian* graph that includes being *connected*: an Eulerian graph is understood as one with all vertices connected by an Eulerian circuit.

$$
\begin{aligned}
\{ham\} &\rightarrow \{conn\}\\
\{eul\} &\rightarrow \{conn\}\\
\{for\} &\rightarrow \{bip, plan\}\\
\{comp\} &\rightarrow \{conn\}\\
\{disc, bip, cbip\} &\rightarrow \{for, plan\}\\
\{conn, comp, eul\} &\rightarrow \{ham\}\\
\{conn, bip, tree, for, plan, ham\} &\rightarrow \{comp, cbip, eul\}\\
\{conn, bip, tree, for, plan, eul\} &\rightarrow \{comp, cbip, ham\}\\
\{conn, bip, cbip, plan, ham\} &\rightarrow \{eul\}\\
\{conn, bip, comp\} &\rightarrow \{cbip, tree, for, plan\}
\end{aligned}
$$

These ten implications constitute the canonical basis relative to the background implications. Every implication that is valid in the context can be deduced from them and the four background implications. The overall canonical basis consists of all implications in the relative basis and additionally of the first and the fourth background implications and the two implications {*tree*} → {*conn, bip, for, plan*} and {*conn, bip, for, plan*} → {*tree*}. In this example, the cardinality of the relative canonical basis is just the difference between the cardinalities of the background implications and the overall canonical basis, but, in general, it may be larger. The concept lattice resulting from this exploration is shown in Figure 4.17.

During the exploration, there are some implications that are contradicted by only one counterexample (up to isomorphism). If we want to determine the general structure of graph theory without bothering with pathological cases, we may confirm some implications that are true for *almost all* graphs and keep in mind the exceptions. For our example, we shall regard as exceptions all graphs that are (up to isomorphism) unique in having exactly their attributes. More generally, we assume that the formal context $\mathbb{K} = (G, M, I)$ under exploration has a finite subcontext $\mathbb{X} = (X, M, I \cap X \times M)$ of **exceptions** ($X \subseteq G$). We want to find the implication theory of

$$\mathbb{K} \div \mathbb{X} := (G \setminus X, M, I \cap ((G \setminus X) \times M)),$$

as well as identify a subcontext $\mathbb{Y}$ of $\mathbb{X}$ that contains counterexamples for all implications valid in $\mathbb{K} \div \mathbb{X}$, but not in $\mathbb{K}$.

It seems reasonable to suggest that the general knowledge about the domain should be separated from the knowledge of which objects may be considered exceptions for a particular application. Therefore, to take care of exceptions, we need interactive input from the **application expert** in addition to the interactive input we get from the domain expert in the basic version of attribute exploration. For a given object $g \in G$, the application expert decides whether $g \in X$, i.e., whether g is an exception.

Back to our example, the beginning of the exploration dialogue remains unchanged. The first difference appears with the question:

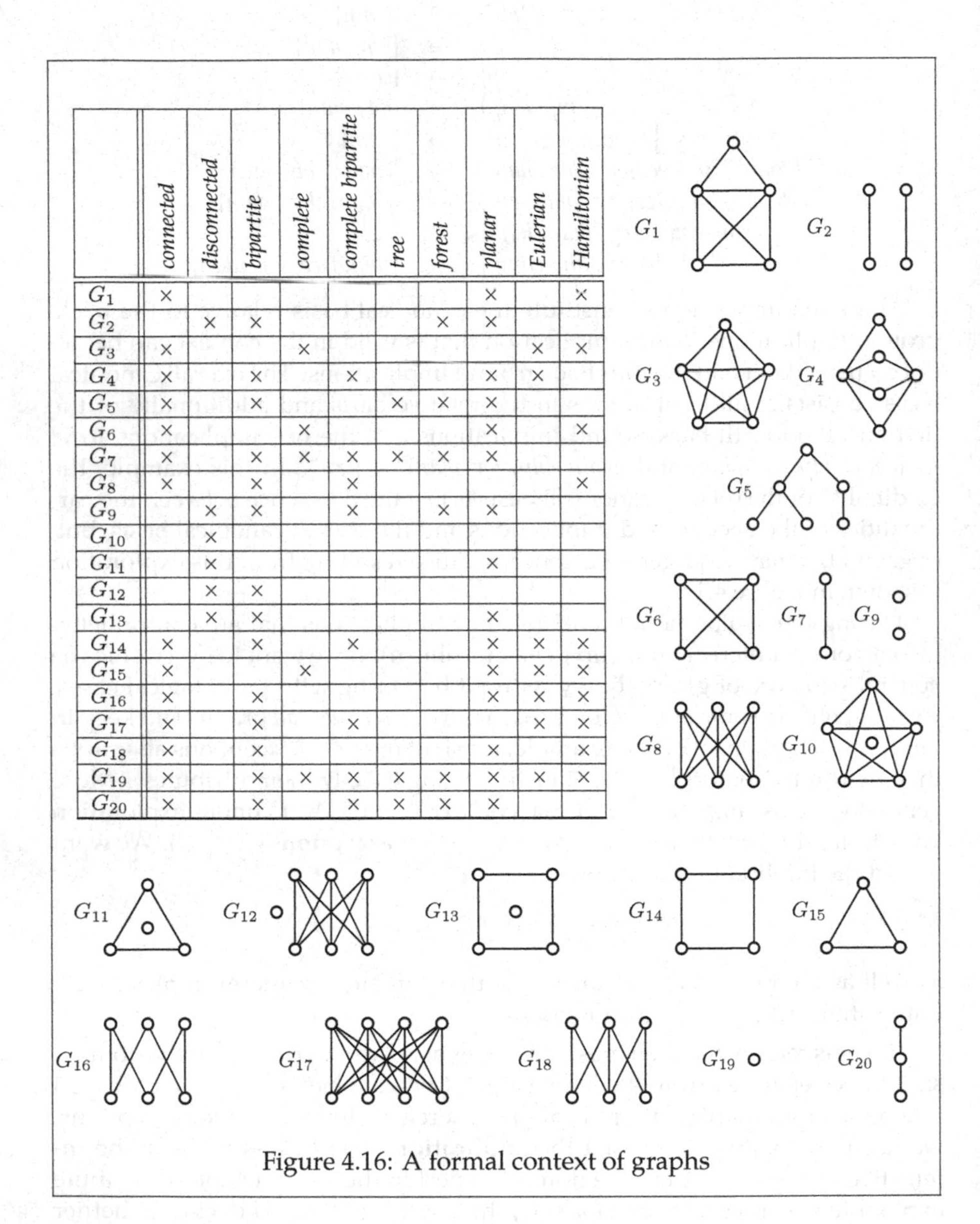

	connected	disconnected	bipartite	complete	complete bipartite	tree	forest	planar	Eulerian	Hamiltonian
G_1	×							×		×
G_2		×	×				×	×		
G_3	×			×					×	×
G_4	×		×		×			×	×	
G_5	×		×			×	×	×		
G_6	×			×				×		×
G_7	×		×	×	×	×	×	×		
G_8	×		×		×					×
G_9		×	×		×		×	×		
G_{10}		×								
G_{11}		×						×		
G_{12}		×	×							
G_{13}		×	×					×		
G_{14}	×		×		×			×	×	×
G_{15}	×			×				×	×	×
G_{16}	×		×					×	×	×
G_{17}	×		×		×				×	×
G_{18}	×		×					×		×
G_{19}	×		×	×	×	×	×	×	×	×
G_{20}	×		×		×	×	×	×		

Figure 4.16: A formal context of graphs

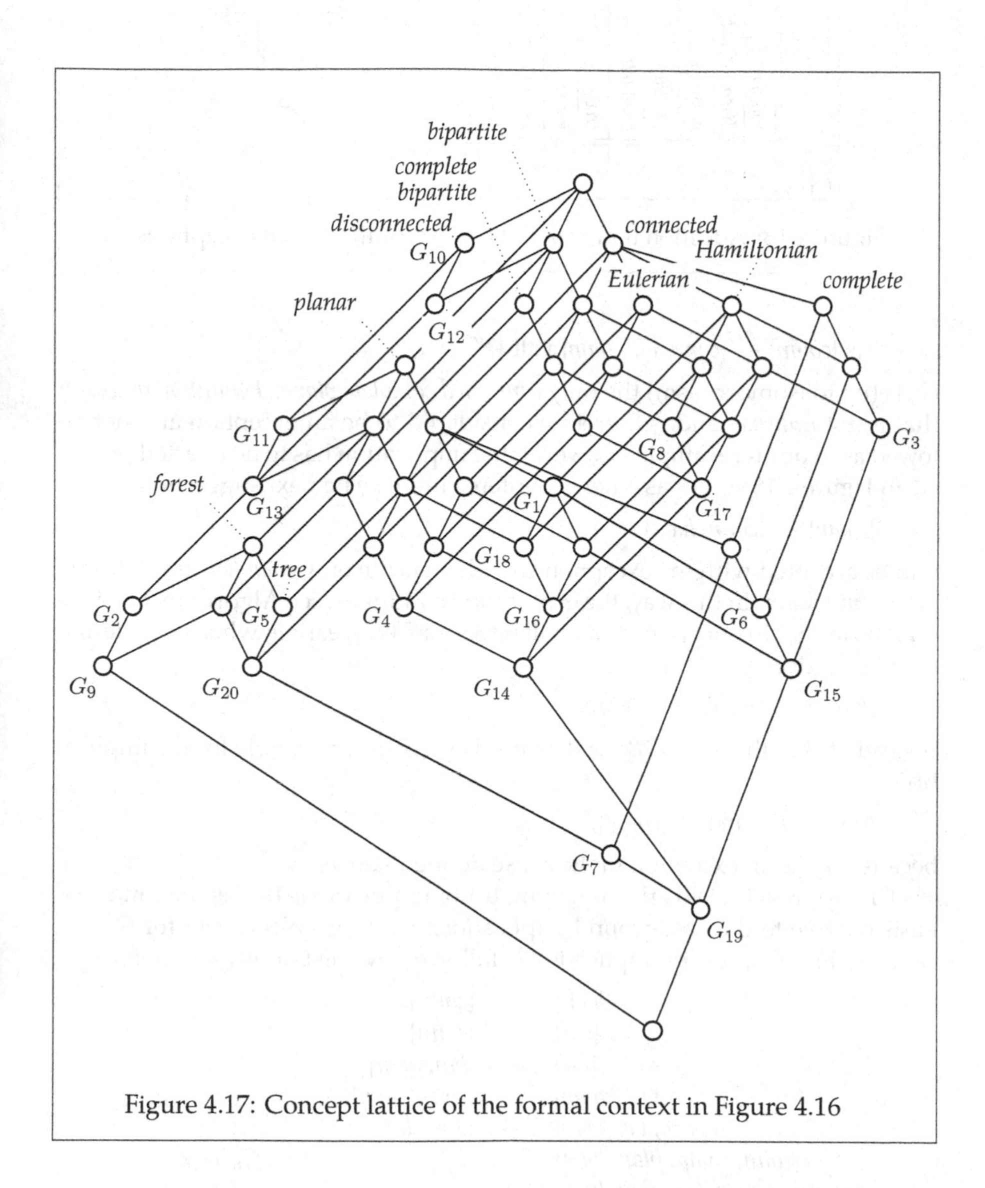

Figure 4.17: Concept lattice of the formal context in Figure 4.16

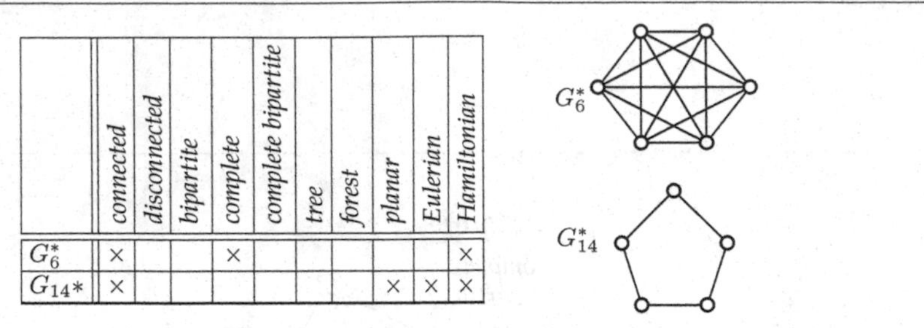

	connected	disconnected	bipartite	complete	complete bipartite	tree	forest	planar	Eulerian	Hamiltonian
G_6^*	×			×						×
$G_{14}*$	×							×	×	×

Figure 4.18: Additional graphs for the exploration with exceptions

"Is $\{comp\} \to \{conn, eul, ham\}$ valid?"

G_6 is (up to isomorphism) the only *connected complete planar Hamiltonian* graph that is not *Eulerian* and is therefore considered to be an exception and not allowed as a counterexample. However, the implication has to be rejected: graph G_6^* in Figure 4.18 serves as a new counterexample. The next suggestion

$$\{comp\} \to \{conn, ham\}$$

will be accepted with the exception G_7, which is the only *complete* graph that is not *Hamiltonian*. In this way, the exploration continues, see Algorithm 20. Since G_7 is no longer in our context, we will have to add G_{20} earlier, when considering the implication

$$\{bip, cbip, for, plan\} \to \{disc\}.$$

Instead of G_{14}, the graph G_{14}^* will be used as a counterexample for the implication

$$\{conn, eul, ham\} \to \{comp\},$$

because G_{14} is an exception in the sense defined above.

This approach yields the following list of implications that is the canonical basis relative to the background implications for all graphs except for G_6, G_7, G_{14}, G_{15}, and G_{19}. Every implication is followed by a list of its exceptions.

$$\begin{array}{rcll}
\{ham\} & \to & \{conn\} & \\
\{eul\} & \to & \{conn\} & \\
\{for\} & \to & \{bip, plan\} & \\
\{comp\} & \to & \{conn, ham\} & (G_7) \\
\{disc, bip, cbip\} & \to & \{for, plan\} & \\
\{conn, comp, plan, ham\} & \to & \bot & (G_6, G_{15}, G_{19}) \\
\{conn, bip, tree, for, plan, ham\} & \to & \bot & (G_{19}) \\
\{conn, bip, tree, for, plan, eul\} & \to & \bot & (G_{19}) \\
\{conn, bip, cbip, plan, ham\} & \to & \bot & (G_{14}, G_{19}) \\
\{conn, bip, comp, ham\} & \to & \bot & (G_{19})
\end{array}$$

Algorithm 20 ATTRIBUTE EXPLORATION with support for exceptions and background implications

Input: Possibly empty subcontexts $\mathbb{E} = (E, M, J = I \cap E \times M)$ of $\mathbb{K} \div \mathbb{X}$ and $\mathbb{Y} = (Y, M, I \cap Y \times M)$ of $\mathbb{X}$, where $\mathbb{K} = (G, M, I)$, M finite, is the domain context and $\mathbb{X}$ is its finite subcontext of exceptions; a set $\mathcal{L}$ of implications over M that hold in $\mathbb{K} \div \mathbb{X}$.

Interactive input: {Domain expert} Upon request, confirm that $A \to A^{JJ}$ holds in $\mathbb{K} \div \mathbb{Y}$ or give an object $g \notin Y$ with intent g^I such that $A \subseteq g^I$, but $A^{JJ} \not\subseteq g^I$.
{Application expert} Upon request, check if object g is in the context $\mathbb{X}$ of exceptions.

Output: The canonical basis of $\mathbb{K} \div \mathbb{X}$ relative to the input $\mathcal{L}$ and possibly enlarged subcontexts $\mathbb{E}$ and $\mathbb{Y}$ such that $\mathbb{E}$ satisfies precisely all the implications of $\mathbb{K} \div \mathbb{Y}$ and these are the same as the implications of $\mathbb{K} \div \mathbb{X}$.

```
A := ∅
while A ≠ M do
    while A ≠ A^{JJ} do
        if A → A^{JJ} holds in K ÷ Y then                {Domain expert}
            L := L ∪ {A → A^{JJ}}
            exit while
        else
            obtain some object g ∈ A^I \ (A^{JJI} ∪ Y) {Domain expert}
            if g is an exception then               {Application expert}
                extend Y with g
            else
                extend E with g
    A := NEXT CLOSURE(A, M, L)
return L, E, Y
```

The sixth implication indicates that there exist (up to isomorphism) only three *complete planar Hamiltonian* graphs: G_6, G_{15}, and G_{19}. That G_{19} is the only *Hamiltonian tree* and the only *Eulerian tree* is expressed by the seventh and eighth implications, respectively. It is also the only *Hamiltonian bipartite complete* graph (the last implication).

The resulting concept lattice is shown in Figure 4.19. The implications valid in this lattice are exactly those that are valid for all graphs except for G_6, G_7, G_{14}, G_{15}, and G_{19}. The concept lattice of these exceptions is shown in Figure 4.20. In particular, one can see in the diagram that all exceptions are *connected planar* graphs.

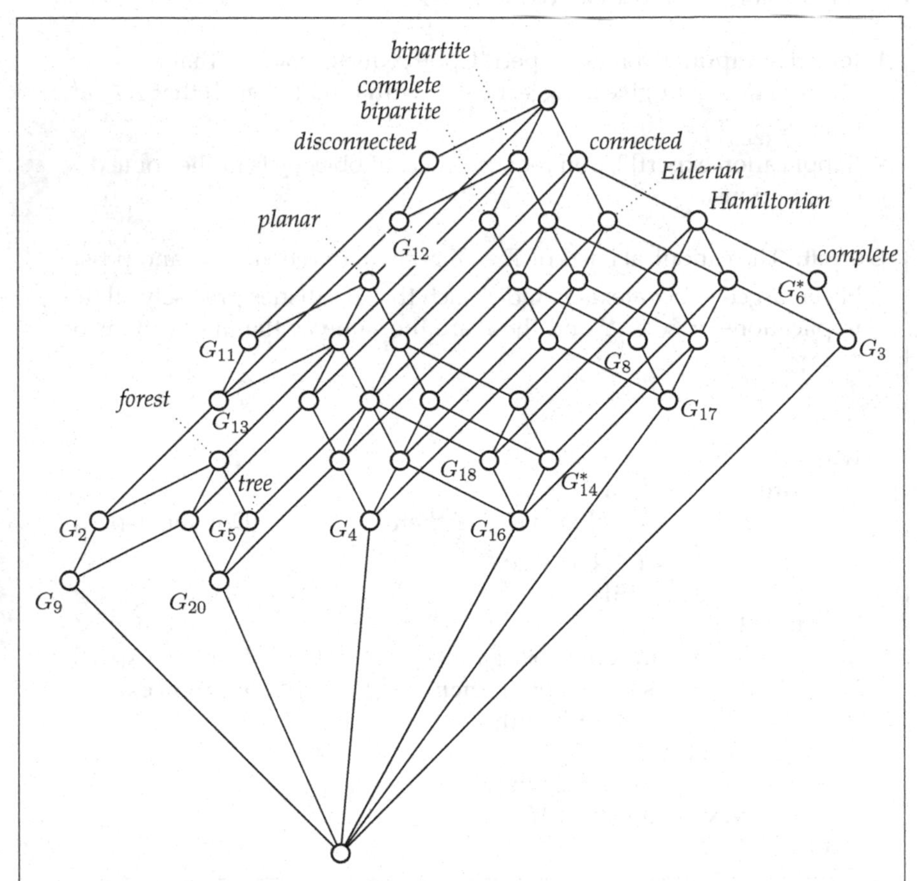

Figure 4.19: Concept lattice resulting from the exploration with exceptions. The labels for the reducible objects G_1 and G_{10} are omitted

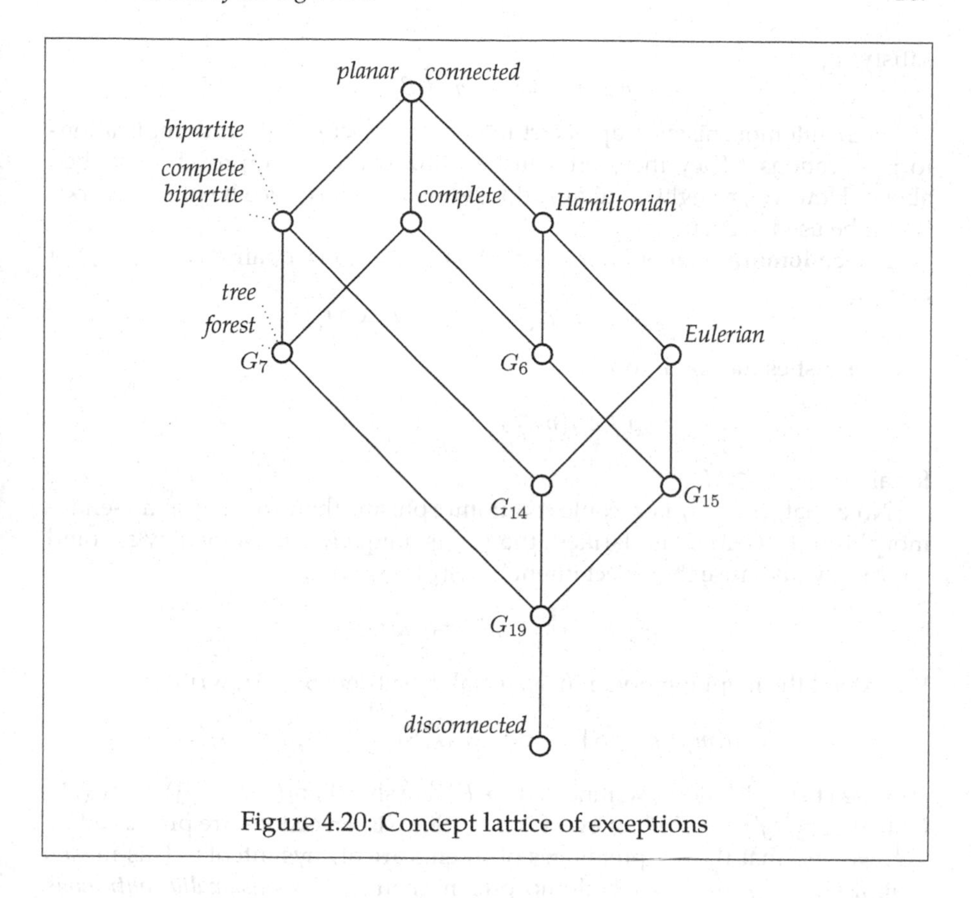

Figure 4.20: Concept lattice of exceptions

4.3.3 Exploration under symmetry

A form of background knowledge that seamlessly integrates with our method occurs when the presence of certain implications or of counterexamples implies that of other ones. Then, if implications or examples are added to the exploration knowledge base, the implied ones should automatically be added, too. There are other possible causes than just implication inference.

Symmetry is a typical instance of this. If it is known that the domain under investigation has certain symmetries, then, whenever a new implication is added to the basis, all its "isomorphic copies" must be added as well, and likewise for counterexamples. We make this more precise now, using context automorphisms, which were defined on page 67 as pairs (α, β) of mappings

$$\alpha : G \to G, \quad \beta : M \to M$$

satisfying

$$g\ I\ m \iff \alpha(g)\ I\ \beta(m).$$

Such automorphisms map object intents to object intents and implications to implications. They therefore can be utilized for the approach described above. However, a slight generalization, which we introduce next, also works. It will be used in Section 6.1.

A **c-endomorphism** of a formal context (G, M, I) is a pair $\varepsilon := (\varepsilon_G, \varepsilon_M)$ of mappings

$$\varepsilon_G : G \to G, \qquad \varepsilon_M : M \to M,$$

which satisfies the condition

$$g\ I\ \varepsilon_M(m) \iff \varepsilon_G(g)\ I\ m$$

for all $g \in G, m \in M$.

Note that, if (α, β) is a context automorphism, then (α^{-1}, β) is a c-endomorphism. If (G, M, I) is clarified, then ε_G is uniquely determined by ε_M (and conversely), because the object intent of $\varepsilon_G(g)$ is given as

$$\varepsilon_G(g)' = \{m \in M \mid \varepsilon_M(m) \in g'\}.$$

We extend the mapping notation, as usual, to subsets $S \subseteq M$, writing

$$\varepsilon_M(S) := \{\varepsilon_M(m) \mid m \in S\} \quad \text{and} \quad \varepsilon_M^{-1}(S) := \{m \in M \mid \varepsilon_M(m) \in S\},$$

and even to implications, writing $\varepsilon(A \to B)$ as a shorthand for $\varepsilon_M(A) \to \varepsilon_M(B)$. Note that $\varepsilon_M^{-1}(g') = \varepsilon_G(g)'$ for every $g \in G$. Since intersections are preserved by ε_M^{-1}, we find that the ε_M-preimages of intents are always intents. This means that, if $\varepsilon := (\varepsilon_G, \varepsilon_M)$ is a c-endomorphism, then ε_G is *extensionally continuous* and ε_M is *intensionally continuous* in the sense of [110], Definition 89. This implies that c-endomorphisms preserve implications, as the following propositions show:

Proposition 29 *If ε is a c-endomorphism, then $\varepsilon_G(g)'$ respects $A \to B$ iff g' respects $\varepsilon(A \to B)$.*

Proof $\varepsilon_G(g)'$ does not respect $A \to B$

$$\begin{aligned}
&\iff A \subseteq \varepsilon_G(g)' \text{ and } B \nsubseteq \varepsilon_G(g)' \\
&\iff \forall_{m\in A}\ \varepsilon_G(g)\ I\ m \text{ and } \exists_{m\in B}\ \varepsilon_G(g)\ \not{I}\ m \\
&\iff \forall_{m\in A}\ g\ I\ \varepsilon_M(m) \text{ and } \exists_{m\in B}\ g\ \not{I}\ \varepsilon_M(m) \\
&\iff \varepsilon_M(A) \subseteq g' \text{ and } \varepsilon_M(B) \nsubseteq g' \\
&\iff g' \text{ does not respect } \varepsilon(A \to B).
\end{aligned}$$

□

Proposition 30 *If $A \to B$ holds in (G, M, I) and ε is a c-endomorphism, then $\varepsilon(A \to B)$ holds in (G, M, I) as well.*

Proof Suppose that $\varepsilon(A \to B)$ does not hold. Then there must be some object $g \in G$ such that g' does not respect $\varepsilon(A \to B)$. By the preceding proposition, $A \to B$ cannot hold either, because it is not respected by $\varepsilon_G(g)'$. □

These results can be very useful for an actual attribute exploration, provided that both the universe of the exploration and the context of admissible examples are known to share some automorphisms or, more generally, c-endomorphisms. Then whenever an implication $A \to B$ is confirmed, we also know that all its images $\varepsilon(A \to B)$ must hold, and these can automatically be added to the implication list of the exploration. Likewise, when a counterexample g is given, we may automatically add all its endomorphic copies to our supply of examples. This avoids asking the domain expert questions that have already been answered in an "isomorphic setting".

A simplification "modulo symmetry" may not only reduce the workload of the expert; it can also reduce the computational work. An alternative to automatically adding endomorphic implications and counterexamples is to modify the algorithms accordingly. An instance of this was already discussed in Section 2.3.4. More of the mathematical and algorithmic foundations were investigated in [49, 54].

4.3.4 Partially given examples

In its basic version, attribute exploration requires full specification of the given examples. In other words, if the algorithm asks whether an implication $A \to B$ holds and the user gives a counterexample, he or she must input all attributes of this example, even those that occur neither in A nor in B. For instance, recall the exploration of *pairs of squares* in Section 4.2.2. One of the questions asked during exploration was

"overlap, common vertex $\to$ parallel, common segment, common edge ?".

When providing a counterexample ( ), we also had to specify if this example has the attribute "disjoint", although this did not occur in the question.

Modifications of the algorithm that allow incompletely specified examples have been studied by several authors (see [56, 170, 124, 58]). Their approaches involve Kleene logic or versions of modal logic.

For our purposes here, a much simpler version, which was first presented in [95], suffices. In the exploration knowledge base, we replace the single subcontext of the domain (G, M, I) by two contexts, $\mathbb{E}_+ := (E, M, J_+)$ and $\mathbb{E}_? :=$

$(E, M, J_?)$ with $E \subseteq G$ and $J_+ \subseteq I \cap (E \times M) \subseteq J_?$, and use their semantics to handle incompleteness. Some readers may prefer to view this context pair as one thing and call $(\mathbb{E}_+, \mathbb{E}_?)$ a **partial formal context**. For an object $g \in E$, we write g^+ instead of g^{J_+} and $g^?$ instead of $g^{J_?}$.

For each object $g \in E$, we have three sets of attributes:

g^+,	the attributes g is known to have;
$g^? \supseteq g^+$,	the attributes g may have; and
$g^- := M \setminus g^?$,	the attributes g is known not to have.

Note that each of the last two sets determines the other.

The interactive procedure is modified as follows:

1. Instead of giving a full specification of an example e, that is, instead of giving an object intent e^I, the user can give *two* disjoint sets e^+ and e^-. They express that the user knows an example e with

$$e^+ \subseteq e^I \quad \text{and} \quad e^I \cap e^- = \emptyset.$$

   In other words, by giving this pair of sets[8], the user expresses that there is an example e that has all attributes from e^+, but no attribute from e^-. The example e is inserted into the context $\mathbb{E}_+$ with intent e^+ and into the context $\mathbb{E}_?$ with intent

$$e^? := M \setminus e^-.$$

   For e to qualify as a counterexample for an implication $A \to B$, it is necessary that $A \subseteq e^+$, but $B \nsubseteq e^?$; i.e., according to what the user knows, e has all attributes from A, but there is at least one attribute from B that e does not have (see Proposition 31).

2. For any given $A \subseteq M$, we replace A^{JJ} by

$$A^{+?} := \bigcap \{e^? \mid e \in E, A \subseteq e^+\}.$$

   We do not claim that the modified operator $A \mapsto A^{+?}$ is a closure operator.

3. Any modification of the list $\mathcal{L}$ of accepted implications leads to a modification of $\mathbb{E}_+$ and $\mathbb{E}_?$ as follows. For each $e \in E$, the object intent e^+ is replaced by its closure under accepted implications:

$$e^+ \leftarrow \mathcal{L}(e^+).$$

   As an additional consistency check, we may test if $e^+ \subseteq e^?$ still holds after this modification.

[8]Such pairs have been called **pods** (partial object descriptions) in [21].

Moreover, we successively remove from $e^?$ elements m that do not satisfy the condition

$$\mathcal{L}(e^+ \cup \{m\}) \subseteq e^?,$$

until all elements of $e^?$ satisfy this condition.

We summarize this procedure in Algorithm 21. This version supports background implications, but does not support exceptions.

We prove now that, starting from an empty list $\mathcal{L}$ of implications, we eventually obtain the canonical basis of the explored context (G, M, I). This algorithm will suggest an implication $A \to A^{+?}$ that is valid in (G, M, I) if and only if A is a pseudo-intent of (G, M, I) and $A^{+?} = A^{II}$.

The algorithm may suggest implications that do not hold in the explored context and that therefore must be refuted by means of a (partial) counterexample. In contrast to what we had so far, this algorithm may suggest implications that are inconsistent with previously given answers. This happens, for example, when $A^{+?} \neq \mathcal{L}(A^{+?})$. In this case, although we know that the implication $A \to A^{+?}$ does not hold, it must be asked, because A can still be the premise of a valid implication $A \to B \subset A^{+?}$, but without an additional counterexample from the user we cannot say what this B might look like. If a counterexample is erroneously not given and the implication is accepted instead, it may occur that $\mathcal{L}(e^+) \not\subseteq e^?$. This can be tested in the additional consistency check mentioned in the third modification above.

If $(\mathbb{E}_+, \mathbb{E}_?)$ is a partial formal context with $\mathbb{E}_+ := (E, M, J_+)$ and $\mathbb{E}_? := (E, M, J_?)$, then any formal context (E, M, J) with $J_+ \subseteq J \subseteq J_?$ is called a **realizer** of $(\mathbb{E}_+, \mathbb{E}_?)$. An implication $A \to B$ is **refuted** by $(\mathbb{E}_+, \mathbb{E}_?)$ if it holds in no realizer of $(\mathbb{E}_+, \mathbb{E}_?)$.

Proposition 31 *An implication $A \to B$ is refuted by a partial formal context $(\mathbb{E}_+, \mathbb{E}_?)$ (defined as above) if and only if there is an object $e \in E$ such that $A \subseteq e^+$, but $B \not\subseteq e^?$.*

Proof Assume that $A \to B$ is not refuted by the partial context $(\mathbb{E}_+, \mathbb{E}_?)$. Then there is a realizer (E, M, J) that respects $A \to B$. Take an arbitrary object $e \in E$. Either $A \not\subseteq e^J$ or $B \subseteq e^J$. Since (E, M, J) is a realizer of $(\mathbb{E}_+, \mathbb{E}_?)$, we have $A \not\subseteq e^+ \subseteq e^J$ or $B \subseteq e^J \subseteq e^?$. It immediately follows that there is no object e such that $A \subseteq e^+$, but $B \not\subseteq e^?$.

Conversely, assume that there is no object e such that $A \subseteq e^+$, but $B \not\subseteq e^?$. We can build a suitable realizer (E, M, J) as follows:

$$e^J := \begin{cases} e^? & \text{if } A \subseteq e^+, \\ e^+ & \text{otherwise.} \end{cases}$$

Consider an arbitrary object $e \in E$. If $A \subseteq e^+$, then $B \subseteq e^? = e^J$. Otherwise, $A \not\subseteq e^+ = e^J$. Consequently, every object intent e^J respects the implication $A \to B$, the implication holds in (E, M, J), and it is not refuted by $(\mathbb{E}_+, \mathbb{E}_?)$. □

Algorithm 21 ATTRIBUTE EXPLORATION with partial examples

Input: A partial formal context $(\mathbb{E}_+, \mathbb{E}_?)$, where $\mathbb{E}_+ = (E, M, J_+)$ and $\mathbb{E}_? = (E, M, J_?)$ are possibly empty subcontexts of $\mathbb{K} = (G, M, I)$, M finite, with $J_+ \subseteq I \cap (E \times M) \subseteq J_?$; a set $\mathcal{L}$ of implications over M that hold in $\mathbb{K}$.

Interactive input: $\{*\}$ Upon request, confirm that $A \to A^{+?}$ holds in $\mathbb{K}$ or give an object $g \in G$ and two sets $C, D \subseteq M$ such that $A \subseteq C \subseteq g^I \subseteq D$, but $A^{+?} \not\subseteq D$.

Output: The canonical basis of $\mathbb{K}$ relative to the input $\mathcal{L}$ and possibly enlarged partial context $(\mathbb{E}_+, \mathbb{E}_?)$, which refutes precisely all the implications invalid in $\mathbb{K}$.

$A := \emptyset$

while $A \neq M$ **do**

  while $A \neq A^{+?}$ **do**
    if $A \to A^{+?}$ holds in $\mathbb{K}$ **then** $\{*\}$
      $\mathcal{L} := \mathcal{L} \cup \{A \to A^{+?}\}$
      for all $e \in E$ **do**
        $J_+ := J_+ \cup \{e\} \times \mathcal{L}(e^+)$
        while there is $m \in e^?$ such that $\mathcal{L}(e^+ \cup \{m\}) \not\subseteq e^?$ **do**
          $J_? := J_? \setminus \{(e, m)\}$
      exit while
    else
      obtain some object $e \in A^I \setminus A^{+?I}$ $\{*\}$
      obtain C and D such that $A \subseteq C \subseteq e^I \subseteq D \not\supseteq A^{+?I}$ $\{*\}$
      extend $\mathbb{E}_+$ with e and $e^+ = C$
      extend $\mathbb{E}_?$ with e and $e^? = D$

  $A :=$ NEXT CLOSURE$(A, M, \mathcal{L})$

return $\mathcal{L}$, $(\mathbb{E}_+, \mathbb{E}_?)$

Proposition 32 *For any given $A \subseteq M$, the set $A^{+?}$ is the largest among all sets B such that the implication $A \to B$ is not refuted by $(\mathbb{E}_+, \mathbb{E}_?)$.*

Proof For an arbitrary object $e \in E$, if $A \subseteq e^+$, then $e \in A^+$ and $A^{+?} \subseteq e^?$. Therefore, $A \to A^{+?}$ is not refuted by $(\mathbb{E}_+, \mathbb{E}_?)$ due to Proposition 31. Conversely, if (E, M, J) is some realizer of $(\mathbb{E}_+, \mathbb{E}_?)$ not refuting $A \to B$, then each e^J must respect $A \to B$. In particular, $B \subseteq e^J \subseteq e^?$ must hold for all $e \in E$ with $A \subseteq e^+ \subseteq e^J$, which proves that $B \subseteq A^{+?}$. □

Proposition 32 should not be overinterpreted: it does not express that the implication $A \to A^{+?}$ is compatible with the information present. The reason is that we have considered *all* possible realizers and not only those that are models of the confirmed list $\mathcal{L}$ of implications. Restricting to such models gives a more complicated result.

Proposition 33 *For any given $A \subseteq M$ and any $m \in A^{+?}$, there is a realizer of $(\mathbb{E}_+, \mathbb{E}_?)$ that is a model of $\mathcal{L} \cup \{A \to m\}$.*

Proof Consider the formal context (E, M, J) defined as follows:

$$e^J := \begin{cases} \mathcal{L}(e^+ \cup \{m\}) & \text{if } A \subseteq e^+, \\ e^+ & \text{otherwise.} \end{cases}$$

Modification 3 in the above list of modifications ensures that all intents of this formal context are $\mathcal{L}$-closed and that $\mathcal{L}(e^+ \cup \{m\}) \subseteq e^?$ for all $e \in A^+ \subseteq m^?$. Moreover, each intent containing A also contains m. □

Loosely speaking, this says that, although the known data may prove that the implication

$$A \to A^{+?}$$

cannot hold, it is still reasonable to ask about it, because the data does not suffice to remove any element from the conclusion.

Proposition 34 *An implication $A \to B$ valid in the domain (G, M, I) is suggested by the algorithm if and only if A is a pseudo-intent of (G, M, I) and $B = A^{II}$.*

Proof We prove this by induction. The basis case, when $A = \emptyset$, is obvious. Assume that the proposition is proved for all subsets of M of size at most k.

Let A be a pseudo-intent of (G, M, I) of size $k+1$. For any implication $C \to D$ that is in the list $\mathcal{L}$ generated by the algorithm, if $C \subset A$, then C is a pseudo-intent of (G, M, I) and $D = C^{II}$ by the induction hypothesis. Therefore, A is

$\mathcal{L}$-closed and the implication $A \to A^{+?}$ will be suggested by the algorithm. If, at this stage, $A^{+?} \neq A^{II}$, the implication $A \to A^{+?}$ does not hold in (G, M, I) and a counterexample must be provided by the user. This will necessarily make $A^{+?}$ smaller and the implication $A \to A^{+?}$ (with the same premise, but a smaller conclusion) will be suggested again. The process continues in this way until no counterexample is possible, which happens when $A^{+?} = A^{II}$. Then, the implication $A \to A^{II}$ will be suggested and accepted.

Conversely, suppose that a valid implication $A \to B$, where $|A| = k + 1$, has been suggested. The algorithm suggests only implications of the form $A \to A^{+?}$. Therefore, $B = A^{+?}$. From $A^{II} \subseteq A^{+?} = B$ (which holds due to Proposition 32) and $A \to B$ being a valid implication (i.e., $B \subseteq A^{II}$), it follows that $B = A^{II}$. It remains to show that A is a pseudo-intent of (G, M, I). Suppose that it is not. Then, since A is not closed either, there must be a pseudo-intent $P \subset A$ of (G, M, I) such that $P^{II} \not\subseteq A$. By the induction hypothesis, the list $\mathcal{L}$ of implications contains $P \to P^{II}$. (Note that, if $P \to P^{II}$ is not in $\mathcal{L}$ by the time when $A \to B$ is considered, it has no chance of getting there later, since implications are added in the lectic order of their premises and $P \subset A$.) This means that A is not $\mathcal{L}$-closed and the implication $A \to B$ cannot be suggested by the algorithm. □

Proposition 35 *Upon termination of Algorithm 21, $\mathbb{E}_+$ and $\mathbb{K}$ have the same implications.*

Proof It follows from Proposition 34 that, upon termination of Algorithm 21, $\mathcal{L}$ is the canonical basis of $\mathbb{K}$. Since e^+ is $\mathcal{L}$-closed for all $e \in E$, the implications from $\mathcal{L}$ are valid in $\mathbb{E}_+$. There cannot be any $A \subseteq M$ with $A = A^{II} \neq A^{++}$, because such a set would be $\mathcal{L}$-closed and the algorithm would ask for examples refuting query $A \to A^{+?}$ until eventually $A = A^{+?}$. But $A \subseteq A^{++} \subseteq A^{+?}$ then yields a contradiction to $A \neq A^{++}$. Thus, all intents of $\mathbb{K}$ must be closed in $\mathbb{E}_+$. It follows that all implications of $\mathbb{E}_+$ are valid in $\mathbb{K}$. □

4.4 A case study in mathematics

There are so many substantial exploration results for mathematical domains, that we can mention only a few of them here. Several students of U. Hebisch have completed explorations as part of their theses. Most remarkable among these is the dissertation of P. Kestler [132]. Kestler studies universal algebras of type (2, 1, 0) as objects and 70 short identities as attributes. It took him several years to answer about ten thousand queries proposed by the exploration algorithm. For each query he had to find a counterexample or to provide a proof. The resulting concept lattice has 24 143 elements. Kestler's results were verified by A. Revenko [183], who pursued exactly the same exploration, but replaced the human domain expert by computer software. Each query was proposed to a "theorem prover" (Prover9) as well as an "example generator" (Mace4, plus a program which can construct infinite examples) [160] and was answered this way. Revenko's fully automatized exploration took about two weeks. His thesis contains a second example for another problem from universal algebra (p-clones), where he again verifies a tedious investigation by means of a fully automatic attribute exploration with modifications.

A very early example is the exploration of lattice properties done by S. Reeg and W. Weiß. For their "Diplomarbeit" (master thesis, supervised by R. Wille) they investigated 50 properties of finite lattices (such as distributivity, being complemented, etc.), gave proofs for all valid implications and counterexamples for all others. Figure 6.1 of [110] shows a small part of their findings with only nine of the 50 attributes. L. Kwuida, C. Pech, and H. Reppe [148] classified generalizations of Boolean algebras (such as Okham algebras, Stone algebras, or ortholattices). More recently, V. Glatz [112] presented an exploration of model-theoretic notions such as ω-categoricity, model completeness, etc. Revenko and Kuznetsov [184] explore properties of functions on ordered sets.

In this section, we discuss in detail an exploration that is very similar to what Kestler and Revenko did, but much smaller: short identities for universal algebras with two unary operations.

4.4.1 Simple mathematical structures: bi-unars

Attribute exploration of mathematical domains is often easier than "real life" examples for several reasons. The interpretation of mathematical attributes is usually clear and no matter of discussion. Sometimes some trivial examples can be generated automatically or at least with ease, giving a start for the exploration. It is not to be expected that attribute exploration can solve hard mathematical problems. Its usefulness lies in organizing the very easy ones, which is often tedious to do otherwise.

The example we present here is very simple and mathematically not very important. Its purpose is to demonstrate the potential of the method.

We investigate **bi-unars**. These are algebraic structures consisting of a (finite) base set S together with two (unary) mappings f and g. Figure 4.21 shows an example of a bi-unar.

As attributes we choose very small equations that such structures may or may not satisfy. The smallest product terms that can be built from the operation symbols are

$$\text{id},\quad f,\quad g,\quad f\circ f,\quad f\circ g,\quad g\circ f,\quad g\circ g,$$

A bi-unar (S, f, g) on the base set $S := \{1, 2, 3\}$ with unary maps

$$f : S \to S, \quad g : S \to S$$

given by

s	1	2	3
$f(s)$	2	3	1

and

s	1	2	3
$g(s)$	3	1	2

,

is abbreviated as (231)(312).

Figure 4.21: A bi-unar on the base set $S := \{1, 2, 3\}$ and its shorthand notation

and the possible equations combining these[9] are

$$\begin{array}{lllll} \mathrm{id} = f, & \mathrm{id} = g, & \mathrm{id} = f \circ f, & \mathrm{id} = f \circ g, & \mathrm{id} = g \circ f, \\ \mathrm{id} = g \circ g, & f = g, & f = f \circ f, & f = f \circ g, & f = g \circ f, \\ f = g \circ g, & g = f \circ f, & g = f \circ g, & g = g \circ f, & g = g \circ g, \\ f \circ f = f \circ g, & f \circ f = g \circ f, & f \circ f = g \circ g, & f \circ g = g \circ f, & f \circ g = g \circ g, \\ g \circ f = g \circ g. & & & & \end{array}$$

These are the 21 attributes for our exploration, while the objects are all possible bi-unars. The bi-unar in Figure 4.21, for example, satisfies the equations

$$\mathrm{id} = f \circ g, \quad \mathrm{id} = g \circ f, \quad f = g \circ g, \quad g = f \circ f, \quad f \circ g = g \circ f,$$

and none of the other 16 identities.

Although these equations are rather trivial, their combinations are not obviously classified. Indeed, the canonical basis for this data set consists of 112 implications. But this can be simplified drastically using the above methods.

4.4.2 Generating examples

The number of mappings from an n-element set to itself is n^n. The number of bi-unars on such a set (which are essentially pairs of maps) is therefore n^{2n}. Thus there are 16 bi-unars on two elements, 729 on three, 65 536 on four, and 9 765 625 on five elements.

As a starting point for an exploration we use the 729 three-element bi-unars. It is an easy exercise to generate them with a small computer program and

[9] id $= f$ is short for $x = f(x)$, etc.

to check the identities. The result is a formal context with 729 objects, 21 attributes, having 110 formal concepts and a canonical basis consisting of 103 implications. We may object-reduce this context without changing the implications, and Figure 4.22 shows a reduced version. We have omitted the bi-unars on two elements, but it can be seen that these are implicitely covered by the 729 examples which were generated. One might extend the computer search to larger examples, but that does not seem very promising. The first problem to attack is that the number of implications is too large.

4.4.3 Using background knowledge and symmetries

With a naive application of attribute exploration, the next step would be to check the 103 implications of the canonical basis one by one for their validity in all bi-unars. This is not an inviting task, and giving 103 proofs seems too much for such a trivial problem.

There is obvious background knowledge that can be used here: equational theories are equivalence relations! In simpler terms, we know that

$$a = b \quad \text{and} \quad b = c \quad \text{together imply} \quad a = c,$$

and that this must be true for every choice of a, b, c from the seven terms id, f, g, $f \circ f$, $f \circ g$, $g \circ f$, $g \circ g$ used in the equations. From this, we get

$$\binom{7}{3} \cdot 3 = 105$$

background implications, such as

$$f \circ g = g \circ f, \quad f \circ g = g \circ g \quad \rightarrow \quad g \circ f = g \circ g.$$

The number of background implications is even larger than the size of the canonical basis!

Using these background implications, we can compute the relative canonical basis as it was introduced in Section 4.3.2. As it turns out, this basis consists of 40 implications. This is better, but is still too much work to do. There is an almost trivial symmetry of the attribute set that can be used for further simplification. This symmetry is induced by interchanging f and g. What is meant precisely will become clearer a little later. But first, we mention that this reduces the "relative" canonical basis to 23 implications.

The lectically first relative pseudo-intent is

$$\{f \circ f = g \circ f, \quad f \circ f = g \circ g, \quad g \circ f = g \circ g\}.$$

It is, however, not orbit-maximal, since interchanging f and g in all terms gives the attribute set

$$\{f \circ f = f \circ g, \quad f \circ f = g \circ g, \quad f \circ g = g \circ g\},$$

(123)(321)	(123)(111)	(231)(312)	(321)(123)	(321)(132)	(321)(222)	(321)(321)	(321)(311)	(321)(111)	(221)(223)	(221)(222)	(221)(121)	(121)(122)	(121)(131)	(121)(221)	(211)(213)	(111)(123)	(111)(132)	(111)(222)	(111)(131)	(111)(321)	(111)(111)	
×	×																					$id = f$
			×													×						$id = g$
×	×		×	×	×	×	×	×														$id = ff$
		×				×																$id = fg$
		×				×																$id = gf$
×			×	×		×									×	×	×			×		$id = gg$
						×															×	$f = g$
×	×											×	×	×		×	×	×	×	×	×	$f = ff$
			×						×							×	×	×	×	×	×	$f = fg$
			×								×	×				×	×		×		×	$f = gf$
×		×																	×		×	$f = gg$
		×	×							×											×	$g = ff$
×	×				×					×		×		×							×	$g = fg$
×	×				×			×		×			×					×			×	$g = gf$
	×		×		×			×	×	×	×	×				×		×			×	$g = gg$
						×				×	×					×	×	×	×	×	×	$ff = fg$
						×			×	×		×			×	×	×		×		×	$ff = gf$
×			×	×		×				×									×		×	$ff = gg$
×	×	×	×		×	×				×						×	×		×		×	$fg = gf$
	×				×	×	×			×		×	×						×		×	$fg = gg$
	×				×	×		×		×				×				×	×		×	$gf = gg$

Figure 4.22: Examples of bi-unars on three elements and the small identities they fulfill

which is lectically larger (for the linear order on the attributes shown in Figure 4.22). Only the second pseudo-intent will therefore be used as a premise of an implication to be asked by the exploration algorithm: do the identities

$$f \circ f = f \circ g, \quad f \circ f = g \circ g, \quad f \circ g = g \circ g$$

together imply

$$f \circ f = g \circ f, \quad f \circ g = g \circ f, \quad \text{and} \quad g \circ f = g \circ g?$$

The answer is "no", and we can provide a counterexample: The bi-unar on $S := \{1, 2, 3, 4\}$ given by

s	1	2	3	4
$f(s)$	2	2	4	2

and

s	1	2	3	4
$g(s)$	2	2	1	1

has the property that $f \circ f, g \circ g$, and $f \circ g$ all map every element to 2, while $g \circ f$ does not. The bi-unar therefore satisfies all three equations of the premise, but none of the suggested conclusion. Actually, these three identities are the only ones of our list that are satisfied by that example.

As a consequence, we extend our list of examples (Figure 4.22) not only by this bi-unar (S, f, g, id), but also by its automorphic copy (S, g, f, id).

The next orbit-maximal pseudo-intent to be found by the algorithm is the singleton set $f = g \circ g$, and the question asked is

$$\text{does } f = g \circ g \text{ imply } f \circ g = g \circ f?$$

Obviously, the answer is "yes", because from $f = g \circ g$ we obtain

$$f \circ g = g \circ g \circ g = g \circ f.$$

The implication list therefore must be extended by this implication, but not only by this one. Its automorphic image

$$g = f \circ f \quad \to \quad f \circ g = g \circ f$$

must also be included.

4.4.4 The exploration completed

- The next problem posed by the exploration algorithm is to decide whether

  $$f = g \circ g, \quad f \circ f = f \circ g, \quad f \circ f = g \circ f, \quad \text{and} \quad f \circ g = g \circ f$$

  together imply

  $$f = f \circ f, \quad f = f \circ g, \quad f = g \circ f, \quad f \circ f = g \circ g, \quad f \circ g = g \circ g, \quad g \circ f = g \circ g.$$

  The answer is "no", a counterexample is, in shorthand notation,

  $$(1112)(1123)$$

- Do
$$f = g \circ f \quad \text{and} \quad f \circ g = g \circ g$$
together imply
$$f = f \circ f \quad \text{and} \quad f \circ f = g \circ f?$$
This is indeed true and requires a short proof:

From $f = g \circ f$ we obtain $f \circ f = f \circ g \circ f$, which, using $f \circ g = g \circ g$, can be written as $f \circ f = g \circ g \circ f$ This may be shortened using the first identity:
$$f \circ f = g \circ g \circ f = g \circ f = f.$$

- Do
$$f = g \circ f \quad \text{and} \quad f \circ f = g \circ g$$
together imply
$$f = f \circ g \quad \text{and} \quad f \circ g = g \circ f?$$
This is an interesting point in this exploration. So far, all questions were almost trivial to answer. But this one, though easy, is somewhat of a riddle. This is typical for such explorations: they lead the user to non-trivial problems, those that cannot easily be answered automatically. Can you prove that, whenever f and g are mappings satisfying $f = g \circ f$ and $f \circ f = g \circ g$, they must also satisfy $f = f \circ g$? Or can you give a counterexample?

We do not disclose the answer here and leave the problem to the reader. But we show how the exploration algorithm continues after the correct answer was provided.

We have still not reached the end of this exploration. Actually, we are not even very close. We give some less interesting steps in small print and in slightly abbreviated form:

- The next question asked is if
$$f = g \circ f \quad \text{and} \quad g = f \circ g$$
together imply
$$f = f \circ f, \quad g = g \circ g, \quad f \circ f = g \circ f \quad \text{and} \quad f \circ g = g \circ g.$$
This is indeed true, and the proofs are easy. We show only the first one, since the remaining ones are similar:
$$f = g \circ f = f \circ g \circ f = f \circ f.$$
Note that in this case only one implication will be added, since this one is self-symmetric.

- Do
$$f = f \circ g \quad \text{and} \quad g \circ f = g \circ g$$
together imply
$$f = f \circ f, \quad \text{and} \quad f \circ f = f \circ g?$$
Yes, they do, since
$$f \circ f = f \circ g \circ f = f \circ g \circ g = f \circ g = f.$$
- Do
$$f = f \circ g \quad \text{and} \quad f \circ f = g \circ g$$
together imply
$$f = g \circ f, \quad \text{and} \quad f \circ g = g \circ f?$$
Again, yes. Here is a proof:
$$f = f \circ g = f \circ g \circ g = f \circ f \circ f = f \circ f \circ g \circ f =$$
$$= g \circ g \circ g \circ f = g \circ f \circ f \circ f = g \circ f \circ g \circ g = g \circ f \circ g = g \circ f.$$
This proof is very similar to the one that we did not disclose, four steps above.
- Do
$$f = f \circ g \quad \text{and} \quad g = g \circ f$$
together imply
$$f = f \circ f, \quad g = g \circ g, \quad f \circ f = f \circ g, \quad \text{and} \quad g \circ f = g \circ g?$$
The answer is "yes", and the proofs are trivial.
- Do
$$f = f \circ g, \quad f = g \circ f, \quad g = f \circ f, \quad \text{and} \quad f \circ g = g \circ f$$
together imply
$$g = g \circ g \quad \text{and} \quad f \circ f = g \circ g?$$
Again, the answer is "yes": $f = f \circ g$ and $g = f \circ f$ give $f = f \circ f \circ f$, from which we derive $f \circ f = f \circ f \circ f \circ f$, i.e., $g = g \circ g$.
- Does $f = g$ imply $f \circ f = f \circ g = g \circ f = g \circ g$? Obviously, "yes".
- Does $\text{id} = f \circ g$ imply $\text{id} = g \circ f$ and $f \circ g = g \circ f$? The answer is "no", but counterexamples have to be infinite. Since we have restricted ourselves to finite base sets, the answer is "yes".
- Do
$$f \circ g = \text{id} = g \circ f \quad \text{and} \quad f \circ f = g \circ g$$
imply $f = g$? "No", a counterexample is $(2341)(4123)$.
- Do
$$f \circ g = \text{id} = g \circ f \quad \text{and} \quad f = g \circ g$$
imply $g = f \circ f$? "Yes", obviously.
- Do
$$f \circ g = \text{id} = g \circ f \quad \text{and} \quad f = f \circ f$$
imply all other identities? "Yes", because these conditions imply $f = \text{id} = g$.

- Do

$$\mathrm{id} = f \circ f, \quad g = g \circ f, \quad \text{and} \quad f \circ g = g \circ g$$

imply

$$g = f \circ g, \quad g = g \circ g, \quad f \circ g = g \circ f, \quad \text{and} \quad g \circ f = g \circ g?$$

The answer is "yes", the proofs are easy.

- The answer to the next question, which asks if

$$\mathrm{id} = f \circ f \text{ and } g = f \circ g \quad \text{imply} \quad g = g \circ f \text{ and } f \circ g = g \circ f,$$

is "no", since there is a counterexample: $(2134)(3434)$. We find that the example satisfies the two equations $\mathrm{id} = f \circ f$ and $g = f \circ g$ of the premise and none of the two equations $g = g \circ f$ and $f \circ g = g \circ f$ of the suggested conclusion. But what about the other 17 identities? Do we really have to check them all? Would it make much difference if we input this as a *partial* example as described in Section 4.3.4 above? In the $+-?$ notation this would be

```
??+?????????+-????-??
```

However, the saturation process described there would complete this to

```
?-+----?----+-?----??
```

because, for example, the implications collected so far enforce that

$$\mathrm{id} = g, \quad \mathrm{id} = f \circ f, \text{ and } g = f \circ g \quad \text{imply} \quad g = g \circ f.$$

The symmetry between f and g allows us to also add the partial example

```
?????+??-+????????-??,
```

which gets completed to

```
-?---+-?-+----???----.
```

- Do

$$\mathrm{id} = f \circ f, \quad g = f \circ g, \quad \text{and} \quad g \circ f = g \circ g$$

imply

$$g = g \circ f, \quad g = g \circ g, \quad f \circ g = g \circ f, \quad \text{and} \quad f \circ g = g \circ g?$$

"Yes". Adding this implication and its isomorphic copy allows the algorithm to further refine the two partial examples from the previous step to

```
?-+----?----+-?----?-
```

and

```
-?---+-?-+----?-?----.
```

- Do

$$\mathrm{id} = f \circ f, \quad g = f \circ g, \quad f \circ g = g \circ g, \quad \text{and} \quad g = g \circ g$$

imply

$$g \circ f = g \circ g, \quad g = g \circ f, \quad \text{and} \quad f \circ g = g \circ f?$$

"No", a counterexample is (1243)(1212). We enter it as

```
??+?????????+-+???-+-,
```

but it is automatically completed to

```
?-+----?----+-+----+-.
```

- Do

$$\mathrm{id} = f \circ f, \quad f = g \circ g, \quad \text{and} \quad f \circ g = g \circ f$$

imply

$$f = f \circ f = g \circ g = \mathrm{id} \quad \text{and} \quad g = f \circ g = g \circ f?$$

"No", a counterexample is (2143)(3421). We enter it as

```
-?+??-?-??+?--???-+??,
```

and it gets completed to

```
-?+-------+?------+--.
```

- Do

$$\mathrm{id} = f \circ f \quad \text{and} \quad f = g \circ f$$

imply

$$f = f \circ g = g \circ f \quad \text{and} \quad g = g \circ g = f \circ f = \mathrm{id}?$$

"Yes", since $g = g \circ \mathrm{id} = g \circ f \circ f = f \circ f = \mathrm{id}$.

- Do

$$\mathrm{id} = f \circ f \quad \text{and} \quad f = f \circ g$$

imply

$$f = f \circ g = g \circ f \quad \text{and} \quad g = g \circ g = f \circ f = \mathrm{id}?$$

"Yes", since $g = \mathrm{id} \circ g = f \circ f \circ g = f \circ f = \mathrm{id}$.

- Does

$$\mathrm{id} = f \circ f = f \circ g = g \circ f$$

imply

$$g \circ g = \mathrm{id} = f \circ f = f \circ g = g \circ f \quad \text{and} \quad f = g?$$

"Yes", since $f = f \circ \mathrm{id} = f \circ f \circ g = \mathrm{id} \circ g = g$.

- Does

$$\mathrm{id} = f$$

imply

$$\mathrm{id} = f = f \circ f \quad \text{and} \quad g = f \circ g = g \circ f?$$

"Yes", obviously.

4.5 Further reading

The articles [92] and [93] are often cited as the first publications on attribute exploration. But Rudolf Wille suggested the development of such methods from the very beginning of Formal Concept Analysis, and [92] only offered the first algorithmic realization of Wille's abstract idea.

A method similar to attribute exploration, but not at all related to Formal Concept Analysis, seems to be behind S. Fajtlowicz's "conjecture making" GRAFFITI software [84].

It was D. Borchmann who suggested an abstract view of attribute exploration [52].

More about the "Components of musical experience" can be found in [223].

The example of applying attribute exploration to the construction of a mandatory access control model presented in Section 4.2.4 is from [173]. Dau and Knechtel discuss how attribute exploration can be used in the context of role-based access control [69]. Their approach combines attribute exploration with Description Logics (see Section 6.2.3) and triadic formal contexts (see Section 6.4.3), which capture various permissions granted to roles with respect to document types.

Partially given examples are treated in a much different way by Li, Mei, and Lv [152]. They use a many-valued context in which all attribute values are from $\{+, ?, -\}$ to represent incomplete data, define approximate concepts, and discuss knowledge acquisition on these grounds.

4.6 Exercises

4.1 Start attribute exploration on the context in Exercise 2.1 from Chapter 2 (p. 82). The first question will be: "entertainment → walking?" Suppose you know of someone who enjoys entertainment (cinema, theater, etc.), but doesn't like walking — and you know nothing about his or her other interests. So, you reject the implication by adding the following counterexample:

	car	*hobby*	*guests*	*sewing*	*entertainment*	*walking*
A	?	?	?	?	×	

What will the next question be? Suppose that you answer "yes". How does the context change?

4.2 For a vertex $v \in V$ of the graph (V, E) from Exercise 2.4 in Chapter 2 (p. 83), define $\phi(v)$ as the set consisting of v, all friends of v, and their friends. Build a formal context (V, V, I), where $v\ I\ w \iff w \in \phi(v)$; in other words, a cross between object v and attribute w means that $w = v$, or w is a friend of v, or w is a friend of a friend of v. Perform attribute exploration on this context assuming that you are aware of a slightly larger fragment of the underlying social network:

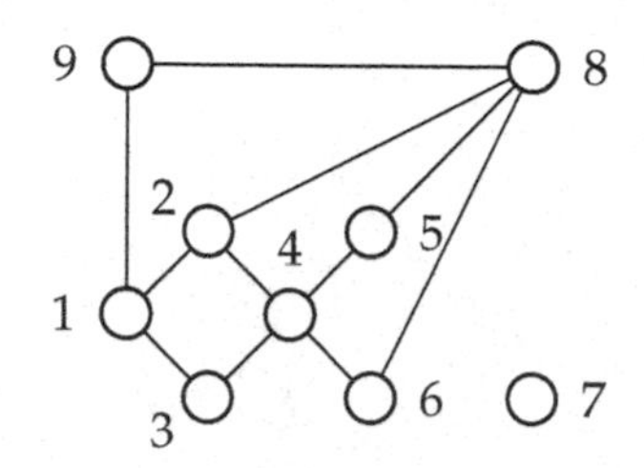

4.3 (a) For loopless directed graphs, explore the attributes *strongly connected, rooted, weakly connected, disconnected, acyclic, transitive, tournament.*

Definitions: A *loopless directed graph* (V, A) consists of a set V of vertices together with an irreflexive relation $A \subseteq V \times V$. A sequence $v_0, \ldots, v_n$ of vertices is called a *directed path* (of length $n \geq 0$) from v_0 to v_n if $(v_{i-1}, v_i) \in A$ for all $i \in \{1, \ldots, n\}$. It is an *undirected path* if $(v_{i-1}, v_i) \in A$ or $(v_i, v_{i-1}) \in A$ for all $i \in \{1, \ldots, n\}$. The graph is *strongly connected* if it contains, for any two $v, w \in V$, a directed path from v to w. It is *rooted* if there is some vertex from which there is a directed path to every other vertex. It is *weakly connected* if any two vertices are connected by an undirected path, *disconnected*, if it is not weakly connected, *acyclic*, if every directed path from a vertex to itself has length 0, *transitive*, if $(v, w), (w, x) \in A$ implies $(v, x) \in A$, and is a *tournament*, if for every two distinct vertices $v, w \in V$ exactly one of $(v, w), (w, v)$ is an element of A.

(b) Draw a line diagram of the concept lattice.

(c) Are any of the counterexamples "pathological", i. e., unique up to isomorphism?

(d) Take a look at the resulting canonical basis. Which of the implications are obviously true? Determine the basis relative to the obvious implications.

Chapter 5

Exploring data tables

In this chapter we extend attribute exploration by allowing for "non-implicational" background knowledge. This is propositional background knowledge that is not as "harmless" as the background implications of Section 4.3.2.

It is needed for two reasons. One is that real-word data often come in the form of data tables with attributes that have *values*, rather than as a binary relation. Formal Concept Analysis handles such data tables as *many-valued contexts*. They are transformed into the usual formal contexts by means of *conceptual scaling*. The scaling process induces patterns on the derived contexts, and attribute exploration must take these patterns into account. This can be done using propositional background knowledge.

A second reason is the use of *compound attributes*, which are derived from composing other attributes by means of the logical connectives *"and"*, *"or"*, or *"not"*. The knowledge that some attributes logically depend on other ones can be utilized for implication inference via propositional background knowledge.

The presence of propositional background knowledge increases the computational complexity of implication inference, see Section 5.3.

Attribute exploration of arbitrary propositional theories (not just implications) is studied in Section 5.2. Implications can indeed be replaced by the more general *cumulated clauses*. This yields a generalization which is, at first glance, very similar to attribute exploration. A closer inspection shows that many nice properties of the basic theory are lost. The computational complexity is much worse, and the "basis" is no longer of minimal size. The general method is nevertheless useful for organizing background knowledge.

5.1 Background clauses and why we need them

5.1.1 An unpleasant example

We know from Theorem 7 (p. 96) and from Proposition 17 (p. 97) that the canonical basis is non-redundant and of minimal cardinality. Sometimes this is hard to believe. Consider, as an example, the following little table. It lists all possible outcomes of a driving test. Its logical content is rather obvious: to pass a driver's license exam, it is necessary and sufficient to pass both the theory part and the practice part, see Figure 5.1. The logic of this example can be expressed

	theory	*driving*	*license*
1	*pass*	*pass*	*pass*
2	*pass*	*fail*	*fail*
3	*fail*	*pass*	*fail*
4	*fail*	*fail*	*fail*

Figure 5.1: Possible outcomes of a driving test

by a single equivalence, i.e., by two implications:

$$\textit{theory: pass} \;\wedge\; \textit{driving: pass} \iff \textit{license: pass}.$$

We should obtain the same result when computing the canonical basis. The table is not a formal context, but it can easily be transformed into the one in Figure 5.2, which obviously expresses the same.

	theory		*driving*		*license*	
	pass	*fail*	*pass*	*fail*	*pass*	*fail*
1	×		×		×	
2	×			×		×
3		×	×			×
4		×		×		×

Figure 5.2: Possible outcomes of a driving test as a formal context

However, computing the canonical basis yields a completely different result, see Figure 5.3. We obtain *eight* implications. Note that this is an irredundant system of minimal cardinality! Something must be wrong!

There is a simple explanation: the logical expression "test passed iff both parts are passed" is sufficient only if we presuppose that *"fail"* is the negation

driving: fail	$\to$	*license: fail*
theory: fail	$\to$	*license: fail*
license: fail, driving: pass	$\to$	*theory: fail*
license: fail, theory: pass	$\to$	*driving: fail*
driving: pass, theory: pass	$\to$	*license: pass*
license: pass	$\to$	*driving: pass, theory: pass*
license: fail, theory: fail, driving: pass, driving: fail	$\to$	$\perp$
license: fail, theory: fail, theory: pass, driving: fail	$\to$	$\perp$

Figure 5.3: The canonical basis for Figure 5.2

of *"pass"*. In other words, we intuitively use as *background knowledge* that the inferences

$$\text{"}\textit{pass}\text{ and }\textit{fail} \;\to\; \perp\text{"} \qquad \textit{and} \qquad \text{"}\top \;\to\; \textit{pass}\text{ or }\textit{fail}\text{"}$$

hold for all three parts of the test. Note that the second condition cannot be expressed by an implication.

Such situations occur often, in particular, when many-valued attributes are used. We shall discuss this later in more detail, but it becomes apparent that we need a version of attribute exploration that supports using non-implicational background knowledge.

Before we come to this, we should recall why we have concentrated on implications so far. In principle, one could develop an exploration using arbitrary propositional formulas. But this would give rise to several problems. One is comprehensibility: implications are intuitively much easier than other propositional formulas. Another one is computational complexity: implication inference is easy, whereas general propositional inference is co$\mathcal{NP}$-hard. We make this more precise in the next section.

5.1.2 Clauses of a formal context

To handle attribute exploration with background knowledge, we make use of the rich theory of propositional logic. However, since we are interested in the attribute logic of formal contexts, our language will be slightly modified. What is called a "propositional variable" in logic will be just an "attribute" for us, and we prefer to speak of *sets* of simple expressions rather than the formulas obtained as their *conjunctions*. For our purposes here we need only two elementary types of formulas, *implications* and *clauses*, and later a common generalization of both, called *cumulated clauses*.

Let M be a fixed set of attributes. *Implications* over M were defined in Chapter 3 as pairs $A \to B$ of subsets $A, B \subseteq M$. An implication $A \to B$ *holds* in a formal context (G, M, I) iff for all $g \in G$

$$A \subseteq g' \quad \text{implies} \quad B \subseteq g',$$

or verbally, iff

every object that has all attributes from A also has all attributes from B.

In propositional logic, an implication corresponds to a conjunction of Horn clauses of a special form. A **Horn clause** over a finite set M of propositional variables is a disjunction of variables from M and their negations with at most one unnegated variable, called *positive literal*. A *definite Horn clause* contains exactly one positive literal. A **Horn sentence** is a conjunction of Horn clauses.

A Horn sentence consisting of definite Horn clauses that have exactly the same negated variables $\neg p_1 \vee \cdots \vee \neg p_n$ can equivalently be represented by propositional implication

$$p_1 \wedge \cdots \wedge p_n \to q_1 \wedge \cdots \wedge q_m,$$

where $p_i, q_i \in M$. We usually write the above implication as

$$\{p_1, \ldots, p_n\} \to \{q_1, \ldots, q_m\}.$$

Thus an implication $A \to B$ as defined in Section 3.1 corresponds to a conjunction of definite Horn clauses sharing the negated variables:

$$\bigwedge_{p \in B} (p \vee \bigvee_{q \in A} \neg q).$$

Note that our definition of implication covers only definite Horn clauses.

A Horn clause is a special case of a **clause**, which is a disjunction of variables and their negations with no restriction on the number of positive literals. For our purposes, we introduce a clause as a pair of subsets $A, B \subseteq M$, denoted by

$$A \multimap B,$$

and say that clause $A \multimap B$ **holds** in a formal context (G, M, I) iff for all $g \in G$

$$A \subseteq g' \quad \text{implies} \quad B \cap g' \neq \emptyset,$$

or verbally, iff

every object that has all attributes from A has at least one attribute from B.

The definitions of clauses and implications are similar, but clauses are much more expressive. Indeed, *every* propositional formula is logically equivalent to a conjunction of clauses ("conjunctive normal form"). Therefore, families of clauses suffice to describe all possible kinds of propositional background knowledge. Note that for any subset $A \subseteq M$ the models of the clause

$$A \multimap (M \setminus A)$$

are all subsets except A. So in order to describe a family $\mathcal{F}$ of subsets we may simply use clauses of this form to rule out all sets not belonging to $\mathcal{F}$.

The models of a family of implications always form a closure system, while *every* collection of subsets of M is the set of models of some family of clauses over M. For finite M, this can be formulated as follows: a formal context is determined by its implications *up to reduction* and by its clauses *up to clarification*.

Another difference is that implications are much easier to handle than clauses, even than "short" clauses. We have seen in Proposition 16 (p. 92) and Section 3.1.3 that it can easily and efficiently be decided if a given implication follows from a given list of implications. Deciding if a given clause follows from a given list of clauses is known to be hard (co$\mathcal{NP}$-complete).

It is even co$\mathcal{NP}$-complete to decide if a given *implication* can be inferred from a given list of clauses or from a list consisting of clauses and implications. This is bad news for attribute exploration with background clauses. Recall that the algorithm systematically generates implications from the known examples and checks, before posing these as problems to the expert, if they can be derived from the confirmed logic. If each such test amounts to solving a co$\mathcal{NP}$-complete problem, performance may become a serious issue.

Moreover, it is also difficult to simplify a given list of clauses. There is very little hope for a theorem like that of Duquenne and Guigues but for clauses. We shall actually present a similar result below, but due to these complexity issues it will be much less useful than that for implications.

5.1.3 Cumulated clauses

We need formulas that generalize both implications and clauses.

Definition 16 A **cumulated clause** over a set M is an expression of the form

$$A \multimap \{B_1, B_2, \ldots\},$$

where A and all the B_i are subsets of M. A subset $S \subseteq M$ is a **model** of (or, synonymously, **respects**) the cumulated clause $A \multimap \{B_1, B_2, \ldots\}$ iff

$$A \not\subseteq S \quad \text{or} \quad B_i \subseteq S \text{ for some } i.$$

$\Diamond$

Note that the family $\{B_1, B_2, \ldots\}$ is allowed to be empty. The cumulated clause $A \multimap \{\}$ is usually noted down as

$$A \multimap \bot.$$

It has no model containing the set A. If one of the B_i is empty, then the clause is trivial, because it is respected by *all* sets.

Readers with a background in propositional logic will probably prefer a version in their usual syntax, which is, at least for short formulas, sometimes better readable, see, e.g., Figure 5.7. We indicate it briefly. An *implication* $A \to B$ stands for *"conjunction implies conjunction"*, i.e., for $\bigwedge A \to \bigwedge B$. A *clause* $A \multimap B$ stands for *"conjunction implies disjunction"*, i.e., for $\bigwedge A \to \bigvee B$. A *cumulated clause* stands for *"conjunction implies disjunction of conjunctions"*, i.e., for $\bigwedge A \to \bigvee_i \bigwedge B_i$. They are often written in this form when calculations are done manually. The representation by sets is convenient for computer programming.

Due to the distributive law, we can find sets C_j such that

$$\bigvee_i \bigwedge B_i = \bigwedge_j \bigvee C_j,$$

and then the cumulated clause $\bigwedge A \to \bigvee_i \bigwedge B_i$ is equivalent to $\bigwedge A \to \bigwedge_j \bigvee C_j$, which in turn is equivalent to $\bigwedge_j (\bigwedge A \to \bigvee C_j)$, i.e., to a conjunction of clauses. This explains the name "cumulated clause". It also makes evident that cumulated clauses generalize both implications and clauses. It was already stated that every propositional formula is equivalent to a conjunction of clauses (and thus of cumulated clauses).

Note that a single cumulated clause may be exponential in size compared to the set M of attributes.

5.1.4 Contextual attribute logic

Implications and (cumulated) clauses are instances of **compound attributes** (see [109]), by which we mean additional attributes that can be obtained from the attribute set M using the symbols $\neg$, $\bigwedge$, and $\bigvee$ iteratively.[1] It is then straightforward to extend the incidence relation to compound attributes; the definition of when an object *has* a given compound attribute follows the standard definitions of propositional logic. We must be careful when stating that two compound attributes "mean the same": we call such attributes

- **extensionally equivalent** over (G, M, I) if they have the same extent in (G, M, I) and

[1] As extreme cases, we obtain the attribute $\bot$ (which no object has) and the attribute $\top$ (which every object has). Other types of compound attributes will be introduced in Section 6.2

- **globally equivalent** if they have the same extent in every context with the attribute set M.

Global equivalence can be tested in one particular formal context, as the next proposition shows:

Proposition 36 *Two compound attributes of an attribute set M are globally equivalent if and only if they are extensionally equivalent in the context*

$$(\mathfrak{P}(M), M, \ni).$$

For this reason, $(\mathfrak{P}(M), M, \ni)$ is called the **test context** over M.

The test context gives another, purely semantic, view on background knowledge that is often helpful in understanding what is going on. It is usually not very practical though.

Without background knowledge, the examples can be arbitrary. More precisely, their intents may be arbitrary subsets of the attribute set M. We may think of the context $(\mathfrak{P}(M), M, \ni)$, the rows of which correspond to the $2^{|M|}$ bit vectors of length $|M|$, as the context from which the examples are drawn.

The presence of background knowledge can be interpreted as a change of the test context. All examples must be models of the background knowledge. Therefore, if $\mathcal{F}$ denotes the family of background clauses and $\text{Mod}(\mathcal{F})$ the set of its models, then the background knowledge makes it necessary to draw all examples from the **relative test context** $(\text{Mod}(\mathcal{F}), M, \ni)$, which is a subcontext of $(\mathfrak{P}(M), M, \ni)$, see Figure 5.12 (p. 211) for an example. The lattice of possible closure systems (and, dually, of their implication theories) is no longer the extent lattice of $(\mathfrak{P}(M), \text{Imp}(M), \models)$ as discussed in Section 3.4.4 (p. 117), but is isomorphic to that of the subcontext

$$(\text{Mod}(\mathcal{F}), \text{Imp}(M), \models),$$

where $\text{Imp}(M)$ is the set of all implications over M, and thus is a $\bigvee$-subsemilattice of the lattice of all closure systems on M.

The test context also gives a *semantic* interpretation of what it means that an implication follows from a list $\mathcal{L}$ of implications in the presence of a set $\mathcal{F}$ of background knowledge clauses: $A \to B$ follows from $\mathcal{L}$ iff $A \to B$ belongs to the concept intent of $(\text{Mod}(\mathcal{F}), \text{Imp}(M), \models)$ generated by $\mathcal{L}$. Theoretically, this gives us an algorithm: store the context $(\text{Mod}(\mathcal{F}), \text{Imp}(M), \models)$ in the computer and use its derivation operators for implication inference. This would make the algorithmic considerations of the subsequent subsections inessential. It seems however feasible only for rather small attribute sets (less than 20 attributes), for exceptional cases where the background knowledge is so rich that the test context becomes small, or for the case of plain scaling, where the relative test context has the form of a semi-product, see Proposition 40.

Exploration with compound attributes can be organized using the methods of this section. One simply adds the definitions of the compound attributes to the background knowledge and then follows the instructions of Section 5.3.4, which are condensed in Algorithm 28.

Suppose, for example, that c is a compound attribute defined as

$$c := m \wedge \neg n.$$

We add this definition to the background knowledge, but first translate it into a conjunction of cumulated clauses, namely

$$c \to m, \quad n \wedge c \to \bot, \quad \text{and } m \to n \vee c.$$

It will be explained in Section 5.2 how these cumulated clauses can be found. To check that their conjunction is indeed equivalent to $c = m \wedge \neg n$, consider Figure 5.4.

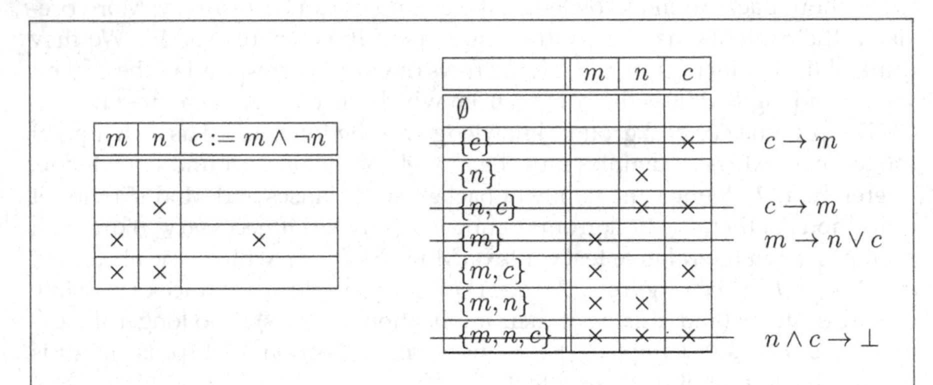

m	n	$c := m \wedge \neg n$
	×	
×		×
×	×	

	m	n	c	
$\emptyset$				
~~$\{c\}$~~			×	$c \to m$
$\{n\}$		×		
~~$\{n, c\}$~~		×	×	$c \to m$
~~$\{m\}$~~	×			$m \to n \vee c$
$\{m, c\}$	×		×	
$\{m, n\}$	×	×		
~~$\{m, n, c\}$~~	×	×	×	$n \wedge c \to \bot$

Figure 5.4: The left table shows how the incidences of $c := m \wedge \neg n$ depend on those of m and n. On the right, is the test context for $\{m, n, c\}$. The lines that are not models of $c = m \wedge \neg n$ are crossed, and a cumulated clause that is not respected is indicated for each crossed line

5.2 A generalized canonical basis

Having said that there is no hope for an equivalent to the theorem of Duquenne and Guigues for clauses, we are now going to prove one. But, as already stated, it will be much less useful. However, there is a natural application for it, which will be studied in the last section of this chapter. For the time being, the task is to make implication inference easy if there is only "a little" propositional background knowledge.

5.2.1 Pseudo-models

We first generalize the notion of a pseudo-closed set, which was introduced in Definition 11 (p. 95) for closure systems. Here is a reformulation of the recursive definition for a closure system $\mathcal{C}$:

Proposition 37 *A finite set $P \subseteq M$ is pseudo-closed w.r.t. a closure system $\mathcal{C}$ iff*

- *$P \notin \mathcal{C}$ and*
- *for every pseudo-closed set Q that is properly contained in P, there is a closed set C with $Q \subseteq C \subseteq P$.*

This indeed is easy to generalize. Let $\mathcal{D}$ be an arbitrary family of subsets of M, not necessarily a closure system.

Definition 17 A finite set $P \subseteq M$ is called a **pseudo-model** of $\mathcal{D}$ iff

- $P \notin \mathcal{D}$ and
- for every pseudo-model Q that is properly contained in P, there is $D \in \mathcal{D}$ with $Q \subseteq D \subseteq P$. ◊

The definition reads easier if one allows calling the elements of $\mathcal{D}$ "models". This seems irritating at first glance, because we so far have no formulas of which they could be models. But later $\mathcal{D}$ will be given as $\mathcal{D} := \text{Mod}(\mathcal{F})$, the set of models of the background knowledge. With this, we may reformulate as follows:

- *A set $P \subseteq M$ is pseudo-closed iff it is not closed and there is a closed set between P and every pseudo-closed proper subset of P.*
- *A set $P \subseteq M$ is a pseudo-model iff it is not a model and there is a model between P and every pseudo-model that is a proper subset of P.*

Of course, this is a proper generalization. The pseudo-models of a closure system are precisely its pseudo-closed sets.

Example 13 Consider the small graph in Figure 5.5. It has, as we shall see later, 37 induced connected subgraphs, including the empty graph. These will be the "models" in this example. Thus let $M := \{1,2,3,4,5,6\}$ be the set of vertices and let $\mathcal{D}$ consist of all (vertex sets of) induced connected subgraphs, like $\{1,2,4,6\}$ or $\{1,3,4,5\}$. Note that the family $\mathcal{D}$ is not a closure system, since the intersection of connected subgraphs is not necessarily connected. For example, the intersection of the two connected subgraphs mentioned above

$$\{1,4\} = \{1,2,4,6\} \cap \{1,3,4,5\}$$

induces a disconnected graph.

What are the pseudo-models of $\mathcal{D}$? According to Definition 17, no element of $\mathcal{D}$ is a pseudo-model. Since subgraphs with less than two elements are automatically connected, each pseudo-model must have at least two elements. Two vertices connected by an edge also form an element of $\mathcal{D}$; so, as candidates for pseudo-models, we must consider sets consisting of two non-adjacent vertices. There are nine of them:

$$\{1,4\},\{1,5\},\{1,6\},\{2,3\},\{2,5\},\{2,6\},\{3,5\},\{3,6\},\{5,6\},$$

and each of these is indeed a pseudo-model. This is true for a trivial reason: they contain no proper subsets that are pseudo-models. Therefore, the condition of Definition 17 is automatically satisfied.

These nine sets are the only pseudo-models of $\mathcal{D}$. Indeed, consider any set D containing more than two vertices. If D is a pseudo-model, it must be disconnected. Let v and w be any two disconnected vertices of D. There is no edge between v and w; therefore, $\{v, w\}$ is a pseudo-model contained in D. For D to be a pseudo-model, it must also contain a superset of $\{v, w\}$ that is a model and, thus, induces a connected subgraph. Hence, v and w cannot be disconnected in D, which leads to a contradiction.

Example 14 To give an example of slightly less trivial pseudo-models, let us now assume that $\mathcal{D}$ consists of all vertex sets of *disconnected* induced subgraphs of the graph in Figure 5.5. In this case, pseudo-models are among connected induced subgraphs, the empty graph being the smallest pseudo-model. Every larger pseudo-model must then contain some model, that is, a disconnected induced subgraph. It is easy to see that none of the connected one- or two-element graphs contains such a subgraph. Therefore, non-empty pseudo-models must contain at least three vertices each. The subgraph induced by $\{1,2,3\}$ is one such pseudo-model: it is connected, but it contains a disconnected subgraph induced by $\{2,3\}$. In fact, all the nine connected induced three-element subgraphs

$$\{1,2,3\},\{1,2,4\},\{1,3,4\},\{2,3,4\},\{2,4,5\},\{2,4,6\},\{3,4,5\},\{3,4,6\},\{4,5,6\}$$

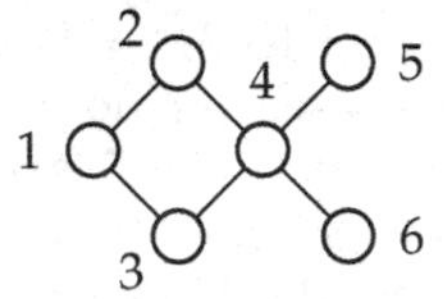

Figure 5.5: A small graph with 37 induced connected subgraphs

are pseudo-models, since the graph does not contain three-element cliques and, thus, among any three vertices, there are two that induce a disconnected subgraph. Obviously, none of the four-element subgraphs is a pseudo-model: each properly contains a three-element pseudo-model, but no larger model. It is not hard to check that the five-element subgraphs are not pseudo-models, either. The same is true about the graph itself: it contains a pseudo-model $\{1, 2, 4\}$, which cannot be extended to a disconnected induced subgraph.

5.2.2 The "canonical basis" of cumulated clauses

For any pseudo-model P of a given set $\mathcal{D}$ of subsets of M, there is a naturally associated cumulated clause, which we abbreviate to $\alpha_P^{\mathcal{D}}$:

$$\alpha_P^{\mathcal{D}} := \quad P \multimap \{D \in \mathcal{D} \mid P \subseteq D\}.$$

If M is finite, a simpler, but equivalent form is used:

$$P \multimap \{D \in \mathcal{D} \mid D \text{ is minimal w.r.t. } P \subseteq D\}.$$

We usually do not distinguish this in notation and use the simpler form whenever possible.

The meaning of this cumulated clause is easy to understand: a set S is a model of $\alpha_P^{\mathcal{D}}$ if it satisfies the following:

if $P \subseteq S$, then there is a model D between P and S.

Theorem 10 *The subsets of M that are models of all cumulated clauses in*

$$\{\alpha_P^{\mathcal{D}} \mid P \text{ is a pseudo-model of } \mathcal{D}\}$$

are exactly the elements of $\mathcal{D}$. This set of cumulated clauses is irredundant.

The **Proof** is simple: S is a model of all the $\alpha_P^{\mathcal{D}}$ if and only if S satisfies the following condition:

If $P \subseteq S$ is a pseudo-model, then there is a model $D \in \mathcal{D}$ between P and S.

If $S \in \mathcal{D}$, then this is obviously satisfied. If $S \notin \mathcal{D}$, but still satisfies the above condition, then S fulfills the conditions for being a pseudo-model and therefore must be a model of $\alpha_S^{\mathcal{D}}$ (i.e., there must be a model $D \in \mathcal{D}$ between S and S), which is impossible. If one of the cumulated clauses is removed, say $\alpha_P^{\mathcal{D}}$, then the set of models changes, since P is a model of the remaining ones and $P \notin \mathcal{D}$. □

$$\begin{array}{lllllllll}\{1,4\} & \multimap & \{\{2\},\{3\}\}, & \{2,3\} & \multimap & \{\{1\},\{4\}\}, & \{3,5\} & \multimap & \{\{4\}\}, \\ \{1,5\} & \multimap & \{\{2,4\},\{3,4\}\}, & \{2,5\} & \multimap & \{\{4\}\}, & \{3,6\} & \multimap & \{\{4\}\}, \\ \{1,6\} & \multimap & \{\{2,4\},\{3,4\}\}, & \{2,6\} & \multimap & \{\{4\}\}, & \{5,6\} & \multimap & \{\{4\}\}.\end{array}$$

Figure 5.6: The basis of cumulated clauses for the connected subgraphs of the graph in Figure 5.5

We have achieved a "basis" result for cumulated clauses, and, yes, it smoothly generalizes the result of Duquenne and Guigues. The generalized basis is indeed sound, complete, irredundant, and even invariant under context automorphisms. And in the case when the set family $\mathcal{D}$ is a closure system, the pseudo-models are pseudo-closed sets and the cumulated clauses $\alpha_P^{\mathcal{D}}$ are the implications that form the canonical basis.

But an important property of the canonical implication basis is not shared in general by that of cumulated clauses: it is not necessarily small. The canonical basis always has the smallest possible number of implications to form a basis. The analogue for cumulated clauses is not true.

Nevertheless, we shall use that basis for handling propositional background knowledge. There may be other roads to the same goal.

Example 15 We have found nine pseudo-models in Example 13. The (simplified[2]) cumulated clauses $\alpha_P^{\mathcal{D}}$ are shown in Figure 5.6 and translated to propositional formulas in Figure 5.7. These nine cumulated clauses are straightfor-

$$\begin{array}{lllllllll}1\wedge 4 & \to & 2\vee 3, & 2\wedge 3 & \to & 1\vee 4, & 3\wedge 5 & \to & 4 \\ 1\wedge 5 & \to & (2\wedge 4)\vee(3\wedge 4), & 2\wedge 5 & \to & 4, & 3\wedge 6 & \to & 4 \\ 1\wedge 6 & \to & (2\wedge 4)\vee(3\wedge 4), & 2\wedge 6 & \to & 4, & 5\wedge 6 & \to & 4.\end{array}$$

Figure 5.7: The basis of cumulated clauses written as propositional formulas

ward to interpret and express the expected. The first one states that

any connected subgraph containing vertices 1 and 4
must contain vertex 2 or vertex 3,

[2]We write $D \setminus P$ instead of D.

and the one below that one stands for

any connected subgraph containing vertices 1 and 5
must contain vertices 2 and 4 or vertices 3 and 4.

These nine conditions are indeed characteristic for the connected subgraphs of the graph in Figure 5.5.

To understand that the generalized basis is not necessarily minimal in size, have a look at Figure 5.8.

$$\begin{array}{rclrcl} a & \to & b \vee d, & b & \to & a \vee c, \\ c & \to & b \vee d, & d & \to & a \vee c, \\ a \wedge c & \to & \bot, & b \wedge d & \to & \bot. \end{array}$$

a d b c

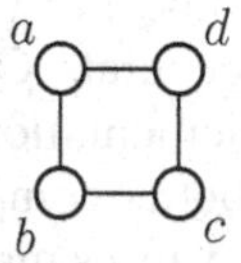

Figure 5.8: The models of the six cumulated clauses on the left are the edges of the graph and the empty set

The pseudo-models of this family of sets are all one- and all three-element subsets. Thus, there are eight pseudo-models and, consequently, eight cumulated clauses in the basis, as shown in Figure 5.9. Thus, Proposition 17 (p. 97) does not generalize to cumulated clauses.

$$\begin{array}{rclrcl} a & \to & (a \wedge b) \vee (a \wedge d), & b & \to & (a \wedge b) \vee (b \wedge c), \\ c & \to & (b \wedge c) \vee (c \wedge d), & d & \to & (a \wedge d) \vee (c \wedge d), \\ a \wedge b \wedge c & \to & \bot, & b \wedge c \wedge d & \to & \bot, \\ c \wedge d \wedge a & \to & \bot, & d \wedge a \wedge b & \to & \bot. \end{array}$$

Figure 5.9: The "basis" of cumulated clauses for the example given in Figure 5.8 has eight elements. It is equivalent to the set of six cumulated clauses in Figure 5.8

On the other hand: even when the basis is short in terms of the number of cumulated clauses, it is often not very intuitive. A typical example is when the models are mutually incomparable with respect to set inclusion, as it is the case for the "driving test" example (see Figure 5.12 below for the relative test context). Then, the basis contains one cumulated clause with the empty set as the premise and the disjunction of all models as the conclusion, followed by other clauses ensuring that there are no larger models. The background knowledge that is actually used in Section 5.3.3 is equivalent, but consists of a larger number of cumulated clauses.

5.2.3 Finding models of cumulated clauses

In Chapter 3, methods were developed for working with implications, and these will now further be generalized to cumulated clauses. An important difference is that we have no closure operator $T \mapsto \mathcal{L}(T)$ available. For any given subset $T \subseteq M$, there may be many models minimal w.r.t. containing T. As a substitute, we define $\mathcal{L}(T)$ to be the lectically smallest model containing T with respect to some linear order on M. In this chapter, the set M is therefore always assumed to be finite and linearly ordered.

It is indeed possible to mimic the algorithmic strategy that we have used for implications. But we should keep in mind that handling clauses is more complex in general. Our approach cannot overcome this. There are highly efficient implementations of propositional logic calculi, and it may well be that these show better computational results than the algorithms presented here. Background clauses may also be interpreted as *constraints*, and finding models is closely related to the **constraint satisfaction problem**, for which very efficient implementations are also available.

We will now present a rather simple algorithm that computes, for a given set $\mathcal{L}$ of cumulated clauses and a given subset $T \subseteq M$, a model of $\mathcal{L}$ containing T if such a model exists. For efficiency reasons, the algorithm divides the set of cumulated clauses into two parts, one of which contains only "cumulated Horn clauses", i.e., implications or expressions of the form $A \multimap \bot$. We denote the set of such clauses by $\mathcal{L}_{\mathcal{H}}$. The closure $\mathcal{L}_{\mathcal{H}}(T)$ can efficiently be computed using Algorithm 15 (p. 94).[3] The remaining cumulated clauses are denoted by $\mathcal{L}_{\mathcal{N}} := \mathcal{L} \setminus \mathcal{L}_{\mathcal{H}}$. In the pseudocode, see Algorithm 22, we explicitly specify $<$, the order on M, as a parameter of the algorithm. This will be useful later.

It is rather obvious that the algorithm terminates, because at every step some set in $\mathcal{X}$ is replaced by strictly larger sets or is altogether removed from $\mathcal{X}$ (e.g., if a clause $A \multimap \bot$ with $A \subseteq S$ is found). Since M is finite, this cannot go on indefinitely. So we obtain some result $\mathcal{L}(T)$.

Theorem 11 *If T is not contained in any model of $\mathcal{L}$, then $\mathcal{L}(T) = \bot$. Otherwise, $\mathcal{L}(T)$ is the lectically smallest model of $\mathcal{L}$ containing T.*

Proof The proof relies on the following proposition:

Proposition 38 *All elements of $\mathcal{X}$ contain T as a subset. If there are models containing T, then $\mathcal{X}$ always contains a subset of the lectically smallest such model.*

Proof This is clearly true upon initialization, when the only element of $\mathcal{X}$ is $\mathcal{L}_{\mathcal{H}}(T)$, unless $\mathcal{L}_{\mathcal{H}}(T) = \bot$, in which case there are no models containing T.

[3] Technically, Algorithm 15 has to be modified slightly to handle clauses of the form $A \multimap \bot$: at the beginning, the algorithm adds $\bot$ to M and then halts by returning $\bot$ whenever $\bot$ is added to the closure being computed.

Algorithm 22 First T-model$(T, \mathcal{L}, <)$

Input: A subset T of a finite set M, a set $\mathcal{L}$ of cumulated clauses over M, and a linear order $<$ on M.
Output: The lectically smallest w.r.t. $<$ model of $\mathcal{L}$ containing T if it exists; $\bot$, otherwise.

Split $\mathcal{L}$ into the Horn part $\mathcal{L}_{\mathcal{H}}$ and the remaining cumulated clauses $\mathcal{L}_{\mathcal{N}}$

if $\mathcal{L}_{\mathcal{H}}(T) = \bot$ **then**
    $\mathcal{X} := \emptyset$
else
    $\mathcal{X} := \{\mathcal{L}_{\mathcal{H}}(T)\}$
while $\mathcal{X} \neq \emptyset$ **do**
    let S be the lectically smallest element of $\mathcal{X}$
    if there is $A \multimap \{B_1, B_2, \ldots\} \in \mathcal{L}_{\mathcal{N}}$ of which S is not a model **then**
        $\mathcal{X} := \mathcal{X} \setminus \{S\}$
        for all B_i such that $\mathcal{L}_{\mathcal{H}}(S \cup B_i) \neq \bot$ **do**
            $\mathcal{X} := \mathcal{X} \cup \{\mathcal{L}_{\mathcal{H}}(S \cup B_i)\}$
    else
        return S
return $\bot$

Now, suppose that Y is the smallest model containing T and, at some step of the algorithm, $\mathcal{X}$ contains a set $Z \subseteq Y$. If Z is not the smallest element of $\mathcal{X}$, then it is unchanged and will be contained in $\mathcal{X}$ at the next step as well. If Z is the smallest element of $\mathcal{X}$, then it is either equal to Y (and the algorithm terminates), or $\mathcal{X}$ is modified according to some $A \multimap \{B_1, B_2, \ldots\}$ of which Z is not a model. But Y is a model and must therefore contain one of the sets $\mathcal{L}_{\mathcal{H}}(Z \cup B_i)$, which then belongs to $\mathcal{X}$. □

To conclude the proof of Theorem 11, consider the situation when the algorithm terminates with $\mathcal{L}(T) \neq \bot$. The output is a model containing T and also the lectically smallest element of $\mathcal{X}$. Any subset of the smallest model $Y \supseteq T$ is lectically smaller or equal to Y and therefore comes before any other model containing T. Since such a set is a member of $\mathcal{X}$, it must be the output. □

Example 16 Let us apply this algorithm to find the lectically smallest superset of {1, 6} that is a model of the set of cumulated clauses from Figure 5.6, i.e., the lectically smallest (vertex set of a) connected subgraph induced by vertices 1 and 6. First, we divide the set of cumulated clauses into the set $\mathcal{L}_\mathcal{H}$ of implications and the set $\mathcal{L}_\mathcal{N}$ containing the remaining (non-Horn) cumulated clauses:

$$\mathcal{L}_\mathcal{H} = \left\{ \begin{array}{l} \{2,5\} \to \{4\}, \\ \{2,6\} \to \{4\}, \\ \{3,5\} \to \{4\}, \\ \{3,6\} \to \{4\}, \\ \{5,6\} \to \{4\} \end{array} \right\} \quad \mathcal{L}_\mathcal{N} = \left\{ \begin{array}{l} \{1,4\} \multimap \{\{2\},\{3\}\}, \\ \{1,5\} \multimap \{\{2,4\},\{3,4\}\}, \\ \{1,6\} \multimap \{\{2,4\},\{3,4\}\}, \\ \{2,3\} \multimap \{\{1\},\{4\}\} \end{array} \right\}$$

Then, we set $\mathcal{X} = \{\mathcal{L}_\mathcal{H}(\{1,6\})\} = \{\{1,6\}\}$ and continue with $S = \{1,6\}$. At this point, S is not a model of $\mathcal{L}_\mathcal{N}$, since it is not a model of the cumulated clause $\{1,6\} \multimap \{\{2,4\},\{3,4\}\}$. We replace S in $\mathcal{X}$ with $\mathcal{L}_\mathcal{H}(\{1,2,4,6\})$ and $\mathcal{L}_\mathcal{H}(\{1,3,4,6\})$. Now, $\mathcal{X} = \{\{1,2,4,6\},\{1,3,4,6\}\}$, and the algorithm terminates by selecting $S = \{1,3,4,6\}$, the lectically smallest element of $\mathcal{X}$, which is also a model of $\mathcal{L}_\mathcal{N}$.

To see how dividing the set of cumulated clauses into two parts helps avoid unnecessary work, consider how the calculation of the lectically smallest model containing $\{2,3,5\}$ may proceed if implications are not separated from other cumulated clauses, that is, when $\mathcal{L}_\mathcal{H} = \emptyset$ and $\mathcal{L}_\mathcal{N}$ is the set of all cumulated clauses from Figure 5.6. The set $\{2,3,5\}$ is not a model of $\{2,3\} \multimap \{\{1\},\{4\}\}$; having selected this clause in the **if** condition of Algorithm 22, we must deal with two extensions of the initial set: $\{1,2,3,5\}$ and $\{2,3,4,5\}$. But the first extension will lead us along the wrong path, since any suitable model must contain 4 because of the implication $\{2,5\} \to \{4\}$ (as well as $\{3,5\} \to \{4\}$). Applying implications before other clauses forces us to extend the current set with what must be there prior to considering optional elements.

5.2.4 All models and the basis

We can use Theorem 11 to find, for any given set A, the lectically next model of $\mathcal{L}$ if such a model exists. Let

$$M := \{m_1 < m_2 < \ldots < m_n\}.$$

Define for $A \subsetneq M$ and $m_i \in M \setminus A$

$$A \oplus m_i := \mathcal{L}((A \cap \{m_1, \ldots, m_{i-1}\}) \cup \{m_i\}).$$

Theorem 12 *The lectically smallest model of $\mathcal{L}$ larger than A, if such a model exists, is $A \oplus m_i$, where m_i is the largest element of M satisfying $A <_i A \oplus m_i$.*

Example 17 Algorithm 23, All Models, can be applied to find the 37 connected subgraphs of the small graph in Figure 5.5 using the basis of cumulated clauses shown in Figure 5.6. The table in Figure 5.10 has 37 rows below the top row, each of them encoding a connected graph.

Algorithm 23 ALL MODELS(M, $\mathcal{L}$, $<$): Generating all models of $\mathcal{L}$

Input: A set $\mathcal{L}$ of cumulated clauses over a finite set M and a linear order $<$ on M.
Output: All models of $\mathcal{L}$ in lectic order.

```
A := FIRST MODEL(L, <)                         {see Algorithm 24}
while A ≠ ⊥ do
    output A
    A := NEXT MODEL(A, M, L, <)                {see Algorithm 25}
```

Algorithm 24 FIRST MODEL($\mathcal{L}$, $<$)

Input: A set $\mathcal{L}$ of cumulated clauses over a finite set M and a linear order $<$ on M.
Output: The lectically first model A of $\mathcal{L}$ if it exists.

```
return FIRST T-MODEL(∅, L, <)                  {see Algorithm 22}
```

Algorithm 25 NEXT MODEL(A, M, $\mathcal{L}$, $<$)

Input: A subset A of a finite set M, a set $\mathcal{L}$ of cumulated clauses over a M, and a linear order $<$ on M.
Output: The lectically next model of $\mathcal{L}$ after A.

```
for all m ∈ M in reverse order do
    if m ∈ A then
        A := A \ {m}
    else
        B := FIRST T-MODEL(A ∪ {m}, L, <)      {see Algorithm 22}
        if B ≠ ⊥ and B \ A contains no element < m then
            return B
return ⊥
```

1	2	3	4	5	6
					×
				×	
			×		
			×		×
			×	×	
			×	×	×
		×			
		×	×		
		×	×		×
		×	×	×	
		×	×	×	×
	×				
	×		×		
	×		×		×
	×		×	×	
	×		×	×	×
	×	×	×		
	×	×	×		×

1	2	3	4	5	6
	×	×	×	×	
	×	×	×	×	×
×					
×		×			
×		×	×		
×		×	×		×
×		×	×	×	
×		×	×	×	×
×	×				
×	×		×		
×	×		×		×
×	×		×	×	
×	×		×	×	×
×	×	×			
×	×	×	×		
×	×	×	×		×
×	×	×	×	×	
×	×	×	×	×	×

Figure 5.10: The 37 connected induced subgraphs of the graph in Figure 5.5, as generated by the ALL MODELS algorithm

Theorem 13 *Let $\mathcal{D}$ be a family of subsets of the (finite, linearly ordered) set M and let $P_1, \ldots, P_n$ be some pseudo-models of $\mathcal{D}$. If there is a lectically smallest set that respects the cumulated clauses $\alpha^{\mathcal{D}}_{P_i}$ for all $i \in \{1, \ldots, n\}$, but is not in $\mathcal{D}$, then it is a pseudo-model of $\mathcal{D}$. If no such set exists, then there are no pseudo-models other than the P_i.*

Proof If there is a pseudo-model other than $P_1, \ldots, P_n$, then it must respect all $\alpha^{\mathcal{D}}_{P_i}$ and is not in $\mathcal{D}$. It therefore has the prescribed property. Let P be the smallest set with the property. Any pseudo-model Q properly contained in P is lectically smaller and thus must be one of the P_i. Therefore, P respects $\alpha^{\mathcal{D}}_{Q}$ for any pseudo-model $Q \subsetneq P$ and is not in $\mathcal{D}$. So P is a pseudo-model. □

Theorem 13 suggests an algorithm for finding all pseudo-models of any given family $\mathcal{D}$ of subsets of M. This includes, of course, finding the pseudo-models of the family

$$\mathcal{D} := \{g' \mid g \in G\}$$

of all object intents of a formal context (G, M, I), and, building on that, the "basis" of cumulated clauses as introduced in Theorem 10. This is the job of

$$
\begin{array}{rcl}
ce & \to & pa \wedge cs \wedge cv \\
cs & \to & pa \\
pa & \to & di \vee ov \vee cv \vee cs \\
pa \wedge cv \wedge cs & \to & ce \\
ov \wedge pa \wedge cv & \to & cs \wedge ce \\
di \wedge cv & \to & \bot \\
di \wedge pa \wedge cs & \to & \bot \\
di \wedge ov & \to & \bot
\end{array}
$$

Figure 5.11: Cumulated clauses for the pairs of squares

Algorithm 29 (p. 231) to be discussed later.

Theorem 13 also makes it possible to compute the pseudo-models of a given list $\mathcal{L}$ of cumulated clauses, since D is a model of $\mathcal{L}$ iff $\mathcal{L}(D) = D$. The minimal models containing a given pseudo-model P of $\mathcal{L}$ can be computed using a modification of Algorithm 22: whenever a model D containing P is found, the clause $D \multimap \bot$ is added to $\mathcal{L}$ and the algorithm is invoked again. It will then only search for models not containing D.

Applying the technique that is implicit in Theorem 13 to the twelve pairs of squares (in Figures 4.3 and 4.5) yields the basis of cumulated clauses that is shown in Figure 5.11, noted down as propositional logic formulas.

The basis is identical to the canonical basis for the implications (which can be read from the last column of Figure 4.4), except for one proper cumulated clause stating that two parallel squares must be disjoint or overlapping, or have a vertex or a line segment in common. This clause expresses the fact that the attribute concept of *parallel* in Figure 4.6 is not an object concept even in the full context of all pairs of squares. The top element and the attribute concepts of *common vertex* and of *overlap* are object concepts in the formal context of Figure 4.3. They are join-reducible and thus are not labeled in the diagram.

5.3 Making inference feasible

It is not our aim to generalize attribute exploration through replacing implications by cumulated clauses. That would not be an intuitive and efficient method, we believe. The generalization we are aiming at is more modest: we assume that we have some background knowledge about possible counterexamples and try to include this knowledge in the exploration process. The strategy is to express this background knowledge by cumulated clauses. Then, in order to check if a new implication follows from given ones, it must be asked if it can be inferred from the given implications and the background clauses.

5.3.1 Complexity considerations

The crucial question concerning the efficiency of all these computations is how many steps are required in the computation of $\mathcal{L}(T)$ (Algorithm 22). In general, this cannot be easy, because it includes the well-known *satisfiability problem*, which is a generic $\mathcal{NP}$-complete problem: $\mathcal{L}$ is satisfiable iff $\mathcal{L}(\emptyset) \neq \perp$.

Let $\mathcal{L}_{\mathcal{N}}$ denote the "non-Horn" part of $\mathcal{L}$:

$$\mathcal{L}_{\mathcal{N}} := \left\{ \bigwedge A_i \to \bigvee_{t \in T_i} \bigwedge B_t \;\middle|\; i \in \{1, \ldots, N\} \right\},$$

where $N := |\mathcal{L}_{\mathcal{N}}|$ and the indexing is chosen in such a way that

$$|T_1| \geq |T_2| \geq \ldots \geq |T_N|.$$

It was shown in [101] that the total number of sets generated by Algorithm 22 is bounded by

$$\sigma(\mathcal{L}_{\mathcal{N}}) := 1 + |T_1| + |T_1| \cdot |T_2| + \ldots + |T_1| \cdot |T_2| \cdots |T_N|.$$

$$\tau(\mathcal{L}_{\mathcal{N}}) := N + |T_1| + \ldots + |T_N|$$

is the total number of sets occurring in $\mathcal{L}_{\mathcal{N}}$. With these parameters, we get

Theorem 14 ([101]) *Constructing $\mathcal{L}(T)$ by Algorithm 22 requires at most*

$$O\left(\sigma(\mathcal{L}_{\mathcal{N}}) \cdot (\tau(\mathcal{L}_{\mathcal{N}}) + |\mathcal{L}_{\mathcal{H}}| \cdot |M|) \cdot |M|\right)$$

bit operations.

Let us remark that, for fixed $\mathcal{L}_{\mathcal{N}}$, this is

$$O\left(|\mathcal{L}_{\mathcal{H}}| \cdot |M|^2\right),$$

which with a pinch of salt may be called *linear in the size of $\mathcal{L}_{\mathcal{H}}$*.

The applications at which we are aiming will have a small amount of non-implicational background knowledge, and it will typically remain constant during an exploration. For such a setting, Algorithm 22 is quite efficient, according to Theorem 14.

We should however clarify that computational efficiency is not the main purpose of our presentation. For problems of propositional calculus, there are powerful software implementations available, so-called **SAT solvers** for problems in conjunctive normal form (CNF), and computer algebra systems (**CAS**) for Boolean polynomials. Alonso-Jiménez et.al. [5] have studied the second possibility and suggest to utilize Boolean derivatives as well. Indeed, it may well be that implementations using such software perform better than the algorithms presented here.

The NEXT CLOSURE algorithm is not the theoretically fastest one for generating formal concepts, but due to its simplicity it is extremely flexible and can be adapted to many other settings. Its generalization to cumulated clauses should have similar properties: useful for building a unified theory, but not necessarily the fastest in every instance.

5.3.2 Implication inference with background knowledge

We have already mentioned that inferring an implication from a set of implications and clauses is co$\mathcal{NP}$-hard. This is however not true when the "clauses" part is fixed and only the "implications" part varies, because the problem can be reduced to computing the $\mathcal{L}(\cdot)$-operator. Here goes the trick.

An implication $A \to B$ follows from a list $\mathcal{L}$ of cumulated clauses iff all models of $\mathcal{L}$ that contain A also contain B. Therefore, $B \subseteq \mathcal{L}(A)$ is a necessary condition (which is true by convention if $\mathcal{L}(A) = \bot$), but not always sufficient. There may be lectically larger models of $\mathcal{L}$ containing A but not B. However, this cannot happen if the linear order of the attribute set is such that sets not containing B come before those containing B in the lectic order. If B is an initial segment in the linear order, then this is indeed the case. This was already discussed in Section 2.3.1 (p. 52).

Proposition 39 *If B is an initial segment in the linear order of M, then the implication $A \to B$ follows from a list $\mathcal{L}$ of cumulated clauses iff $B \subseteq \mathcal{L}(A)$.*

As a consequence, it can be decided with a single invocation of the $\mathcal{L}(\cdot)$-operator if an implication $A \to B$ follows from $\mathcal{L}$ (see Algorithm 26). The complexity of implication inference thus is the same as the complexity of computing $\mathcal{L}(\cdot)$. The linear order of M that induces the lectic order on the subsets may however change between such invocations. This may make the implementation tedious, but has little effect on the computation cost.

Example 18 To give a very small example, let

$$M := \{a < b < c < d\}$$

Algorithm 26 FOLLOWS($A \to B, M, \mathcal{L}$)

Input: A set $\mathcal{L}$ of cumulated clauses and an implication $A \to B$ over a finite set M.
Output: **true** if $A \to B$ follows from $\mathcal{L}$, **false** otherwise.

let $<$ be a linear order on M such that $b < m$ for all $b \in B$ and $m \notin B$
return $B \subseteq$ FIRST T-MODEL($A, \mathcal{L}, <$) {see Algorithm 22}

and consider a set $\mathcal{B}$ of two background cumulated clauses, one of which is actually an implication, given by

$$\mathcal{B} := \{\, \emptyset \multimap \{\{a\},\{b\}\}, \quad \{c,d\} \to \{a,b\} \,\}.$$

Let the Horn part consist of a single implication (brackets omitted)

$$\mathcal{H} := \{b \to d\},$$

and let $\mathcal{L} := \mathcal{B} \cup \mathcal{H}$.

What is $\mathcal{L}(\{c\})$? In the call to Algorithm 22 (FIRST T-MODEL), we set $T := \{c\}$ and obtain

$$\mathcal{X} := \{\mathcal{L}_{\mathcal{H}}(\{c\})\} = \{\{c\}\}.$$

There is a cumulated clause of which $\{c\}$ is not a model, namely,

$$\emptyset \multimap \{\{a\},\{b\}\}.$$

Applying it as prescribed by the algorithm, we replace $\{c\}$ by $\mathcal{L}_{\mathcal{H}}(\{a,c\})$ and $\mathcal{L}_{\mathcal{H}}(\{b,c\})$, from which we get the new value of $\mathcal{X}$:

$$\mathcal{X} = \{\{a,c\},\{a,b,c,d\}\},$$

since $\mathcal{L}_{\mathcal{H}}$ contains $\{c,d\} \to \{a,b\}$, as well as $b \to d$. The lectically smallest element of $\mathcal{X}$ is now a model of $\mathcal{L}$, and the algorithm terminates with the result that

$$\mathcal{L}(\{c\}) = \{a,c\}$$

is the lectically smallest model of $\mathcal{L}$ containing c. Since a is first in the order $a < b < c < d$, we also get implication inference:

$$c \to a \quad \text{follows from} \quad b \to d$$

with respect to the given background knowledge.

For any $A \subseteq M$, there is always a unique maximal $B \subseteq M$ such that $A \to B$ follows from the list $\mathcal{L}$ of cumulated clauses: this B is simply the intersection of all models of $\mathcal{L}$ containing A as a subset. When $\mathcal{L} = \mathcal{H} \cup \mathcal{B}$, where $\mathcal{H}$ contains only implications, we may say that $\mathcal{L}$ induces the following closure operator mapping $A \subseteq M$ to its *$\mathcal{H}$-closure under background knowledge $\mathcal{B}$*:

$$A \mapsto \bigcap_{A \subseteq B = \mathcal{L}(B)} B.$$

Algorithm 27 computes the $\mathcal{H}$-closure of set $T \subseteq M$ under background knowledge $\mathcal{B}$ maintaining two approximations of the closure: the lower bound S and the upper bound U initialized to T and M, respectively. The algorithm computes the lectically first model containing S and removes elements not contained in this model from the upper bound U. It then adds to S elements of the updated U that form the initial segment in the linear order of M. If S is still different from U, the algorithm changes the order on M by putting all elements of U to the front and computes the lectically first model of S with respect to this new order. The computation proceeds in this way until the two approximations, S and U, coincide. The algorithm terminates at some point, since S is always a subset of U and U becomes smaller at each iteration — except, probably, the last one, but in this case, U forms the initial segment of the linear order on M and, therefore, all its elements are added to the lower bound S, which becomes equal to the upper bound U.

Algorithm 27 BACKGROUND CLOSURE($T, M, \mathcal{H}, \mathcal{B}$)

Input: A subset T of a finite set M, a set $\mathcal{H}$ of implications (Horn clauses), and a set $\mathcal{B}$ of cumulated clauses over M.
Output: The $\mathcal{H}$-closure of T under background knowledge $\mathcal{B}$.

```
Let < be an arbitrary linear order on M
if FIRST T-MODEL(S, H ∪ B, <) = ⊥ then                 {see Algorithm 22}
    return M

S := T
U := M
while S ≠ U do
    let < be a linear order on M such that u < m if u ∈ U and m ∉ U
    U := U ∩ FIRST T-MODEL(S, H ∪ B, <)                {see Algorithm 22}
    S := S ∪ {u ∈ U | u < m for all m ∈ M \ U}
return S
```

Example 19 We enrich the background knowledge of the previous example to

$$\mathcal{B} := \{\ \emptyset \multimap \{\{a\},\{b\}\},\quad \emptyset \multimap \{\{c\},\{d\}\},\quad \{c,d\} \to \{a,b\},\quad \{a,b\} \to \{c,d\}\ \}.$$

Obviously, $c \to a$ still follows from $b \to d$ under $\mathcal{B}$. We put $\mathcal{H} := \{c \to a\}$ and run Algorithm 27 to find the $\mathcal{H}$-closure of $\{b\}$ under background knowledge $\mathcal{B}$. We set $S := \{b\}$ and $U := \{a,b,c,d\}$ and compute the FIRST T-MODEL for S. In Algorithm 22 implementing this computation, $\mathcal{X} := \{\{b\}\}$ initially, but $\{b\}$ is not a model of

$$\emptyset \multimap \{\{c\},\{d\}\}.$$

Thus, $\mathcal{X}$ is updated to

$$\mathcal{X} = \{\mathcal{L}_{\mathcal{H}}(\{b,c\}), \mathcal{L}_{\mathcal{H}}(\{b,d\})\} = \{\{b,d\},\{a,b,c,d\}\}$$

where $\mathcal{L} = \mathcal{H} \cup \mathcal{B}$ and $\mathcal{L}_{\mathcal{H}}$ consists of the implications contained in $\mathcal{L}$. Although $\{b,d\}$ is a model of $\mathcal{L}$, we cannot infer $b \to d$ at this point, because d is not initial in the order $a < b < c < d$. We update $U := \{b,d\}$ keeping S unchanged and compute the FIRST T-MODEL for S again, but now with a different order on M: $b < d < a < c$. Again, the result is $\{b,d\}$. Therefore, U is unchanged, $S := \{b,d\}$, and the algorithm terminates. Thus,

$$b \to d \quad \text{follows from} \quad c \to a$$

with respect to the given background knowledge.

Note that Algorithm 27 can easily be optimized. In the previous example, we computed the model $\{b,d\}$ twice, which is clearly unnecessary. Instead, after updating U, we could simply have removed $\{b,d\}$ from $\mathcal{X}$ and continue with $\mathcal{X} = \{\{a,b,c,d\}\}$ and the modified order on M. However, this optimization requires merging Algorithm 22 with Algorithm 27 or, on the contrary, decomposing Algorithm 22 into smaller subprocedures, which would make things too technical for our purposes here.

5.3.3 No canonical basis

With respect to the background knowledge of Example 19, the two implications $c \to a$ and $b \to d$ imply each other and therefore generate the same set of implications. They clearly cannot be replaced by simpler ones. This shows that there is no such thing as a "canonical" basis when background knowledge is present.

It is worthwhile to take a closer look at the "driving test" example from Section 5.1.1, where the canonical basis surprisingly consists of eight implications, while intuitively only two were expected (because of tacitly made assumptions). The background knowledge in that case is that *pass* and *fail* are negations of each other, i.e.,

$$pass \wedge fail \to \bot, \qquad \top \to pass \vee fail.$$

This holds for all three aspects of the test (theory, driving, and license), which amounts to six clauses. These have only eight models, shown in Figure 5.12 as the rows of the *relative test context*. The first four rows are identical to those in Figure 5.2 (p. 188). There are exactly 32 implications of the form $A \to b$

	theory		*driving*		*license*	
	pass	*fail*	*pass*	*fail*	*pass*	*fail*
1	×		×		×	
2		×	×			×
3	×			×		×
4		×		×		×
5		×	×		×	
6	×			×	×	
7		×		×	×	
8	×		×			×

Figure 5.12: The relative test context for the "driver's licence" example

which hold in the "driving test" context (Figure 5.2) but not in this relative test context. Combining those having the same premise yields 17 implications. These are listed in Figure 5.13.

It is shown in Figure 5.14 which of the rows 5 to 8 of Figure 5.12 are models of which implications from Figure 5.13. It requires at least two implications to rule out all four "unwanted" attribute combinations. One of them must be implication 1, the other one may arbitrarily be chosen from implications 4, 12, 16, 17. More concretely, a minimal set of implications characterizing the "driving test" with respect to the background knowledge consists of two implications. One of the two is

$$\textit{license:pass} \to \textit{theory:pass, driving:pass};$$

the other may be each of

$$\begin{aligned} \textit{driving:pass, license:fail} &\to \textit{theory:fail},\\ \textit{theory:pass, license:fail} &\to \textit{driving:fail},\\ \textit{theory:pass, driving:pass} &\to \textit{license:pass},\\ \textit{theory:pass, driving:pass, license:fail} &\to \bot. \end{aligned}$$

Again, there is no canonical basis.

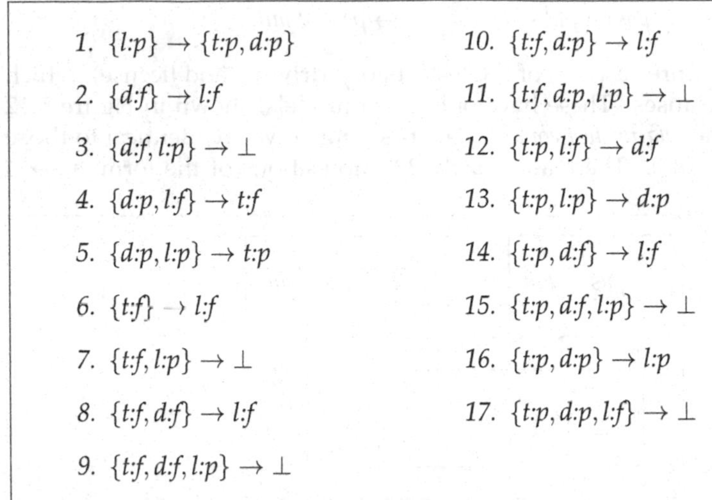

1. $\{l{:}p\} \to \{t{:}p, d{:}p\}$
2. $\{d{:}f\} \to l{:}f$
3. $\{d{:}f, l{:}p\} \to \bot$
4. $\{d{:}p, l{:}f\} \to t{:}f$
5. $\{d{:}p, l{:}p\} \to t{:}p$
6. $\{t{:}f\} \to l{:}f$
7. $\{t{:}f, l{:}p\} \to \bot$
8. $\{t{:}f, d{:}f\} \to l{:}f$
9. $\{t{:}f, d{:}f, l{:}p\} \to \bot$
10. $\{t{:}f, d{:}p\} \to l{:}f$
11. $\{t{:}f, d{:}p, l{:}p\} \to \bot$
12. $\{t{:}p, l{:}f\} \to d{:}f$
13. $\{t{:}p, l{:}p\} \to d{:}p$
14. $\{t{:}p, d{:}f\} \to l{:}f$
15. $\{t{:}p, d{:}f, l{:}p\} \to \bot$
16. $\{t{:}p, d{:}p\} \to l{:}p$
17. $\{t{:}p, d{:}p, l{:}f\} \to \bot$

Figure 5.13: These implications are not valid in Figure 5.12, but they hold in the first four rows

5.3.4 Attribute exploration with background knowledge

The existence of a canonical basis is convenient, but not essential for attribute exploration. Its main virtue is that it is non-redundant and complete, because this guarantees that no superfluous questions are asked and nothing is left open. All this remains true for attribute exploration with background knowledge, as presented in Algorithm 28. What is lost (due to the absence of a canonical basis) is the guarantee of a compact representation: the family of confirmed implications may be larger than necessary (though it remains irredundant).

In practice, there is not much of a difference. In order to explore the implication theory of some "large" and "unknown" formal context (G, M, I), we collect a list $\mathcal{H}$ of implications known to hold in (G, M, I) and a subcontext $(E, M, J := I \cap E \times M)$ of examples. In addition, a list $\mathcal{B}$ of cumulated clauses is given, representing background knowledge.

The generalized algorithm is identical to the basic version, Algorithm 19 (p. 129), except that it scans through sets that are $\mathcal{H}$-closed *under background knowledge* $\mathcal{B}$ (in lectic order). For each such set A, it investigates if this is a potential implication premise. This is the case if the closure A^{JJ} in the context of examples is properly larger than A. The implication $A \to A^{JJ}$ then can neither be inferred from $\mathcal{H} \cup \mathcal{B}$ nor is it refuted by any example of the subcontext. If $A \to A^{JJ}$ can be derived from the available information, then the user will not be bothered. The result is included in the list $\mathcal{H}$, and the algorithm continues. Algorithm 28 implements this. Note the notation: $\phi(\mathcal{H}, \mathcal{B})$ denotes the

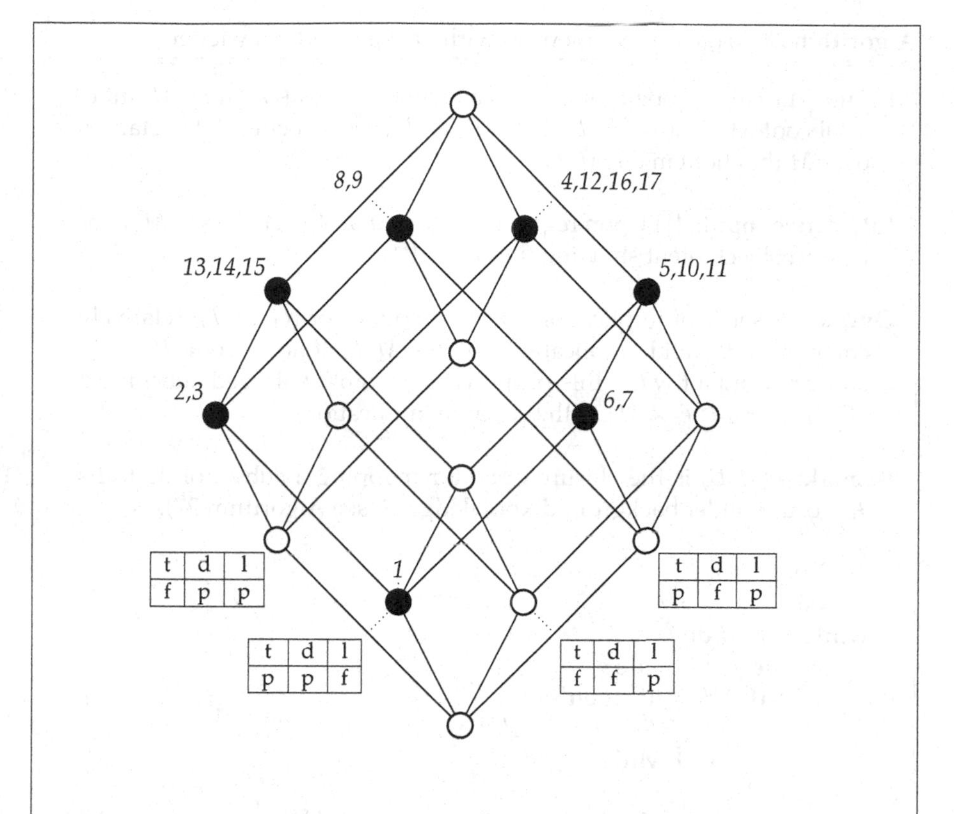

Figure 5.14: The implications from Figure 5.13 and their models among the last four rows of Figure 5.12

closure operator mapping a subset T of M to its $\mathcal{H}$-closure under background knowledge $\mathcal{B}$, which can be implemented by a call to Algorithm 27:

$$\text{BACKGROUND CLOSURE}(T, M, \mathcal{H}, \mathcal{B}).$$

We illustrate the method by a small exploration of comets in our solar system. Before doing so, we mention two basic aspects that occur frequently, not only in this instance.

First we must make precise in what sense our domain expert is supposed to be a "reliable source of information". New comets are discovered regularly, and the knowledge about comets is constantly changing. An "absolute" domain expert who knows the "truth" about comets is therefore not available. Our reference is WIKIPEDIA, where the state of knowledge is extensively (and

Algorithm 28 ATTRIBUTE EXPLORATION with background knowledge

Input: The closure operator, $(\cdot)^{II}$, of a formal context (G, M, I), M finite, a subcontext $(E, M, J := I \cap E \times M)$, and a set $\mathcal{B}$ of cumulated clauses over M that hold in (G, M, I).

Interactive input: $\{*\}$ Upon request, confirm that $A = A^{II}$ in (G, M, I) or give an object intent showing that $A \neq A^{II}$.

Output: A set $\mathcal{H}$ of implications over M sound for (G, M, I), relatively complete (all valid implications of (G, M, I) follow from $\mathcal{H} \cup \mathcal{B}$), and irredundant w.r.t. this property; a possibly enlarged subcontext $(E, M, J := I \cap E \times M)$ with the same implications.

Remark: $\phi(\mathcal{H}, \mathcal{B})$ is the closure operator mapping a subset of M to its $\mathcal{H}$-closure under background knowledge $\mathcal{B}$ (see Algorithm 27).

$\mathcal{H} := \emptyset$
$A := \emptyset$
while $A \neq M$ **do**
  while $A \neq A^{JJ}$ **do**
    if $A^{JJ} = A^{II}$ **then** $\{*\}$
      $\mathcal{H} := \mathcal{H} \cup \{A \rightarrow A^{JJ}\}$
      exit while
    else
      extend E by some object $g \in A^I \setminus A^{JJI}$ $\{*\}$
  $A :=$ NEXT CLOSURE$(A, M, \phi(\mathcal{H}, \mathcal{B}))$
return $\mathcal{L}$, (E, M, J)

quite reliably) documented. So the domain to be explored is not the "reality", but the knowledge as it is presently (Jan. 2016) represented in WIKIPEDIA. More precisely, we even restrict to WIKIPEDIA's list of *numbered comets* (containing those that were observed at least twice). The outcome of an exploration must be understood accordingly.

The second aspect is that the parameters for comets are mostly real-valued. For example, the *Orbital Period*, given in years, is typically between 3 and 200. Our first step therefore must be a transformation of such data into a formal context. The standard method for this, *conceptual scaling*, is discussed in the next section. It is extremely flexible and would even allow to distinguish between all these values. But that would make the exploration much too complex and conceptually ill-structured. For our example, we make do with a simple approach:

we pick a few meaningful intervals and use membership in these intervals as attributes. For example, *Jupiter-family comets* have orbital periods of less than twenty years and low inclinations ($< 30°$); so these values may define derived attributes "*fast*" and "*low inclination*". But why exactly twenty years and $30°$? These bounds are somewhat arbitrarily chosen, and one often would appreciate algorithmic support for the selection of such intervals. We are not aware of such tools, nor of more flexible variants of attribute exploration that would allow changing such choices during an exploration.

Here are some (unprofessional) suggestions for attributes and their definitions:

fast:	orbital period ≤ 20 years
slow:	orbital period ≥ 50 years
not fast:	orbital period > 20 years
not slow:	orbital period < 50 years
low inclination:	inclination $< 30°$
high inclination:	inclination $> 40°$
circular:	eccentricity ≤ 0.75
elliptic:	eccentricity > 0.75
small:	absolute magnitude ≤ 13.5
large:	absolute magnitude > 13.5

The exploration process is as usual; a typical query is

Is it true that every large comet is fast and has low inclination?

This question is answered by a counterexample, such as the comet **206P/Barnard-Boattini** with the following parameters:

orbital period	5.83
inclination	32.9309
eccentricity	0.646433
magnitude	20.3

This comet is *fast*, *not slow*, *circular*, and *large*, and is thus a counterexample to the query (since it does not have the attribute *low inclination*).

It looks as if we are back to plain attribute exploration, except for the mapping that transforms the numerical attribute values into the one-valued attributes. The basic algorithm may indeed be used for this exploration. It results in 18 implications and 14 counterexamples, see Figure 5.15. The concept lattice has 74 elements.

Using the basic algorithm in this example is however quite inefficient. Seven of the 18 implications follow immediately from the definitions of the attributes: *fast* $\rightarrow$ *not slow*, *slow* $\rightarrow$ *not fast*, and the five with $\bot$ as their conclusion — those even without *small* in their premises. These attributes can be used as harmless

	fast	*not slow*	*not fast*	*slow*	*low inclin.*	*high inclin.*	*circular*	*elliptic*	*small*	*large*
1P/Halley			×	×		×		×	×	
2P/Encke	×	×			×			×		×
4P/Faye	×	×			×		×		×	
6P/D'Arrest	×	×			×		×			×
8P/Tuttle	×	×				×		×	×	
177P/Barnard			×	×				×		×
23P/Brorsen-Metcalf			×	×	×			×	×	
95P/Chiron			×	×	×		×		×	
126P/IRAS	×	×				×	×		×	
76P/West-Kohoutek-Ikemura	×	×					×			×
27P/Crommelin		×	×		×			×	×	
28P/Neujmin	×	×			×			×	×	
174P/Echeclus		×	×		×		×		×	
55P/Temple-Tuttle		×	×			×		×	×	

$$
\begin{array}{rcl}
\textit{small, large} & \to & \bot \\
\textit{circular, large} & \to & \textit{fast, not slow} \\
\textit{circular, elliptic} & \to & \bot \\
\textit{high inclination} & \to & \textit{small} \\
\textit{high inclination, circular, small} & \to & \textit{fast, not slow} \\
\textit{low inclination, large} & \to & \textit{fast, not slow} \\
\textit{low inclination, high inclination, small} & \to & \bot \\
\textit{slow} & \to & \textit{not fast} \\
\textit{not fast, large} & \to & \textit{slow, elliptic} \\
\textit{not fast, circular} & \to & \textit{low inclination, small} \\
\textit{not fast, high inclination, small} & \to & \textit{elliptic} \\
\textit{not fast, low inclination} & \to & \textit{small} \\
\textit{not slow, large} & \to & \textit{fast} \\
\textit{not slow, not fast} & \to & \textit{small} \\
\textit{not slow, not fast, slow, small} & \to & \bot \\
\textit{fast} & \to & \textit{not slow} \\
\textit{fast, not slow, elliptic, large} & \to & \textit{low inclination} \\
\textit{fast, not slow, not fast, small} & \to & \bot
\end{array}
$$

Figure 5.15: The formal context of counterexamples for the exploration of comets, and its canonical basis below

background knowledge. The definitions also yield some background information that cannot be expressed by implications:

$$\top \rightarrow \textit{fast} \vee \textit{not fast}, \qquad \top \rightarrow \textit{slow} \vee \textit{not slow},$$
$$\top \rightarrow \textit{circular} \vee \textit{elliptic}, \qquad \top \rightarrow \textit{small} \vee \textit{large}.$$

The exploration can be done using Algorithm 28 with these eleven cumulated clauses as background knowledge. The counterexamples remain the same, but instead of verifying all 18 implications of the canonical basis, the domain expert has to validate only seven implications, see Figure 5.16. The first four of these implications are also contained in the canonical basis, see Figure 5.15, which then continues with

$$\textit{not fast}, \textit{ large} \rightarrow \textit{slow}, \textit{ elliptic}.$$

$$\begin{array}{rcl} \textit{circular}, \textit{ large} & \rightarrow & \textit{fast}, \textit{ not slow} \\ \textit{high inclination} & \rightarrow & \textit{small} \\ \textit{high inclination}, \textit{ circular}, \textit{ small} & \rightarrow & \textit{fast}, \textit{ not slow} \\ \textit{low inclination}, \textit{ large} & \rightarrow & \textit{fast}, \textit{ not slow} \\ \textit{not fast}, \textit{ elliptic}, \textit{ large} & \rightarrow & \textit{slow} \\ \textit{not fast}, \textit{ circular}, \textit{ small} & \rightarrow & \textit{low inclination} \\ \textit{fast}, \textit{ not slow}, \textit{ elliptic}, \textit{ large} & \rightarrow & \textit{low inclination} \end{array}$$

Figure 5.16: With Algorithm 28, only these implications need to be validated. The remaining ones in Figure 5.15 can be derived using the background knowledge

This is not a query of Algorithm 28, because the premise $\{\textit{not fast}, \textit{large}\}$ is not $\phi(\mathcal{H}, \mathcal{B})$-closed, for the following reason: from the background clause

$$\top \rightarrow \textit{circular} \vee \textit{elliptic}$$

it follows that every model must contain either *circular* or *elliptic*. But since *circular* and *large* together imply *fast* (due to the first implication of the basis), no model can contain

not fast, *circular*, and *large*.

Therefore *elliptic* is in the $\phi(\mathcal{H}, \mathcal{B})$-closure of $\{\textit{not fast}, \textit{large}\}$.

5.4 Conceptual scaling of data tables

The examples in the previous section show that attribute exploration works well with *data tables*, where the attributes may have many values, not just one. This is made more precise and practical in the present section, where we formalize the transformation from a data table to a formal context (in 5.4.2). The method, called "plain conceptual scaling", is extremely flexible and versatile. An act of interpretation is required from the user: choosing the *conceptual scales*. This choice affects what formal context is *derived* from the data table and may be used as the domain of an exploration. The background knowledge is then the combined logic of the conceptual scales.

According to this approach, a data table does not have a unique attribute logic, it has many. Which one is used, follows from the conceptual scaling decision. It is not unusual to consider several conceptual scalings for the same data table for different aspects of data analysis.

5.4.1 A warm-up example

Data tables have already occurred in this chapter, see Figures 5.1 and 5.15 (the latter is an already *scaled* data table). As an additional example we investigate the data table in Figure 5.17. It is part of a spectrometric investigation of enamel found in Babylon, in the palace of Nebuchadnezzar II.

Such tables look similar to the cross-tables representing formal contexts except for an essential difference: the attributes, noted in the column heads like Na, Ca, ..., have *values*. These values may be numbers, but often are of a different nature. In this example, these are •, +, −, and *o*. We have *many-valued attributes*, in contrast to the **one-valued attributes** that we have considered so far. We may consider the data table as a *many-valued context*, see Definition 18.

In order to apply the methods that we have developed, we may ask the general question if there is a conceptual structure associated to such a many-valued context. The answer is "yes", but not without the intermediate *conceptual scaling* step, which, at first glance, seems a bit long-winded, but is very useful. The formal contexts for the driving test and for the comets examples were derived through ad hoc conceptual scaling. Now we catch up on a systematic presentation of the method.

A conceptual investigation of the data in Figure 5.17 was done by Bartel [29]. We have shortened the data set considerably (the original data has 19 chemical elements instead of 5) to make it easier to use as an example.

The method of conceptual scaling is to derive a formal context from the given data table and apply Formal Concept Analysis to that formal context. For the example of the enamel data, there is a seemingly natural way to do this: just replace the many-valued attributes Na, Ca, Al, Fe, and Cu by their attribute values

	Na	Ca	Al	Fe	Cu
$T1$	$\bullet$	$+$	$+$	$+$	$+$
$T2$	$+$	$\bullet$	$\bullet$	$+$	$-$
$L1$	$\bullet$	$+$	$+$	$\bullet$	o
$L2$	$+$	$+$	$+$	$+$	o
$L3$	$\bullet$	$+$	$+$	$+$	o
$T3$	$\bullet$	$\bullet$	$\bullet$	$+$	$-$
$L4$	$\bullet$	$+$	$\bullet$	$+$	o
$L5$	$\bullet$	$+$	$\bullet$	$+$	o
$L6$	$+$	$+$	$+$	$+$	o
$T4$	$\bullet$	$+$	$+$	$-$	$\bullet$
$L7$	$\bullet$	$+$	$+$	$+$	$\bullet$
$L8$	$+$	$+$	$+$	$+$	$\bullet$
$L9$	$\bullet$	$-$	$+$	$+$	$\bullet$
$T5$	$\bullet$	$\bullet$	$\bullet$	$+$	$-$
$T6$	$\bullet$	$+$	$+$	$+$	o
$L10$	$\bullet$	$+$	$+$	$+$	o

Figure 5.17: Chemical components of Babylonian enamel, from [90]. The rows stand for ceramic objects (lions and parts of the throne facade). The entries indicate how much sodium, calcium, aluminum, iron, and copper was found. $\bullet$ = rich, $+$ = medium, $-$ = weak, o = very weak

$$\text{Na}:\bullet,\quad \text{Na}:+,\quad \text{Ca}:\bullet,\quad \text{Ca}:+,\quad \ldots,\quad \text{Cu}:-,\quad \text{Cu}:o,$$

and define the incidence in the obvious way. The first two rows of the derived context would then look as follows:

	Na: $\bullet$	Na: $+$	Ca: $\bullet$	Ca: $+$	Ca: $-$	Al: $\bullet$	Al: $+$	Fe: $\bullet$	Fe: $+$	Fe: $-$	Cu: $\bullet$	Cu: $+$	Cu: $-$	Cu: o
T1	×			×			×		×			×		
T2		×	×			×			×				×	

This is known as **nominal scaling** in Formal Concept Analysis, and we have already seen an example of it in Section 5.1.1 (p. 188) for the driving test data table, where we obtained the cross-table in Figure 5.2 as the derived formal context. In that example, the two attribute values "pass" and "fail" are indeed nominal, which means that they are mutually exclusive and exhaustive, and the scaling therefore was appropriate in that case.

But nominal scaling often leads to counterintuitive results when the implicit structure of the attribute values is not nominal. Formal Concept Analysis offers a more flexible method that requires an *interpretation* of how the attribute values are structured. This is expressed by means of a *conceptual scale* (one for each many-value attribute). A conceptual scale is a formal context the objects of which are called *scale values* and the attributes *scale attributes*. For the example of Babylonian enamels, we shall use a **biordinal scale** as shown in Figure 5.18. This is not the only possible choice, and we do not even claim that it is the best interpretation of the given data.[4] Indeed, Bartel uses a different scaling in his much more elaborate analysis [29].

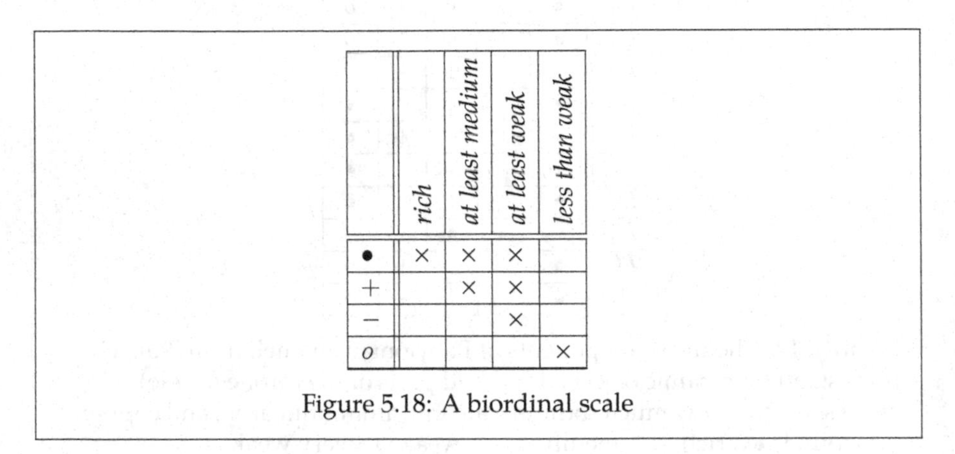

	rich	*at least medium*	*at least weak*	*less than weak*
•	×	×	×	
+		×	×	
−			×	
o				×

Figure 5.18: A biordinal scale

Our biordinal scale interprets •, +, and − as descending grades of presence of the respective element and *o* as its absence. Using this for scaling (a precise definition will be given in the next subsection), we obtain the formal context in Figure 5.19, which, after deleting the (reducible) full and empty columns simplifies to the one in Figure 5.20.

With this procedure, we have obtained a concept lattice for the data table via a conceptual scaling. Its line diagram is shown in Figure 5.21. As a consequence, we now have attribute implications, and some of them can easily be read from the lattice diagram. For example, the enamels rich in calcium (these are T2, T3, and T5) are precisely those that are rich in aluminium and contain at least some copper.

We have therefore reasons to expect that the theory of implications can be extended to (scaled) many-valued contexts, i.e., data tables. Nebuchadnezzar's enamels are however not rich enough in structure to serve as a running example for this. We shall use another data set below. Before, we give the formal definitions of conceptual scaling.

[4]The are several families of *standard scales* often used for conceptual scaling. They correspond to frequently occurring types of data. See [110] for a detailed discussion.

L10	T6	T5	L9	L8	L7	T4	L6	L5	L4	T3	L3	L2	L1	T2	T1	
×	×	×	×		×	×		×	×	×	×		×		×	Na: *rich*
×	×	×	×	×	×	×	×	×	×	×	×	×	×	×	×	Na: *at least medium*
×	×	×	×	×	×	×	×	×	×	×	×	×	×	×	×	Na: *at least weak*
																Na: *less than weak*
		×								×				×		Ca: *rich*
×	×	×		×	×	×	×	×	×	×	×	×	×	×	×	Ca: *at least medium*
×	×	×	×	×	×	×	×	×	×	×	×	×	×	×	×	Ca: *at least weak*
																Ca: *less than weak*
		×						×	×	×				×		Al: *rich*
×	×	×	×	×	×	×	×	×	×	×	×	×	×	×	×	Al: *at least medium*
×	×	×	×	×	×	×	×	×	×	×	×	×	×	×	×	Al: *at least weak*
																Al: *less than weak*
													×			Fe: *rich*
×	×	×	×	×	×		×	×	×	×	×	×	×	×	×	Fe: *at least medium*
×	×	×	×	×	×	×	×	×	×	×	×	×	×	×	×	Fe: *at least weak*
																Fe: *less than weak*
			×	×	×	×										Cu: *rich*
			×	×	×	×									×	Cu: *at least medium*
		×	×	×	×	×				×				×	×	Cu: *at least weak*
×	×						×	×	×		×	×	×			Cu: *less than weak*

Figure 5.19: The derived formal context for the enamel data from Figure 5.17, scaled with the scale in Figure 5.18

	Na: *rich*	Ca: *rich*	Ca: *at least medium*	Al: *rich*	Fe: *rich*	Fe: *at least medium*	Cu: *rich*	Cu: *at least medium*	Cu: *at least weak*	Cu: *less than weak*
T1	×		×			×		×	×	
T2		×	×	×		×			×	
L1	×		×		×	×				×
L2			×			×				×
L3	×		×			×				×
T3	×	×	×	×		×			×	
L4	×		×	×		×				×
L5	×		×	×		×				×
L6			×			×				×
T4	×		×				×	×	×	
L7	×		×			×	×	×	×	
L8			×			×	×	×	×	
L9	×					×	×	×	×	
T5	×	×	×	×		×			×	
T6	×		×			×				×
L10	×		×			×				×

Figure 5.20: The derived formal context, simplified. Its concept lattice is shown in Figure 5.21

5.4.2 Many-valued contexts and their derived contexts

In Formal Concept Analysis, data tables are formalized as *many-valued contexts*.

Definition 18 A **many-valued context** (G, M, W, I) consists of three sets G, M, and W, together with a ternary relation $I \subseteq G \times M \times W$, satisfying the condition

$$(g, m, w_1) \in I \text{ and } (g, m, w_2) \in I \text{ implies } w_1 = w_2.$$

We call the elements of G the **objects**, those of M the **many-valued attributes**, and read$(g, m, w) \in I$ as "the value of the attribute m for the object g is w". This is sometimes abbreviated as $m(g) = w$. ◊

We have seen examples of many-valued contexts in Figures 5.1 (p. 188) and 5.17. The rest of this section will be illustrated by the many-valued context

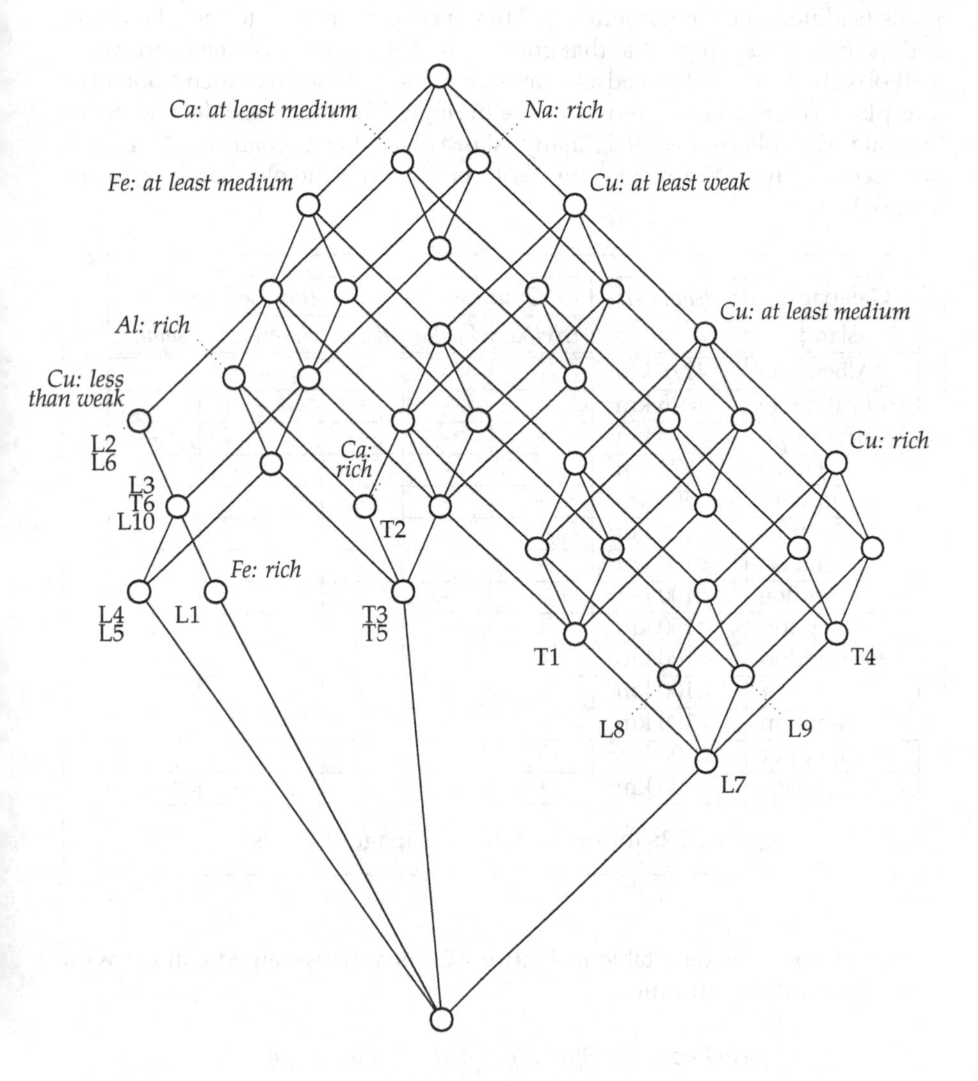

Figure 5.21: The concept lattice of the enamel spectra

in Figure 5.22. It shows some data about tortoises on the Galapagos islands. The elephant tortoises on these islands have developed different shapes of their shells (saddle, dome, intermediate). This may be connected to the island size and the type of cacti (opuntia) that grow on it. The data set is taken from an old textbook [126], where it served as a basis for a first discussion of such a potential interplay. The reader is referred to the biological literature for details. Since the data was collected in 1962, many islands have been renamed: Abingdon has become Pinta, Wenman is now Wolf, etc. We use the old names from the textbook.

Galapagos island	*Island size*	*Opuntia*		*Tortoise type*		
		treelike	*bushy*	*dome*	*intermed.*	*saddle*
Albemarle	4278 km^2	+	-	+	-	-
Indefatigable	1025 km^2	+	-	+	-	-
Narborough	650 km^2	+	-	+	-	-
James	574 km^2	+	-	-	+	-
Chatham	ca. 500 km^2	+	-	-	+	-
Charles	ca. 200 km^2	+	-	-	+	+
Hood	<100 km^2	+	-	-	-	+
Bindloe	<100 km^2	-	+	-	-	-
Abingdon	<100 km^2	+	-	-	-	+
Barringdon	<100 km^2	+	-	-	+	+
Tower	<100 km^2	-	+	-	-	-
Wenman	<100 km^2	-	+	-	-	-
Culpepper	<100 km^2	-	+	-	-	-
Jervis	<100 km^2	+	-	-	+	+

Figure 5.22: Tortoises on the Galapagos islands

We interpret the data table in Figure 5.22 as a many-valued context with three many-valued attributes,

Island size, *Opuntia,* *and* *Tortoise type,*

where the values of "*Island size*" are 4278 km^2, 1025 km^2, etc., the values of "*Opuntia*" are $+-$ and $-+$, and the values of "*Tortoise type*" are $+--$, $-+-$, $--+$, $-++$, and $---$.

In order to carry out a conceptual analysis of this data table, we must decide on conceptual scales and use these to obtain a derived context. This will be done next.

Definition 19 A **conceptual scale** for a many-valued attribute m of a many-valued context (G, M, W, I) is a formal context $\mathbb{S}_m := (G_m, M_m, I_m)$ with the property that

$$\{m(g) \mid g \in G\} \subseteq G_m.$$

The objects of the scale are called **scale values**, and the attributes are **scale attributes**.

◊

A simplified notation is often used when different attribute values have identical scale attributes. Then such scale values are grouped into sets ("object clarification"), and the scale is noted down with these sets as objects. For example, one would tacitly simplify

	even	*odd*
0	×	
1		×
2	×	
3		×

to

	even	*odd*
{0, 2}	×	
{1, 3}		×

.

The first scale in Figure 5.23 is to be interpreted this way.

Definition 20 A conceptual **scaling schema** for a set M of many-valued attributes is a family

$$(\mathbb{S}_m \mid m \in M)$$

of scales for all attributes of M.

◊

To give the data table in Figure 5.22 a conceptual structure, we use the scaling schema that is shown in Figure 5.23.

Definition 21 From a scaling schema $(\mathbb{S}_m \mid m \in M)$ for the many-valued context (G, M, W, I), the **derived context**(G, N, J) is defined through

$$N := \bigcup_{m \in M} \{m\} \times M_m$$

and

$$g\ J\ (m, n) \quad :\Longleftrightarrow \quad n \in M_m \text{ and } m(g)\ I_m\ n.$$

◊

Loosely speaking, the derived (one-valued) context is obtained from the many-valued context by replacing each value by the corresponding row of the respective scale.

Island size	*small*	*not large*	*not small*	*large*
< 100 km^2	×	×		
100–1000 km^2		×	×	
> 1000 km^2			×	×

Opuntia *treelike* | *bushy*		*treelike*	*bushy*
+	−	×	
−	+		×

Tortoise type *dome* | *intermediate* | *saddle*			*dome*	*intermediate or saddle*	*saddle*	*intermediate*
−	−	−				
+	−	−	×			
−	+	−		×	×	
−	−	+		×		×
−	+	+		×	×	×

Figure 5.23: Scales for the tortoises context

	island size: large	*island size: not small*	*island size: not large*	*island size: small*	*opuntia: treelike*	*opuntia: bushy*	*tortoises: dome*	*tortoises: intermediate or saddle*	*tortoises: saddle*	*tortoises: intermediate*
Albemarle	×	×			×		×			
Indefatigable	×	×			×		×			
Narborough		×	×		×		×			
James		×	×		×			×		×
Chatham		×	×		×			×		×
Charles		×	×		×			×	×	×
Hood			×	×	×			×	×	
Bindloe			×	×		×				
Abingdon			×	×	×			×	×	
Barringdon			×	×	×			×	×	×
Tower			×	×		×				
Wenman			×	×		×				
Culpepper			×	×		×				
Jervis			×	×	×			×	×	×

Figure 5.24: The derived context

For the data table of the Galapagos tortoises with the scaling given in Figure 5.23, we obtain the derived context in Figure 5.24 and the concept lattice in Figure 5.25.

For the derived context in Figure 5.24, we make the simple observation that every row is a concatenation of rows of the scales. For example, the two columns for the attributes "opuntia:treelike" and "opuntia:bushy" contain only rows of the form $\begin{array}{|c|c|}\hline . & \times \\ \hline\end{array}$ and $\begin{array}{|c|c|}\hline \times & . \\ \hline\end{array}$, but not $\begin{array}{|c|c|}\hline . & . \\ \hline\end{array}$ or $\begin{array}{|c|c|}\hline \times & \times \\ \hline\end{array}$. Derived contexts for a given scaling schema are thus not arbitrary: they must be built from rows of the scales. Since this will be of importance for implication inference, we give a formal definition.

Definition 22 The **semi-product** of formal contexts

$$\mathbb{S}_m := (G_m, M_m, I_m), \quad m \in M$$

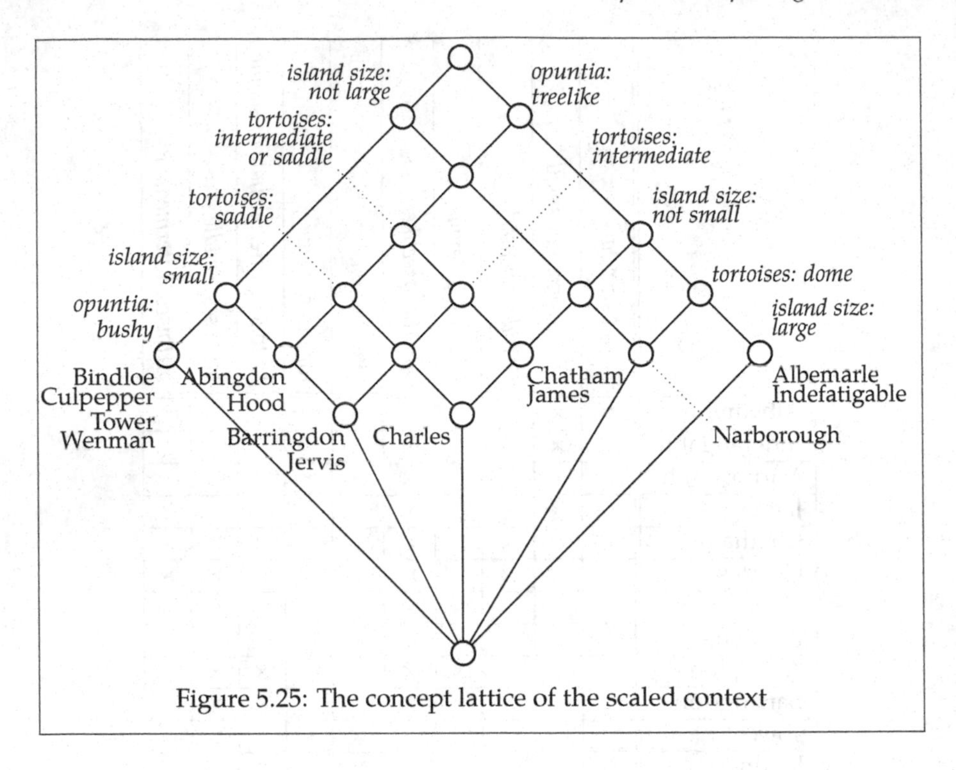

Figure 5.25: The concept lattice of the scaled context

is defined as the formal context[5]

$$(\bigtimes_{m\in M} G_m, \bigcup_{m\in M} \{m\} \times M_m, J),$$

where

$$(g_m \mid m \in M)\ J\ (k,n) :\iff g_k\ I_k\ n.$$

$\Diamond$

To put it simple: the semi-product is obtained by concatenating the rows of the contexts (G_m, M_m, I_m) in all possible ways. The name "semi-product" expresses that the object sets are multiplied and the attribute sets are added.

The intents of the derived context necessarily are intents of the semi-product of the scales. In our example (see Figure 5.23), the scales have 3, 2, and 5 objects (scale values) and 4, 2 and 4 scale attributes respectively. Their semi-product therefore has $3 \cdot 2 \cdot 5 = 30$ objects and $4 + 2 + 4 = 10$ attributes.

Conversely, we have that, given a scaling schema $(\mathbb{S}_m \mid m \in M)$ for a set M of many-valued attributes, every formal context whose rows are rows of the

[5] $\bigtimes_{m\in M} G_m$ denotes the cartesian product of the object sets.

semi-product can be obtained from some many-valued context as a derived context for that scaling schema. This gives the following proposition:

Proposition 40 *The relative test context for a fixed scaling schema is the semi-product of the scales.*

This will be the context from which we will take counterexamples during exploration; it is relative w.r.t. the background knowledge implied by the scales.

For the proof of the following proposition, we need a convenient notation. We will assume that $M = \{1, 2, \ldots, n\}$, so that subsets of the attribute set can be written as follows: we abbreviate

$$\bigcup_{m \in M} \{m\} \times S_m \quad \text{by} \quad (S_1, \ldots, S_n).$$

The derivation of such a set is denoted by

$$(S_1', \ldots, S_n') \quad \text{instead of by} \quad (S_1^{I_1}, \ldots, S_n^{I_n}).$$

Lemma 2 *The non-empty pseudo-models of a semi-product are precisely the sets of the form $(g_1', \ldots, g_{i-1}', P, g_{i+1}', \ldots, g_n')$, where P is a non-empty pseudo-model in $\mathbb{S}_i$ and each g_j' is a minimal object intent in $\mathbb{S}_j$.*

Proof The structure of the proof is as follows: we first show that all sets of the form $(P, g_2', \ldots, g_n')$, where P is a non-empty pseudo-model of $\mathbb{S}_1$, are indeed pseudo-models of the semi-product. This is done in steps: we first consider a potential minimal counterexample, then show that all properly contained pseudo-models must be of the given form, and conclude that the minimal example is not a counterexample. Then we can prove the general case.

Let $g_2', \ldots, g_n'$ be minimal object intents (i.e., minimal models) in $\mathbb{S}_2, \ldots, \mathbb{S}_n$. Let $P \neq \emptyset$ be a pseudo-model of $\mathbb{S}_1$ with the property that, for every non-empty pseudo-model Q of $\mathbb{S}_1$ properly contained in P, the set $(Q, g_2', \ldots, g_n')$ is a pseudo-model of the semi-product. Since P is non-empty, it must contain some object intent g_1'.

Now consider an arbitrary pseudo-model X of the semi-product, which is properly contained in $(P, g_2', \ldots, g_n')$. To prove that $(P, g_2', \ldots, g_n')$ is a pseudo-model of the semi-product, we must show that there is a model between X and $(P, g_2', \ldots, g_n')$. If X is the empty set, $(g_1', g_2', \ldots, g_n')$ is such a model. Let us now assume that X is not empty. Since the g_j' are minimal, X must be of the form $(R, g_2', \ldots, g_n')$ for some set $R \subset P$, and R must contain some object intent of $\mathbb{S}_1$, which w.l.o.g. may be assumed to be g_1. To show that R is a pseudo-model of $\mathbb{S}_1$, let Q be a non-empty pseudo-model of $\mathbb{S}_1$ properly contained in R. By assumption, $(Q, g_2', \ldots, g_n')$ is a pseudo-model of the semi-product. It is properly contained in $(R, g_2', \ldots, g_n')$, which is also a pseudo-model. So there must be a model in between, and consequently there must be an $\mathbb{S}_1$-model between Q and R. This proves that R is a pseudo-model of $\mathbb{S}_1$. We can also conclude that there must be a model between R and P and, therefore, a model of the semi-product between $(R, g_2', \ldots, g_n')$ and $(P, g_2', \ldots, g_n')$. Consequently, $(P, g_2', \ldots, g_n')$ is a pseudo-model of the semi-product, and all sets of this form are.

Finally, let $(P_1, P_2, \ldots, P_n)$ be an arbitrary non-empty pseudo-model of the semi-product. Not all P_i can be models, and there must be some minimal model

$$(g_1', g_2', \ldots, g_n') \subseteq (P_1, P_2, \ldots, P_n).$$

Again, we discuss only the case $i = 1$ and assume that P_1 is not a model of $\mathbb{S}_1$. Then there must be a non-empty pseudo-model P contained in P_1 such that there is no model between P and P_1. As a consequence, there cannot be a model of the semi-product between $(P, g_2', \ldots, g_n')$ and $(P_1, P_2, \ldots, P_n)$. But since both are pseudo-models, they must be equal. □

Theorem 15 *The non-empty pseudo-intents of a semi-product are precisely the sets of the form*

$$(\emptyset'', \ldots, \emptyset'', P, \emptyset'', \ldots, \emptyset''),$$

where P is a non-empty pseudo-intent of the respective factor.

To obtain the corollary from the lemma, note that, for an arbitrary formal context $\mathbb{K} := (G, M, I)$, the pseudo-intents of $\mathbb{K}$ are exactly the pseudo-models of $(\text{Int}(\mathbb{K}), M, \ni)$, where $\text{Int}(\mathbb{K})$ is the system of concept intents of $\mathbb{K}$. This construction commutes with the construction of the semi-product.

To every implication $A \to B$ or clause $A \multimap B$ over M_m there corresponds the *lifted* version over $\bigcup_{m \in M} \{m\} \times M_m$. It is given by $(\{m\} \times A) \to (\{m\} \times B)$ and by $(\{m\} \times A) \multimap (\{m\} \times B)$, respectively. Cumulated clauses can be lifted in the same way. Lifted valid clauses and implications hold in the semi-product.

Proposition 41 *The implications of a semi-product follow from the lifted implications of the factor contexts. Likewise, all valid cumulated clauses of a semi-product follow from the lifted cumulated clauses of its factors contexts.*

This is very useful for obtaining the scaling-induced background knowledge, see below.

5.4.3 Scaling induced background knowledge

Together with the derived formal context, we get its implications. For these, we have a stronger form of implication inference due to the background knowledge obtained from conceptual scaling. Proposition 41 makes explicit how to find this background knowledge in the form of cumulated clauses: one computes compact representations for all scales, lifts them (which requires a simple renaming procedure), and forms the union of the lifted sets of cumulated clauses. For finding compact representations for the individual scales, Algorithm 29 may be used. It is based on Theorem 13.

Algorithm 29 BASIS OF CUMULATED CLAUSES((G, M, I))

Input: A formal context (G, M, I).
Output: The basis of cumulated clauses for the set $\mathcal{D} := \{g' \mid g \in G\}$ of object intents of (G, M, I).

$\mathcal{L} := \emptyset$
$A := \emptyset$
let $<$ be some linear order on M
while $A \neq \bot$ **do**
  $\mathcal{B} :=$ all minimal sets in $\{g' \mid A \subseteq g'\}$
  if $\mathcal{B} \neq \{A\}$ **then**
    $\mathcal{L} := \mathcal{L} \cup \{A \multimap \mathcal{B}\}$
  $A :=$ NEXT MODEL$(A, M, \mathcal{L}, <)$ {See Algorithm 25}
return $\mathcal{L}$

We were warned already that the result of Algorithm 29 is not always the most elegant one. To obtain a more readable but logically equivalent list, one may modify the algorithm by starting with a list $\mathcal{L}$ of some "nice" (valid!) cumulated clauses instead of starting with $\mathcal{L} = \emptyset$.

The output of Algorithm 29 for the scales in Figure 5.23 is shown in Figure 5.26. From these, the twelve "lifted" scales are obtained simply by using the many-valued attribute names as prefixes. The "Opuntia"-scale, for example, contributes

$$\emptyset \multimap \{\{\textit{opuntia: treelike}\}, \{\textit{opuntia: bushy}\}\},$$

$$\{\textit{opuntia: treelike, opuntia: bushy}\} \multimap \bot.$$

The combined set of twelve lifted cumulated clauses is also the basis of the semi-product of the three scales.

5.4.4 The implications of a scaled many-valued context

Using the background knowledge computed in the previous subsection, we now can compute the relative implication basis for the derived formal context in Figure 5.24. This is essentially the same as it was done for the "comets" data

Island size

$$\emptyset \multimap \{\{small, not\ large\}, \{not\ large, not\ small\}, \{not\ small, large\}\}$$

$$\begin{aligned} \{not\ large, not\ small, large\} &\multimap \bot \\ \{small, not\ small, large\} &\multimap \bot \\ \{small, not\ large, large\} &\multimap \bot \\ \{small, not\ large, not\ small\} &\multimap \bot \end{aligned}$$

Opuntia

$$\begin{aligned} \emptyset &\multimap \{\{treelike\}, \{bushy\}\} \\ \{treelike, bushy\} &\multimap \bot \end{aligned}$$

Tortoise type

$$\begin{aligned} \{intermediate\} &\multimap \{interm.\ or\ saddle\} \\ \{saddle\} &\multimap \{interm.\ or\ saddle\} \\ \{interm.\ or\ saddle\} &\multimap \{\{saddle\}, \{intermediate\}\} \\ \{dome,\ interm.\ or\ saddle,\ intermediate\} &\multimap \bot \\ \{dome,\ interm.\ or\ saddle,\ saddle\} &\multimap \bot \end{aligned}$$

Figure 5.26: The bases for the scales in Figure 5.23

in Section 5.3.4, but now on the ground of precise definitions. The canonical basis (without using the background knowledge) consists of 14 implications.

Now we invoke Algorithm 28, use the formal context in Figure 5.24 as the subcontext (E, M, J) and the twelve background implications lifted from Figure 5.26 as the list of cumulated clauses. The algorithm is not run interactively; instead, every query is confirmed, so that no counterexample must be given. The result, slightly simplified, is given in Figure 5.27.

The relative basis of implications is simpler and therefore much better to read. It contains the essential information, without the trivial scaling-induced part. This also makes an exploration using Algorithm 28 much easier for the domain expert.

So what is the result of this data analysis? First of all, we should warn that

tortoises: intermediate or saddle	$\rightarrow$	*island size: not large, opuntia: treelike*
tortoises: dome	$\rightarrow$	*island size: not small, opuntia: treelike*
opuntia: bushy	$\rightarrow$	*island size: small*
island size: small, opuntia: treelike	$\rightarrow$	*tortoises: saddle*
island size: not small, tortoises: intermediate or saddle	$\rightarrow$	*tortoises: intermediate*
island size: large, opuntia: treelike	$\rightarrow$	*tortoises: dome.*

Figure 5.27: The implication basis relative to the scaling induced background knowledge (slightly simplified)

the biology textbook from which we took the data is quite old. Actually, a fourth type of Galapagos tortoises was discovered recently. Figure 5.27 contains a few interesting relations between tortoise type and island size, but it does not encode, surprisingly, a relation which is obvious from Figure 5.22: bushy opuntia grow on islands with no tortoises. That this is not discovered is due to our choice of scales, since tortoise type "none" was not selected as an attribute.

Indeed, if we add *tortoise type: none* as another scale attribute, then the implication basis becomes even simpler. It is shown in Figure 5.28. The new implications now quite clearly express the interplay between tortoise types, opuntia, and island sizes. This demonstrates that a good choice for the conceptual scales is essential.

tortoises: none	$\rightarrow$	*island size: small, opuntia: bushy*
tortoises: intermediate or saddle	$\rightarrow$	*island size: not large, opuntia: treelike*
tortoises: dome	$\rightarrow$	*island size: not small, opuntia: treelike*
island size: small, tortoises: interm. or saddle	$\rightarrow$	*tortoises: saddle*
island size: not small, tortoises: interm. or saddle	$\rightarrow$	*tortoises: intermediate*

Figure 5.28: The same as in Figure 5.27, but with an additional scale attribute "*tortoises: none*"

The practical potential of many-valued attribute exploration was not clear at the time when this book was written. No large application examples comparable to those by Kestler and Revenko (see page 174) had been documented. And some research questions of practical importance, as far as we know, had not been addressed. For example, this one: which results can be saved if during a many-valued exploration the conceptual scaling is altered? Parts of the acquired knowledge are obviously scaling-independent. Counterexamples remain examples, and each confirmed implication excludes the existence of some many-valued examples. But how this can be used in algorithms remains to be worked out.

5.5 Further reading

Dependencies between many-valued attributes were studied by R. Wille in [218]. A popular scaling method different from plain conceptual scaling is "logical scaling", see Prediger [176, 178] and Hereth [123]. More general is "logical concept analysis" by Ferré and Ridoux [87]. "Pattern structures" [102], another variant of Formal Concept Analysis, offer a nice description of functional dependencies as object implications, see Baixeries, Kaytoue, and Napoli [25].

The *inference rules* for implications are different when propositional background knowledge is present. It is possible to make this explicit and to enrich the Armstrong rules (introduced in Section 3.1.2). This was studied in [96], but was never used in algorithms, as far as we know.

H.G. Bartel has several publications related to the material presented in Section 5.4.1, see [27, 28, 30].

An algorithm for computing a "direct" basis of cumulated clauses (in analogy to the basis of proper implications in Section 3.3.4) was given by Krauße [138].

5.6 Exercises

5.1 Let $\mathcal{F}$ consist of all vertex sets of disconnected induced subgraphs of the graph from Exercise 2.3 (assume that the empty graph is connected). List all pseudo-models of size at most 5 w.r.t. $\mathcal{F}$.

5.2 (From [138].) Consider the set $\mathcal{L}$ consisting of the following cumulated clauses on the set $M := \{a, b, c, d, e, f\}$:

$$a \to c \wedge d, \quad b \to c \wedge d, \quad c \to e \wedge f, \quad d \to e \wedge f,$$
$$\top \to e \vee f, \quad a \wedge b \to \bot, \quad c \wedge d \to a \vee b, \quad e \wedge f \to c \vee d \, .$$

Determine all models of $\mathcal{L}$. Draw the concept lattice $\underline{\mathfrak{B}}(\mathrm{Mod}(\mathcal{L}), M, \ni)$, which has ten elements.

5.3 On the right, a diagram of an ordered set $(P, \leq)$ is shown. The (general) ordinal scale for $(P, \leq)$ is the formal context $\mathbb{O}_{(P,\leq)} := (P, P, \leq)$. Write this context as a table and determine its basis of cumulated clauses.
Can you generalize? Which cumulated clauses describe arbitrary ordinal scales?

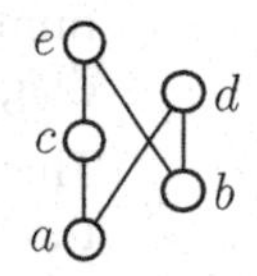

5.4 Given

$$a \to b, \quad d \to e, \quad b \wedge c \to \bot, \quad \top \to b \vee c,$$
$$g \to f, \quad e \wedge g \to \bot, \quad d \wedge f \to \bot, \quad \top \to d \vee f, \quad \top \to e \vee g$$

as background knowledge, do the two implications

$$a, e \to f, \quad b, e \to d$$

together imply $b, e \to a$?

For a proof, give a derivation of the inference. For a disproof, find a model of the background knowledge that respects the two implications, but not the third. Alternatively, show that the background knowledge describes the intents of a semi-product, delete models that do not respect the first two implications, and check if all the remaining ones respect the third.

5.5 (From [28].) Figure 5.29 shows a many-valued context of aromatic molecules.

	m_R	m_{OS}	m_N
Benzene	⬡		
Furan	⬠	O	
Imidazole	⬠		2
1-3-Oxazole	⬠	O	1
Pyrazine	⬡		2
Pyrazole	⬠		2
Pyridine	⬡		1
Pyrimidine	⬡		2
Pyrrole	⬠		1
Thiazole	⬠	S	1
Thiophene	⬠	S	
1-3-5-Triazine	⬡		3

$\mathbb{S}_{m_R}$	5R	6R
⬠	×	
⬡		×

$\mathbb{S}_{m_{OS}}$	O	S
O	×	
S		×

$\mathbb{S}_{m_N}$	≥ 1N	≥ 2N	≥ 3N
1	×		
2	×	×	
3	×	×	×

Figure 5.29: A many-valued context of aromatic molecules, with a scaling schema.

The many-valued attributes describe the ring structure (pentagonal or hexagonal), if the ring contains an oxygen or a sulphur atom, and the number of nitrogen atoms. A conceptual scaling schema is given. Draw a diagram of the concept lattice of the derived context (it has 15 elements).

This many-valued context is the result of an attribute exploration. Find the implication basis relative to the scaling-induced background knowledge.

Chapter 6

More expressive variants of exploration

Several authors have extended the basic method of attribute exploration, as it is presented in the two previous chapters of this book, to more advanced situations, and this last chapter is devoted to such ideas. We start with *rule exploration*, which lifts attribute exploration in a rather natural way to first-order predicate logic, with Horn formulas replacing implications. Next we discuss how attribute exploration can be integrated with *Description Logics*. However, so much substantial research has been contributed to this topic that we can only introduce some first ideas. The third section is devoted to *concept exploration*, which was already mentioned in the very first publication on Formal Concept Analysis. Three more generalizations are sketched in the fourth section, which then concludes by mentioning a few further ones.

In contrast to the previous chapters, we now cannot always illustrate the theory by detailed examples. One reason is that such "use cases" have not yet been worked out. Another one is that, due to the complexity of the methods, realistic examples would go beyond the scope of this book.

6.1 Rule exploration

The use of first-order logic Horn formulas instead of implications has been studied in detail and is actually rather simple. We utilize a brute force approach of "propositionalization" to represent rules by families of (propositional) implications. It is, as we believe, best introduced by means of an example or, better, of two. The first one ("triplets of squares") extends the exploration discussed in Section 4.2.2. It is somewhat uninteresting in its result, but very transparent. A reader who has read this example carefully will probably understand rule exploration even without further definitions.

The second example ("treelike relations") is a little more magical. It combines basic rule exploration with using symmetries and background knowledge obtained from conceptual scaling. The outcome of this exploration is the "discovery" of a non-trivial mathematical theorem, though without a complete proof.

6.1.1 Triplets of squares

We build on one of our running examples, the classification of "pairs of squares", where we asked how two squares can be positioned with respect to six given attributes such as "disjoint" and "having a common vertex".

What about *three* squares? Here are four (out of many) possible configurations of triplets of squares:

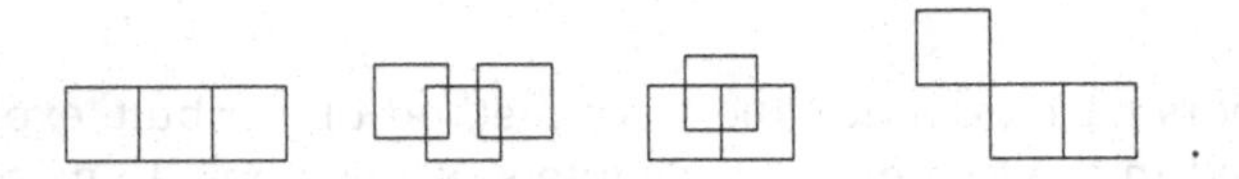

You may guess that a classification of triplets of squares is much more complex than that of pairs. This is indeed the case. In order to fit the example in this book, we therefore restrict to four of the six attributes and only ask if the squares

are *disjoint* or they *overlap*, have a *common vertex* or a *common edge*.

Moreover, we allow only examples where all squares under consideration are of the same size ("unit squares") and have sides parallel to the axes of our coordinate system.

But wait! Being *disjoint* or having a *common edge* are properties of two squares, not of three. We need to specify in each case which two of the three squares are meant. Thus, a typical context row looks like this:

	disjoint			*overlap*			*common vertex*			*common edge*		
	AB	AC	BC	AB	AC	BC	AB	AC	BC	AB	AC	BC
$A \mid B \mid C$		×					×		×	×		×

Instead of four attributes, we now have twelve. Actually, there could have been 36, including *disjoint*(B, A) and *disjoint*(A, A). But obviously all four properties are symmetric: A and B are disjoint iff B and A are, etc. And no square is disjoint to itself, but each square overlaps itself and shares vertices and edges with itself.

The above triplet of squares can be labeled in six different ways. But due to a symmetry, only three different context rows are obtained for this example:

	disjoint			*overlap*			*common vertex*			*common edge*		
	AB	AC	BC	AB	AC	BC	AB	AC	BC	AB	AC	BC
$A \mid C \mid B$	×							×	×		×	×
$A \mid B \mid C$		×					×		×	×		×
$B \mid A \mid C$			×				×	×		×	×	

We gave four ad-hoc examples of triplets of squares at the beginning of this section. If we include the remaining three of them, we get the formal context in Figure 6.1, which may serve as a start for a rule exploration procedure. The fourth triplet gives rise to six lines in this context due to the lack of symmetries. Of course, in an interactive procedure the human user will provide only one copy of each example. Finding the isomorphic ones with different labeling should obviously be left to the computer.

Note that the resulting context has natural context automorphisms as defined in Section 2.3.4 (p. 67). They are induced by the permutations of the labels. For example, interchanging the labels A and B in Figure 6.1 would interchange the columns

2 with 3, 5 with 6, 8 with 9, 11 with 12,

and simultaneously the rows

2 with 3, 5 with 6, 8 with 9, 10 with 11, 12 with 14, 13 with 15,

while the crosses indicating the incidences remain unchanged.

Such automorphisms not only operate on the formal context. They carry over to the conceptual structure, including the implications. We start with an obvious one:

If two squares have a common edge, then they have a common vertex,

or, in short notation,

$$\textit{common-edge} \quad \rightarrow \quad \textit{common-vertex}.$$

In the formal context of Figure 6.1, three instances of this implication hold:

$$\begin{array}{lcl} \textit{common-edge}(A,B) & \rightarrow & \textit{common-vertex}(A,B) \\ \textit{common-edge}(A,C) & \rightarrow & \textit{common-vertex}(A,C) \\ \textit{common-edge}(B,C) & \rightarrow & \textit{common-vertex}(B,C). \end{array}$$

We may introduce variables X, Y and treat the three implications as one *rule*

$$\textit{common-edge}(X,Y) \quad \rightarrow \quad \textit{common-vertex}(X,Y),$$

	disjoint			*overlap*			*common vertex*			*common edge*		
	AB	AC	BC	AB	AC	BC	AB	AC	BC	AB	AC	BC
A C B	×							×	×		×	×
A B C		×					×		×	×		×
B A C			×				×	×		×	×	
A C B	×				×	×						
A B C		×		×		×						
B A C			×	×	×							
C / A B					×	×	×			×		
B / A C				×		×		×			×	
A / B C				×	×				×			×
A / C B	×							×	×			×
B / C A	×							×	×		×	
A / B C		×					×		×			×
C / B A		×					×		×	×		
B / A C			×				×	×			×	
C / A B			×				×	×		×		

Figure 6.1: A formal context based on four ad-hoc triplets of squares

with the interpretation that this rule holds for all **variable assignments**[1]

$$\nu : \{X, Y\} \to \{A, B, C\}.$$

The attribute exploration algorithm applied to the context in Figure 6.1 will find these implications and ask the user if they hold. However, we do not answer the same question several times. The algorithm must be modified to work with *rules* and *generic examples*.

We shall present the details below. For the moment, let us just follow a run of the rule exploration procedure, details of which will be presented later.

0. We start with the examples given in Figure 6.1 and the single rule

$$\textit{common-edge}(X,Y) \quad \to \quad \textit{common-vertex}(X,Y).$$

1. The first rule proposed by the interactive algorithm asks if

$$\textit{common-vertex}(X,Y) \quad \text{and} \quad \textit{common-vertex}(X,Z)$$

together imply

$$\textit{disjoint}(Y,Z).$$

It is easy to provide a counterexample, e.g.,

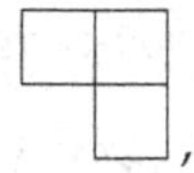

which we add to the formal context.

2. The next rule asked is if

$$\textit{common-edge}(X,Y), \quad \textit{common-edge}(X,Z), \quad \text{and} \quad \textit{common-edge}(Y,Z)$$

together imply all attributes,[2] that is, in essence, if *common-edge*(X,Y), *common-edge*(X,Z) and *common-edge*(Y,Z) cannot occur together. They can however, since we allow squares occupying the same position as in our next example (the "2" indicates that the left square is taken twice):

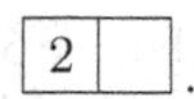

3. Another counterexample, namely,

[1]As a caveat for later, we keep in mind that this includes assignments with $\nu(X) = \nu(Y)$. These pose no problem for the example above.

[2]The premise actually consists of six elements: it also includes the *common-vertex* attributes, which are already known to follow.

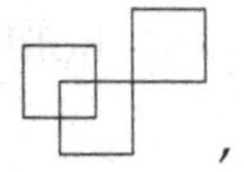

refutes the next suggested rule, which is

$$\mathit{overlap}(X,Y), \mathit{common\text{-}vertex}(X,Z) \quad \rightarrow \quad \mathit{common\text{-}edge}(X,Z).$$

4. Then the rule

$$\begin{gathered}\mathit{overlap}(X,Y), \mathit{common\text{-}vertex}(X,Z), \mathit{common\text{-}vertex}(Y,Z) \\ \rightarrow \\ \mathit{common\text{-}vertex}(X,Y), \mathit{common\text{-}edge}(X,Y), \\ \mathit{common\text{-}edge}(X,Z), \mathit{common\text{-}edge}(Y,Z)\end{gathered}$$

   is disproved by means of the counterexample

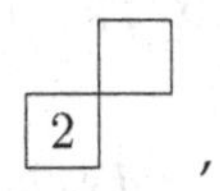

5. but its shorter version

$$\begin{gathered}\mathit{overlap}(X,Y), \mathit{common\text{-}vertex}(X,Z), \mathit{common\text{-}vertex}(Y,Z) \\ \rightarrow \\ \mathit{common\text{-}vertex}(X,Y), \mathit{common\text{-}edge}(X,Y)\end{gathered}$$

   indeed holds true.

6. After another counterexample (2), we encounter the rule

$$\mathit{overlap}(X,Y), \mathit{common\text{-}vertex}(X,Y) \quad \rightarrow \quad \mathit{common\text{-}edge}(X,Y),$$

   which again holds true: two squares (parallel and of equal size) that overlap and share a vertex must be equal and, therefore, also share an edge.[3]

... And so on. Figure 6.2 shows the result of this rule exploration.

The seventeen triplet types in Figure 6.2 all satisfy the nine rules of the given list. Their variable assignments extend the formal context of Figure 6.1 to 56 objects.

The nine rules expand to 42 implications forming a basis of the implication theory. Each implication that holds for all such triplets can be inferred from the

[3] We could and should have used the implications in Figure 4.8 (p. 146) or, even better, the cumulated clauses in Figure 5.11 (p. 205), as background knowledge.

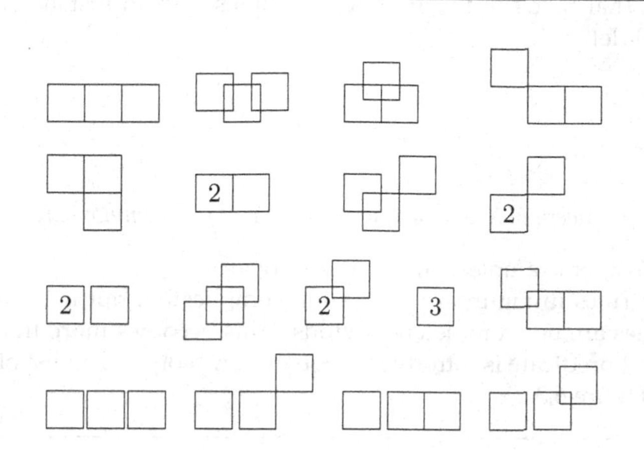

1. $\textit{common-edge}(X,Y) \quad \rightarrow \quad \textit{common-vertex}(X,Y)$
2. $\textit{overlap}(X,Y), \textit{common-vertex}(X,Z), \textit{common-vertex}(Y,Z)$
   $\rightarrow \quad \textit{common-vertex}(X,Y), \textit{common-edge}(X,Y)$
3. $\textit{overlap}(X,Y), \textit{common-vertex}(X,Y) \quad \rightarrow \quad \textit{common-edge}(X,Y)$
4. $\textit{overlap}(X,Y), \textit{common-vertex}(X,Y), \textit{common-vertex}(X,Z)$
   $\rightarrow \quad \textit{common-vertex}(Y,Z)$
5. $\textit{overlap}(X,Y), \textit{common-vertex}(X,Y), \textit{common-edge}(X,Z)$
   $\rightarrow \quad \textit{common-edge}(Y,Z)$
6. $\textit{overlap}(X,Y), \textit{common-vertex}(X,Y), \textit{overlap}(X,Z)$
   $\rightarrow \quad \textit{overlap}(Y,Z)$
7. $\textit{disjoint}(X,Y), \textit{common-vertex}(X,Y) \quad \rightarrow \quad \bot$
8. $\textit{disjoint}(X,Y), \textit{overlap}(X,Z), \textit{common-vertex}(X,Z)$
   $\rightarrow \quad \textit{disjoint}(Y,Z)$
9. $\textit{disjoint}(X,Y), \textit{overlap}(X,Y) \quad \rightarrow \quad \bot$

Figure 6.2: The results of a rule exploration of triplets of parallel unit squares. Rules 4, 5, 6, and 8 have slightly been simplified using Rule 3. See also Figure 6.3

rule set. The formal context is object-reduced (which is not an automatic result of the algorithm), meaning that all 17 examples are needed. Each example refutes a rule that holds in the 16 other examples. As an instance, we look at the tenth triplet,

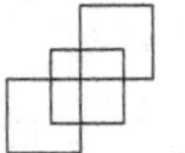

The rule

$$\textit{overlap}(X,Y), \textit{overlap}(X,Z), \textit{common-vertex}(Y,Z) \rightarrow \textit{common-edge}(Y,Z)$$

holds in all 16 other examples, but not in this one.

The list of rules in Figure 6.2 may seem complicated, but it is not. Only two of the rules are not completely obvious. This becomes more transparent if an additional predicate is introduced, equality. A reorganized list of rules is presented in Figure 6.3.

1. $\textit{common-edge}(X,Y) \quad \rightarrow \quad \textit{common-vertex}(X,Y)$
2. $\textit{overlap}(X,Y), \textit{common-vertex}(X,Z), \textit{common-vertex}(Y,Z)$
   $\rightarrow \quad X = Y$
3. $\textit{overlap}(X,Y), \textit{common-vertex}(X,Y) \quad \rightarrow \quad X = Y$
4. $X = Y \quad \rightarrow \quad \textit{overlap}(X,Y), \textit{common-edge}(X,Y)$
5. $X = Y, \textit{common-vertex}(X,Z) \quad \rightarrow \quad \textit{common-vertex}(Y,Z)$
6. $X = Y, \textit{common-edge}(X,Z) \quad \rightarrow \quad \textit{common-edge}(Y,Z)$
7. $X = Y, \textit{overlap}(X,Z) \quad \rightarrow \quad \textit{overlap}(Y,Z)$
8. $X = Y, \textit{disjoint}(X,Z) \quad \rightarrow \quad \textit{disjoint}(Y,Z)$
9. $\textit{disjoint}(X,Y), \textit{common-vertex}(X,Y) \quad \rightarrow \quad \bot$
10. $\textit{disjoint}(X,Y), \textit{overlap}(X,Y) \quad \rightarrow \quad \bot$

Figure 6.3: A more transparent rule set equivalent to that of Figure 6.2 is obtained by introducing equality as an additional predicate

6.1.2 Predicates: attributes of power contexts

Take another look at Figure 6.1. It shows a formal context of a somewhat puzzling nature. Its objects are triplets of squares, so they are *objects composed of parts*. And the attributes are structured too. We need some vocabulary for such cases!

What is new is that the object domain G, which we explore, is structured by one or more relations. In logic, such relations are called **predicates**. For each such predicate, there is a fixed number, the **arity**, indicating how many objects are related in each instance. If p is a k-ary predicate, then

$$p(g_1, \dots, g_k)$$

expresses that the objects $g_1, \dots, g_k$ are in relation p. Another way to write this is

$$(g_1, \dots, g_k) \models p.$$

A k-ary **power context** over a set G is of the form $(T, P, \models)$, where $T \subseteq G^k$ is a set of k-tuples. The attributes $p \in P$ can be considered as k-ary predicate names over G, and $(g_1 \dots, g_k) \models p$ expresses that the tuple $(g_1, \dots, g_k)$ is an instance of the predicate p.

We need however a slightly more general approach. The formal context in Figure 6.1 has triplets as objects, and therefore it is a ternary power context. But its attributes like "disjoint" or "common edge" are essentially binary. This becomes possible through a standard construction from logic, by using *variables*.

As before, we work over a set G of objects, which is structured by relations (also called predicates) of possibly different arities. Let P be the set of the predicate names (also called **predicate symbols**), together with their respective arities. Now fix a set $V := \{v_1, \dots, v_n\}$ of **variables**.

Mappings $\nu : V \to G$ are called **variable assignments**. Each such assignment can be represented by the tuple $(\nu(v_1), \dots, \nu(v_n))$ of its values.

Using variables, we can define new predicates from given ones as follows. Let p be a k-ary predicate symbol, and let $t_1, \dots, t_k$ be variables from V (not necessarily distinct). Then

$$p(t_1, \dots, t_k)$$

is an n-ary (!) predicate over G, which is defined by

$$(\nu(v_1), \dots, \nu(v_n)) \models p(t_1, \dots, t_k) \quad :\iff \quad (\nu(t_1), \dots, \nu(t_k)) \models p.$$

For example, in Figure 6.1, the triplet of squares | A | C | B | has the ternary (!) attribute *disjoint*(A, B) because the depicted variable assignment maps the variable A to the left square, the variable B to the right, and the variable C to the middle one, and indeed, the two squares to which the variables A and B are assigned in this tuple are disjoint.

A k-ary predicate name followed by k variables is called an **atom**. If P is a set of predicate names and V is a set of variables, then P_V denotes the set of all atoms that can be obtained from P and V.

The **FOL power context**[4] over the domain G with predicate names from P and variables from V is

$$(G^V, P_V, \models),$$

where G^V is the set of all $|V|$-tuples of objects, represented by variable assignments $\nu : V \to G$, where P_V is the set of all atoms $p(t_1, \ldots, t_k)$ made of the predicate names in P and the variables in V, and

$$\nu \models p(t_1, \ldots, t_k)$$

indicates that $(\nu(t_1), \ldots, \nu(t_k))$ is an instance of the predicate p.

Such contexts quickly become large, of course, but there is a remedy: variable transformations induce symmetries and endomorphisms in the sense of Section 4.3.3.

A **variable transformation** is any map $\tau : V \to V$ that maps variables to variables. Each such transformation induces a pair of maps $\varepsilon^\tau := (\varepsilon^\tau_G, \varepsilon^\tau_M)$ of the FOL power context as follows:

- If $\nu : V \to G$ is a variable assignment, then $\varepsilon^\tau_G(\nu) := \nu \circ \tau$, i.e.,

$$(\nu(v_1), \ldots, \nu(v_n)) \overset{\varepsilon^\tau_G}{\mapsto} (\nu(\tau(v_1)), \ldots, \nu(\tau(v_n))).$$

- If $p(t_1, \ldots, t_k) \in P_V$ is a predicate name, then

$$\varepsilon^\tau_M(p(t_1, \ldots, t_k)) := p(\tau(t_1), \ldots, \tau(t_k)).$$

Proposition 42 *If $(G^V, P_V, \models)$ is a FOL power context, then $\varepsilon^\tau := (\varepsilon^\tau_G, \varepsilon^\tau_M)$, as defined above, is a c-endomorphism in the sense of Section 4.3.3, i.e., it satisfies*

$$\varepsilon^\tau_G(g) I m \iff g I \varepsilon^\tau_M(m).$$

Proof

$$\begin{aligned}
& (\nu(\tau(v_1)), \ldots, \nu(\tau(v_n))) \models p(t_1, \ldots, t_k) \\
\iff & (\nu(\tau(t_1)), \ldots, \nu(\tau(t_k))) \models p \\
\iff & (\nu(v_1), \ldots, \nu(v_n)) \models p(\tau(t_1), \ldots, \tau(t_k)).
\end{aligned}$$

□

[4]FOL is short for "first-order logic". We assume that the sets G, P, and V are disjoint to avoid misinterpretations.

We give two illustrating examples. First, for triplets of squares, we obtain

$$\begin{array}{|c|c|c|}\hline A & B & C \\ \hline\end{array} \models \mathit{disjoint}(A, C)$$

from

$$\begin{array}{|c|c|c|}\hline A & C & B \\ \hline\end{array} \models \mathit{disjoint}(A, B),$$

using the transformation that exchanges B and C, but keeps A intact.

As a second example, we consider a transformation that is not a bijection:

$$A \mapsto A, \quad B \mapsto B, \quad C \mapsto A.$$

This transformation generates from the predicate $\mathit{disjoint}(A, C)$ the somewhat strange new predicate $\mathit{disjoint}(A, A)$. Both predicates are ternary, since they come from $P_{\{A,B,C\}}$. We ask which triplets of squares are related by this predicate. We obtain that

$$\begin{array}{|c|c|c|}\hline A & B & C \\ \hline\end{array} \models \mathit{disjoint}(A, A)$$

iff

$$\begin{array}{|c|c|c|}\hline AC & B & \\ \hline\end{array} \models \mathit{disjoint}(A, C).$$

The latter is obviously not the case. The same argument works for any other triplet of squares, and $\mathit{disjoint}(A, A)$ is the same as $\bot$: it is never true.

6.1.3 Exploring rules

A rule exploration begins with choosing a set P of predicate names (with arities), a set V of variables[5], and an object domain G where the chosen predicates have an interpretation. The formal context the implication logic of which is to be explored is the FOL power context

$$(G^V, P_V, \models),$$

as defined on page 246.

A **Horn clause** (also called a **rule**) has the form $B \to H$, where both B (the **body**) and H (the **head**) are (possibly empty) sets of atoms. For example, the rule

$$\begin{array}{l} \mathit{overlap}(X, Y), \mathit{common\text{-}vertex}(X, Z), \mathit{common\text{-}vertex}(Y, Z) \to \\ \quad \mathit{common\text{-}vertex}(X, Y), \mathit{common\text{-}edge}(X, Y) \end{array}$$

from Section 6.1.1 has this shape. From the viewpoint of Formal Concept Analysis, it represents an implication holding in the FOL power context $(G^V, P_V, \models)$

[5] Zickwolff [230] uses *typed logic*, where there may be several types of variables.

iff it is respected by all object intents. In first-order logic, it is interpreted as a logical formula universally quantified over all its variables; it is an abbreviation of the first-order sentence

$$\forall X \forall Y \forall Z \big(\mathit{overlap}(X,Y) \wedge \mathit{common\text{-}vertex}(X,Z) \wedge \mathit{common\text{-}vertex}(Y,Z) \rightarrow \\ \mathit{common\text{-}vertex}(X,Y) \wedge \mathit{common\text{-}edge}(X,Y)\big).$$

It would be considered true for the domain G iff it holds true for all possible assignments of the variables to elements of G.

But this is the same! The objects of the FOL power context are just all variable assignments, and an object intent respects a clause iff it is true under the corresponding assignment.

Thus, a **rule exploration** is just an exploration of a FOL power context, and rules are implications of such contexts (see Algorithm 30). The main difference from the general case is the presence of c-endomorphisms, which are heavily used to simplify the exploration procedure.

Algorithm 30 RULE EXPLORATION

Input: A possibly empty subcontext $(E, P_V, \models)$ of a FOL power context $\mathbb{U} := (G^V, P_V, \models)$, where V and P are finite sets of variables and of predicate symbols. The set V^V of all variable transformations is denoted by T.

Interactive input: $\{*\}$ Upon request, confirm for $B, H \subseteq P_V$ that $B \rightarrow H$ holds in $\mathbb{U}$ as a rule or give a variable assignment $\nu : V \rightarrow G$ and its intent $\nu^{\models}$ for which $B \subseteq \nu^{\models}$ and $H \not\subseteq \nu^{\models}$, thus refuting the rule $B \rightarrow H$.

Output: A set $\mathcal{L}$ of rules sound and complete for $\mathbb{U}$ and a possibly enlarged subcontext $(E, P_V, \models)$ with the same rules.

```
ℒ := ∅
B := ∅
while B ≠ P_V do
    while B ≠ H := ⋂{ε^τ_M(e^⊨) | e ∈ E, τ ∈ T, B ⊆ ε^τ_M(e^⊨)} do
        if the rule B → H holds in 𝕌 then                              {*}
            ℒ := ℒ ∪ {B → H}
            exit while
        else
            extend E by some ν : V → G refuting B → H                  {*}
    B := NEXT CLOSURE(B, P_V, {ε^τ(X → Y) | X → Y ∈ ℒ, τ ∈ T})
return ℒ, (E, P_V, ⊨)
```

6.1.4 A problem on evolutionary trees

A fastidious reader, who now takes a fancy to the exploration of rules, may ask if the technique can be enriched by the additional methodology that we have developed in Section 5.4. As another appetizer, we now discuss a more ambitious example that occurred when studying data from comparative linguistics. One of the methods for representing processes with ramifications is the use of *phylogenetic trees*. Here is a tiny example about Indo-European languages from [98].[6]

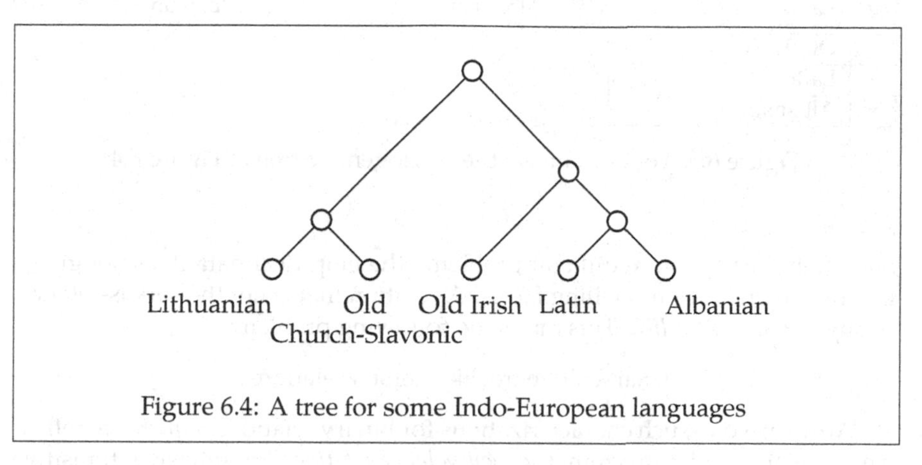

Figure 6.4: A tree for some Indo-European languages

Ideally, such trees are based on data. A natural representation uses a ternary **versus-relation** as it is shown in Figure 6.5. The versus-relation expresses for each choice of three (languages) which two are more closely related. For example, the last entry of Figure 6.5 (the lone box in the fourth row) expresses that Latin and Albanian are closer to each other than they are to Old Irish. This is read as

Latin, Albanian *versus* Old Irish

and abbreviated by

Latin, Albanian | Old Irish

In the tree shown in Figure 6.4, this finds its expression in the fact that Latin and Albanian have a common ancestor node that is not an ancestor of Old Irish.

It is straightforward to read from a (finite) rooted tree its versus-relation. And it is easy to see that any such tree can be reconstructed from its versus-relation (up to non-branching inner vertices). When working with real data,

[6] We shall not discuss here if the tree reflects the true phylogenetic history of these languages, nor if that can be expressed by a tree at all. We also disregard all numerical modelings and only consider the qualitative structure.

×	Lithuanian
×	Old Church-Slavonic
	Old Irish

×	Lithuanian
×	Old Church-Slavonic
	Latin

×	Lithuanian
×	Old Church-Slavonic
	Albanian

	Lithuanian
×	Old Irish
×	Latin

	Lithuanian
×	Old Irish
×	Albanian

	Lithuanian
×	Latin
×	Albanian

	Old Church-Slavonic
×	Old Irish
×	Latin

	Old Church-Slavonic
×	Old Irish
×	Albanian

	Old Church-Slavonic
×	Latin
×	Albanian

	Old Irish
×	Latin
×	Albanian

Figure 6.5: Versus data for the phylogenetic tree in Figure 6.4

one often meets a more difficult problem: the empirical data does not give a tree due to errors or modelling faults. A relation that is not the versus-relation of any tree is *not treelike*. This raises the following problem:[7]

Characterize treelike versus-relations!

We are used to such characterizations for binary relations. *Orders* are reflexive, transitive and antisymmetric; *equivalence relations* are reflexive, transitive and symmetric. We are looking for similar axioms for treelike relations, which are, however, ternary. We are not familiar with properties of ternary relations, nor are these very intuitive. Nevertheless, two properties are quite obvious:[8]

1. $a, b \mid c$ implies $b, a \mid c$;
2. when $a, b \mid c$ holds, then $a, c \mid b$ does *not* hold.

If we restrict to binary trees, in which every vertex has either zero or exactly two successors, we get a third condition:

3. when $a, b \mid c$ and $a, c \mid b$ do not hold, then $b, c \mid a$ holds.

But these conditions are not sufficient for being treelike. How can we come up with further axioms?

We might try rule exploration for treelike relations or, less pathetically, investigate a sufficient number of treelike relations and check what they have in common. There are, up to isomorphism, only two rooted binary trees with four

[7]Treelike versus-relations are well investigated. See the literature cited in [98].

[8]In addition, it is common to define $a, a \mid b$ to hold iff $a \neq b$. We omit these degenerate instances of the relation.

	$\{a,b,c\}$	$\{a,b,d\}$	$\{a,c,d\}$	$\{b,c,d\}$
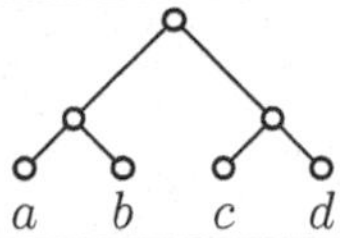	$ab \mid c$	$ab \mid d$	$cd \mid a$	$cd \mid b$
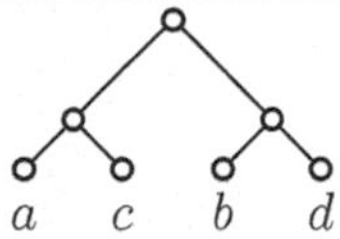	$ac \mid b$	$bd \mid a$	$ac \mid d$	$bd \mid c$
13 more trees with 4 leaves	…	…	…	…

Figure 6.6: A many-valued context of binary trees with four leaves

$\mathbb{S}_{\{x,y,z\}}$	$yz \mid x$	$xz \mid y$	$xy \mid z$	$\neg yz \mid x$	$\neg xz \mid y$	$\neg xy \mid z$
$yz \mid x$	×				×	×
$xz \mid y$		×		×		×
$xy \mid z$			×	×	×	

Figure 6.7: A scale $\mathbb{S}_{\{x,y,z\}}$ for the context of trees

leaves, but, with all their leaf-labelings, we already get 15 examples. These are combined in a many-valued context, a part of which is printed in Figure 6.6. The many-valued attributes correspond to the possible choices of three leaves. The attribute values are the entry in the versus-relation for the respective choice.

Next, a suitable conceptual scaling is required that turns the many-valued context into a one-valued one. It is our choice how to scale. Naturally, we shall use the same scale for all four attributes. Since we expect that good axioms for treelike relations might include negations, we choose a scale that is richer that the nominal scale, see Figure 6.7.

The derived context has 15 objects and 24 attributes. Its canonical basis consists of 88 implications, which would give a rather unpleasant system of axioms.

But many of these 88 implications are "isomorphic" and differ only in the labeling. The automorphism group induced by the 24 permutations of $\{a, b, c, d\}$ groups them together into only six orbits, which we again may read as rules, see Figure 6.8.

So far we have derived six rules that hold in all the 15 examples. We have

1. $\neg(xy \mid z), \neg(yz \mid x) \quad \rightarrow \quad xz \mid y$
2. $xy \mid z \quad \rightarrow \quad \neg(xz \mid y), \neg(yz \mid x)$
3. $xy \mid z, \neg(xy \mid z) \quad \rightarrow \quad \bot$
4. $\neg(xy \mid w), \neg(yz \mid x) \quad \rightarrow \quad \neg(yz \mid w)$
5. $\neg(xz \mid w), yz \mid w, \neg(xz \mid y) \quad \rightarrow \quad yz \mid x$
6. $wx \mid y, \neg(wx \mid z), \neg(wy \mid z), \neg(xy \mid z) \quad \rightarrow \quad wz \mid y, xz \mid y$

Figure 6.8: Rules for the formal context derived from the many-valued context in Figure 6.6 using the scaling from Figure 6.7 (simplified)

1. $\neg(xy \mid z), \neg(yz \mid x) \quad \rightarrow \quad xz \mid y$
2. $xy \mid z \quad \rightarrow \quad \neg(xz \mid y), \neg(yz \mid x)$
3. $xy \mid z, \neg(xy \mid z) \quad \rightarrow \quad \bot$

3* $\top \quad \rightarrow \quad xy \mid z$ or $\neg(xy \mid z)$

Figure 6.9: Background formulas for the scale in Figure 6.7

no guarantee that these rules indeed characterize treelike versus-relations, not even that they hold in *all* such relations. Investigating further examples might perhaps reduce the list. But before we extend our search, we should make full use of the available information. Still unused is the *background information* of the scaling, as described in Chapter 5 of this book. The cumulated clauses can be aggregated to first-order logic formulas using the automorphisms. The result is shown in Figure 6.9.

The first three formulas of the background knowledge are identical to the first three rules. Formulas 3 and 3* remind us that the negation is really meant as negation and not merely as a symbol. We recall from Section 5.1.1 (p. 188) that a modified implication inference is necessary to treat this case. The rules in Figure 6.8 were produced without including background knowledge. Therefore, they may be logically redundant! This could be checked by an algorithm or by hand. Here we demonstrate how the system can be simplified "manually".

Using contraposition, we can slightly simplify Rule 4 to

$$\neg(xy \mid w),\ yz \mid w \quad\rightarrow\quad yz \mid x. \tag{4*}$$

Rule 5, which reads

$$\neg(xz \mid w),\ yz \mid w,\ \neg(xz \mid y) \quad\rightarrow\quad yz \mid x,$$

now follows (apply Rule 4* with y and z interchanged). Rule 6,

$$wx \mid y,\ \neg(wx \mid z),\ \neg(wy \mid z),\ \neg(xy \mid z) \quad\rightarrow\quad wz \mid y,\ xz \mid y,$$

can be derived as well: from Rule 4, we infer

$$yz \mid w,\ \neg(yz \mid x) \quad\rightarrow\quad xy \mid w$$

and, equivalently, by interchanging w with y and x with z,

$$wx \mid y,\ \neg(wx \mid z) \quad\rightarrow\quad wz \mid y. \tag{4**}$$

Note that the premise of (4**) is symmetric in x and w, which implies

$$wx \mid y,\ \neg(wx \mid z) \quad\rightarrow\quad wz \mid y,\ xz \mid y,$$

and thereby Rule 6.

$$wx \mid y,\ \neg(wz \mid y) \quad\rightarrow\quad wx \mid z$$

A single rule for treelike versus-relations

We are left with a single rule! Rules 1, 2, and 3 follow from the preconditions, Rules 5 and 6 from Rule 4. What remains to prove is that this rule indeed characterizes treelike versus-relations. It does, and this can be found as a theorem in the literature (Colonius and Schulze [66]). Rule exploration did not prove this theorem for us, but it showed a way to find it.

6.2 Attribute descriptions

When attributes are not merely "given", but instead are *defined* in some formal way, new possibilities for the exploration method occur. It may then be possible to support the domain expert by logical inference stemming from such definitions. However, this strongly depends on the formalism used for defining attributes.

Particularly useful and well-structured are **Description Logics (DL)**, a family of logical languages that combine expressive power and decidability, see [18] or [188]. Remarkable research was done on melding attribute exploration with Description Logics, starting with pioneering work of F. Baader [17]. Many of the later contributions came from his research group.

The technical terms that are commonly used for Description Logics are different from those for Formal Concept Analysis and sometimes not very compatible. Most of the publications combining attribute exploration with Description Logics use the jargon of Description Logics. In our presentation here, we stay with FCA terminology, for which the necessary generalizations are introduced in the following two subsections.

6.2.1 Binary power context pairs

The universe of an attribute exploration based on Description Logics is a binary **power context pair** $(\mathbb{K}, \mathbb{K}_2)$, where $\mathbb{K} := (G, M, I)$ is, as usual, a formal context with a set G of objects, a set M of attributes, and an incidence relation I. As additional information, we have a binary power context

$$\mathbb{K}_2 := (G \times G, R, I_2).$$

The attributes of $\mathbb{K}_2$ are binary relations on G or, for short, **roles**. The objects of $\mathbb{K}_2$ therefore are pairs of objects from G, and

$$(g, h) \; I \; r$$

expresses that the pair (g, h) is an instance of the role r.

Realistic examples of power contexts quickly become too large to be presented here in detail. We therefore illustrate the notion by two small data sets. The first one almost literally is a toy example, but full of structure and well suited for demonstrating the possibilities. The second is part of a natural language processing example and much closer to real-world applications. Note that both examples are not meant to be explored; they are only examples of binary power context pairs.

Example 20 We consider the 3×3 magic square on the integers 0 to 8 shown in Figure 6.10.

We define a formal context $\mathbb{K} := (G, M, I)$, the objects of which are the nine digits, i.e., $G := \{0, \ldots, 8\}$. As attributes, we take the integral part and the remainder when dividing by three, see the upper part of Figure 6.11.

The product set $G \times G$ in this example has already 81 Elements. If we choose the three roles "being in the *same row*", "being in the *same column*", and "being on the *same diagonal*" (of the magic square), we get the formal context which is shown in clarified form in the lower part of Figure 6.11.

Example 21 The second binary context power pair is from R. Wille's paper [220]. Wille starts with some text, for which he chose a paragraph out of *"The Smithsonian Guide to Historic America"* that informs about Seattle's central business district [155, p. 349]. Here we use only the first sentence of Wille's example:

1	6	5
8	4	0
3	2	7

Figure 6.10: A magic square on the digits $0, \ldots, 8$

"Seattle's central business district, bounded by Yesler Way, Route 5, Stewart Street, and the waterfront, has among its sleek glass monoliths smaller buildings with gargoyles and other lively early-twentieth-century decorations that give it an interesting texture."

Wille first represents this text by a conceptual graph in the sense of J. Sowa. The result is displayed in Figure 6.12. The exact rules for reading conceptual graph diagrams are found in Sowa's book [200]. Here we cite from Wille's paper:

> Each box represents a concept ... together with at least one of its objects ..., while an edge represents at least one instance of a semantic relation. Concept names are written in each box with capitals (T denotes the universal concept). If a box does not contain an object name (written behind the colon), then the existence of an object is assumed for this box to which an individual marker like #437 may be assigned ({*} behind the colon indicates that there exist several objects whose exact number might also be given). Dotted line segments, representing so-called coreference links, indicate that the objects of the joined boxes are the same.
>
> The abbreviations for the used semantic relations translate as follows: PTNT=patient, AGNT=agent, STAT=state, ATTR=attribute, RCPT=recipient, ...

The concept names are the attributes of the formal context $\mathbb{K}$ in Figure 6.13. The semantic relations, understood as roles, become the attributes of the formal context $\mathbb{K}_2$ in Figure 6.14.

6.2.2 Terminological attributes

For binary context power pairs $(\mathbb{K}, \mathbb{K}_2)$, where

$$\mathbb{K} := (G, M, I) \quad \text{and} \quad \mathbb{K}_2 := (G \times G, R, I_2),$$

$\mathbb{K}$	0 (div 3)	1 (div 3)	2 (div 3)	0 (mod 3)	1 (mod 3)	2 (mod 3)
0	×			×		
1	×				×	
2	×					×
3		×		×		
4		×			×	
5		×				×
6			×	×		
7			×		×	
8			×			×

$\mathbb{K}_2$	*same-row*	*same-column*	*same-diagonal*
$(0,0),(2,2),(6,6),(8,8)$	×	×	
$(1,1),(3,3),(4,4),(5,5),(7,7)$	×	×	×
$(1,5),(1,6),(5,1),(5,6),(6,1),(6,5),(0,4),(0,8),(4,0),(4,8),(8,0),(8,4),(2,3),(2,7),(3,2),(3,7),(7,2),(7,3)$	×		
$(1,3),(1,8),(3,1),(3,8),(8,1),(8,3),(0,5),(0,7),(5,0),(5,7),(7,0),(7,5),(2,4),(2,6),(4,2),(4,6),(6,2),(6,4)$		×	
$(1,4),(1,7),(4,1),(4,7),(7,1),(7,4),(3,4),(3,5),(4,3),(4,5),(5,3),(5,4)$			×
$(0,1),(0,2),(1,0),(1,2),(2,0),(2,1),(0,3),(0,6),(3,0),(3,6),(6,0),(6,3),(6,7),(6,8),(7,6),(7,8),(8,6),(8,7),(2,5),(2,8),(5,2),(5,8),(8,2),(8,5)$			

Figure 6.11: The formal contexts $\mathbb{K}$ and $\mathbb{K}_2$ (clarified), derived from the magic square

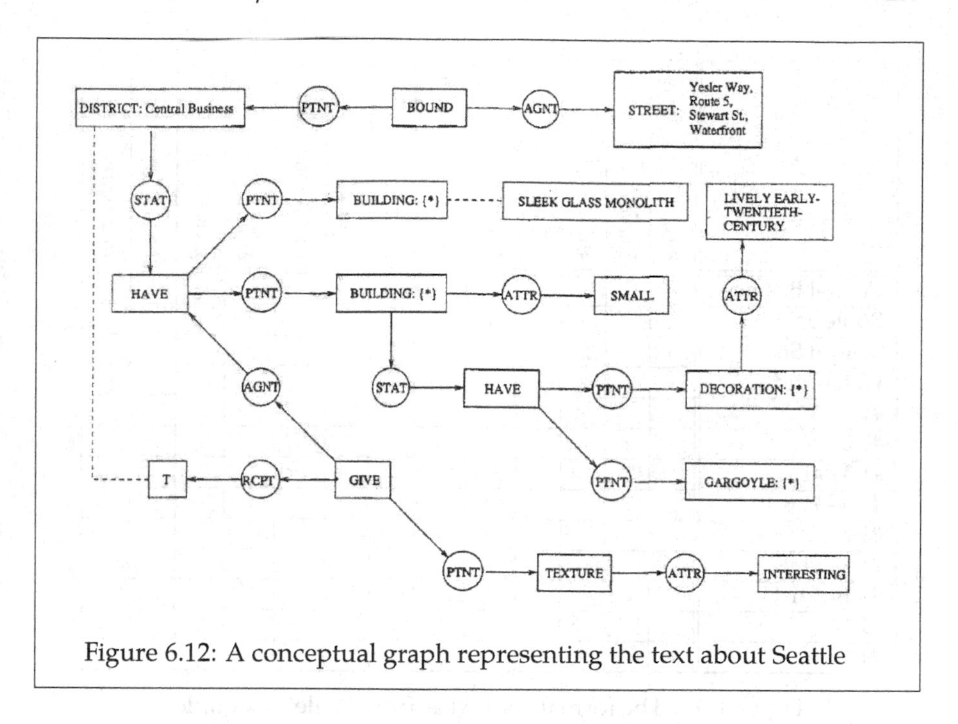

Figure 6.12: A conceptual graph representing the text about Seattle

the attributes of $\mathbb{K}_2$, called *roles*, have as extents binary relations between objects in G. The presence of roles gives the possibility of defining new types of compound attributes, similar to ones which are commonly used in natural language. For example, a sentence like

My house is next to a grocery store.

may be understood as the statement of an attribute:

An attribute of my house is to be next to a grocery store.

This attribute is a compound one, composed from a relational part *(is next to)* and an attribute *(is a grocery store)*, i.e., from a role r and a unary attribute m. It is an example of a *terminological* compound attribute. In the language of Description Logics, it would be abbreviated as

$$\exists r.m.$$

An object g has the terminological attribute $\exists r.m$ iff there is an object h such that (g, h) are in role r and h has m. In the example, it holds that *my house is next to a grocery store* if there is an object h such that *my house is next to* h and h *is a grocery store.*

$\mathbb{K}$	*district*	*bound*	*street*	*have*	*building*	*sleek glass monolith*	*small*	*decoration*	*lively early-twent. c.*	*gargoyle*	*give*	*texture*	*interesting*
Central Business	×												
Route 5			×										
Stewart St.			×										
Waterfront			×										
#1.1			×										
#1.2		×											
#1.3 - #1.4				×									
#1.5 - #1.6					×	×							
#1.7					×		×						
#1.8 - #1.9				×				×	×				
#1.10 - #1.11										×			
#1.12											×		
#1.13												×	×

Figure 6.13: The formal context $\mathbb{K}$ from Wille's example

The availability of terminological attributes greatly enhances the expressive power of the conceptual formalism. We now can build extents like $(\exists r.m)'$, which, provided that the object set consists of houses, comprises

all houses that are next to a grocery store.

As often, there is a price to pay for the better expressive power: both theory and computation become more difficult. This will be discussed below.

Closely related to $\exists$, but somewhat less easy to use, is the attribute constructor

$$\forall r.m.$$

An object g has the terminological attribute $\forall r.m$ iff *all* objects h with $(g, h) I_2 r$ have the attribute m.

There are further possible constructors, see Baader et al. [18]. For a more general approach to terminological attribute logic, see Prediger [177] and Distel [75]. Here we discuss the constructors used for the Description Logic $\mathcal{ALC}$ *("Attributive Language with Complements")*. Context power pairs may be considered as a form of *semantics*. All that is required for the logic formalism is to specify a set M of (unary) attributes together with a set R of roles. This is

$\mathbb{K}_2$	PTNT	AGNT	STAT	RCPT
(#1.1, Central Business)	×			
(#1.1, Yesler Way)		×		
(#1.1, Route 5)		×		
(#1.1, Stewart St.)		×		
(#1.1, Waterfront)		×		
(Central Business, #1.2)			×	
(#1.2, #1.3) - (#1.2, #1.4)	×			
(#1.2, #1.5) - (#1.2, #1.6)	×			
(#1.5, #1.7) - (#1.6, #1.7)			×	
(#1.7, #1.8) - (#1.7, #1.9)	×			
(#1.7, #1.10) - (#1.7, #1.11)	×			
(#1.2, #1.12)		×		
(#1.12, Central Business)				×
(#1.12, #1.13)	×			

Figure 6.14: The formal context $\mathbb{K}_2$ from Wille's example

sometimes called a **signature**. A set of individual object names may also be included [18].

Definition 23 The set $M_{\mathcal{ALC}}$ of $\mathcal{ALC}$ **terminological** attributes over a signature (M, R) is defined as follows:

- $\top \in M_{\mathcal{ALC}}$ and $\bot \in M_{\mathcal{ALC}}$;
- $m \in M_{\mathcal{ALC}}$ for every unary attribute $m \in M$;
- $\neg m \in M_{\mathcal{ALC}}$ for every $m \in M_{\mathcal{ALC}}$;
- $\bigsqcap S \in M_{\mathcal{ALC}}$ and $\bigsqcup S \in M_{\mathcal{ALC}}$ for every finite subset $S \subseteq M_{\mathcal{ALC}}$;
- $\exists r.m \in M_{\mathcal{ALC}}$ and $\forall r.m \in M_{\mathcal{ALC}}$ for every $m \in M_{\mathcal{ALC}}$ and $r \in R$;

and these are all elements of $M_{\mathcal{ALC}}$.

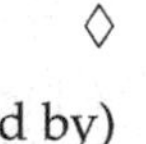

Definition 24 The $\mathcal{ALC}$ **terminological context** derived from (or induced by) a binary power context pair $((G, M, I), (G \times G, R, I_2))$ is the formal context $(G, M_{\mathcal{ALC}}, I_{\mathcal{ALC}})$, where $I_{\mathcal{ALC}}$ is defined as follows:

- $g \; I_{\mathcal{ALC}} \; \top$ for all $g \in G$;

- $g\ I_{\mathcal{ALC}}\ \bot$ for no $g \in G$;
- $g\ I_{\mathcal{ALC}}\ m$ iff $g\ I\ m$, given $m \in M$;
- $g\ I_{\mathcal{ALC}}\ \neg m$ iff $(g, m) \notin I_{\mathcal{ALC}}$;
- $g\ I_{\mathcal{ALC}} \bigsqcap S$ iff $g\ I_{\mathcal{ALC}}\ m$ for all $m \in S$;
- $g\ I_{\mathcal{ALC}} \bigsqcup S$ iff $g\ I_{\mathcal{ALC}}\ m$ for some $m \in S$;
- $g\ I_{\mathcal{ALC}}\ \exists r.m$ iff there is some $h \in G$ such that $(g, h)\ I_2\ r$ and $h\ I_{\mathcal{ALC}}\ m$;
- $g\ I_{\mathcal{ALC}}\ \forall r.m$ iff $h\ I_{\mathcal{ALC}}\ m$ for every h with $(g, h)\ I_2\ r$.

We say that an implication **holds** in a binary power context pair iff it holds in the derived terminological context. ◊

Note that the terminological context contains the original (G, M, I) as a subcontext, since $M \subseteq M_{\mathcal{ALC}}$ and $I = I_{\mathcal{ALC}} \cap (G \times M)$.

There are many Description Logics, and they differ in the constructors that they use. According to Distel [75], a *lightweight* Description Logic is one that does not provide all constructors of $\mathcal{ALC}$, while an *expressive* Description Logic provides more constructors than $\mathcal{ALC}$. The above construction can be made for other Description Logics as well, and we shall use the notation accordingly. Sometimes we do not want to specify which Description Logic is used. In that case, we write

$$(G, M_{\mathrm{DL}}, I_{\mathrm{DL}})$$

for the derived formal context.

The set $M_{\mathcal{ALC}}$ of terminological attributes is infinite even if M is finite and even if $\mathcal{ALC}$ is replaced by a lightweight Description Logic. One may generalize the notion of **global equivalence**, which was defined in Section 5.1.4, to terminological attributes and then identify equivalent ones, but what remains is still infinite. This can be seen from the following simple example, where, as above, the objects are buildings, there is only one attribute $m :=$ *grocery store* and only one role $r :=$ *next to*. Consider the terminological attributes

$$\exists r.m, \quad \exists r.\exists r.m, \quad \exists r.\exists r.\exists r.m, \quad \ldots,$$

which read out as *"my house is next to a grocery store"*, *"my house is next to a building which is next to a grocery store"*, *"my house is next to a building which is is next to a building which is next to a grocery store"*, etc. For any two of these attributes, it is easy to come up with a fictitious city map in which they behave differently, showing that they are not globally equivalent.

Example 20, contd. The formal context $\mathbb{K}$ in Figure 6.11 has exactly 17 extents. These are: the empty set $\emptyset$ and the full set G, all 9 possible singleton sets $\{g\}$, $g \in G$, and the six attribute extents m', $m \in M$. If we allow for disjunction, then we immediately get all the 512 subsets of G as extents of compound attributes, simply as combinations of the singleton sets. Examples of extents of other terminological compound attributes are

$$\begin{aligned} (\exists \text{ same-row.}(2 \text{ (div } 3)))' &= G, \\ (\exists \text{ same-row.}((2 \text{ (div } 3)) \sqcap (2 \text{ (mod } 3)))' &= \{0,4,8\}, \\ (\exists \text{ same-diagonal.}\top)' &= \{1,3,4,5,7\}. \end{aligned}$$

6.2.3 Attribute exploration with Description Logics

Including terminological attributes into attribute exploration considerably changes the rules of the game. It results in a number of difficulties that are not easy to overcome. Remarkable results were obtained in a series of studies, most of them under the guidance of F. Baader. In particular, the Ph.D. theses by Rudolph [186], Sertkaya [194], Distel [75], and Borchmann [50] have greatly advanced the field. Their results are too extensive and manifold to be fully included in this book. We only give a simple version.

We have already mentioned the first problem: the set of terminological attributes becomes infinite even for very simple signatures. The attribute exploration algorithms described in the previous chapters are designed for finite attribute sets and therefore are not immediately applicable.

Several approaches address this problem. Rudolph [186] has derived an exploration algorithm for $\mathcal{FLE}$-attributes[9] of bounded "role depth", a parameter restricting how deep the nesting of quantifiers in a terminological attribute may be. For a given finite signature and a given number n, there are indeed only finitely many $\mathcal{FLE}$-attributes of role depth $\leq n$. Rudolph's algorithm explores these finitely many attributes; the result may then be extended to role depth $\leq n+1$. Rudolph proves that this procedure eventually terminates.

A Description Logic that very naturally extends the FCA language is $\mathcal{EL}^{\perp}$. It only uses propositional conjunction $\sqcap$, $\exists$, and $\perp$ as constructors. Distel [75] shows that each finite $\mathcal{EL}^{\perp}$-model has a finite basis of implications (or "GCIs" in his work). His proof very elegantly makes use of the unusual Description Logic $\mathcal{EL}^{\perp}_{gfp}$, named after its "greatest fixpoint semantics".

Baader et al. [21] generalize attribute exploration to formal contexts with a finite number of terminological attributes, freely chosen by the individual user. Their approach is described in detail below.

A second serious problem comes from what computer scientists may call the "closed world semantics" underlying attribute exploration. More explicitly,

[9] $\mathcal{FLE}$ is a lightweight Description Logic that uses only the constructors $\sqcap$, $\exists$, and $\forall$.

it is the following: for basic attribute exploration, the intermediate knowledge is stored in a formal context (of the "counterexamples" that were provided) and a list of validated implications. That formal context is a subcontext of the universe, i.e., of the formal context to be explored. All implications valid in the universe therefore hold in the context of counterexamples as well.

This is not the case when the universe is a binary power context pair, say,

$$((G, M, I), (G \times G, R, I_2)),$$

the derived terminological context of which is to be explored. If $E \subseteq G$ is a set of "counterexamples", then the terminological context derived from the power context pair of counterexamples

$$((E, M, I \cap E \times M), (E \times E, R, I_2 \cap (E \times E) \times R))$$

usually is *not* a subcontext of the larger one. Implications that hold in general may not hold in the smaller derived context. Figure 6.15 shows an example of a tiny "universe" in order to demonstrate the effect.

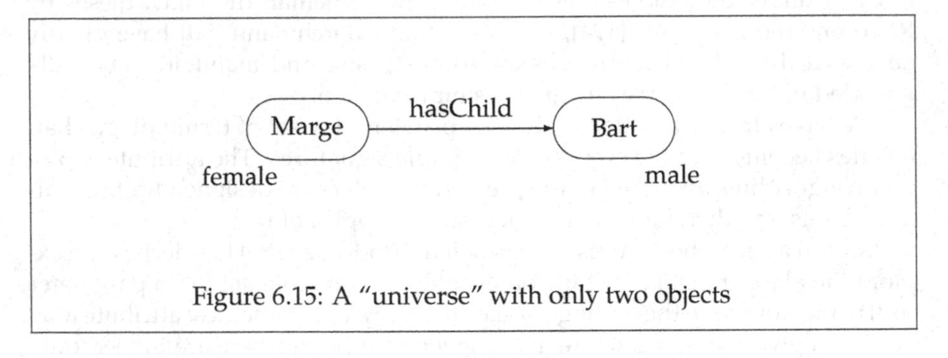

Figure 6.15: A "universe" with only two objects

Here we have only two objects, Marge and Bart; two unary attributes, male and female; and one role, named hasChild. Marge is female; Bart is male; and (Marge, Bart) is the only element of the only role, hasChild.

An implication asked at the very beginning of an exploration could be

Do all objects have the attribute "male"?

There is only one counterexample, "Marge", and when this is given, we have $E = \{\text{Marge}\}$. The intermediate power context pair is

$$((E, M, \{(\text{Marge}, \text{female})\}), (E \times E, R, \emptyset)).$$

Since Bart is not an object in E, it cannot be derived from this singleton cut-out that Marge has the attribute $\exists$ hasChild.male. Thus the implication

$$\text{female} \to \exists\ \text{hasChild.male},$$

which holds in the (derived terminological context of) the universe, does not hold in this cut-out.

One possibility to resolve this problem is to ask the expert to disprove implications not by giving single objects as counterexamples, but to include their role-successors, the successors of these, and so on. Distel [75] has worked this out in his "Model Exploration", where counterexamples have to be given in the form of connected components of the universe.

Another approach is to interpret the expert's answer differently. The expert had given the counterexample with the intention to disprove the implication that was asked. There was no intention to provide information about other attributes, in particular, compound ones. It is therefore natural to use "partially given examples" as introduced in Section 4.3.4. Attribute exploration with partially given terminological description is the topic of Section 6.2.4.

A third problem caused by the enhanced expressive power is that logical inference becomes more difficult. The knowledge base of an attribute exploration consists of some counterexamples and some implications. In the basic version, it is consistent if all these counterexamples respect all implications. With terminological attributes this becomes more demanding for the same reason as in the second problem discussed above: the collection of counterexamples does not necessarily constitute a **model** of the implications, i.e., a binary power context pair in which the implications all hold. It must be checked if a model containing the counterexamples can possibly exist, and this may be difficult. (An exception is Distel's "Model Exploration" method that was mentioned above.)

The task of uncovering knowledge that is only implicitly contained in the knowledge base is called **reasoning**. Fortunately, powerful algorithms, called **reasoners**, have been developed for many Description Logics. They work well in practice, even though the worst-case complexity of the reasoning problems is often scary.

6.2.4 Exploration of terminological attributes

The method presented in this subsection is general in the sense that it does not refer to a particular Description Logic. From a given signature (M, R) of attribute and role names, we derive the set M_{DL} of terminological attributes, depending on the chosen Description Logic.

The knowledge base of an exploration consists of a binary power context pair and a list of implications. Both may initially be trivial, i.e., the implication list and the incidence relations all may be empty.

The binary power context pair is considered as a cut-out of the universe under investigation and therefore as a collection of assertions. It corresponds to the **ABox** in Description Logic language, while the implication list would be called a **TBox**. A **terminological knowledge base**[10] $(\mathcal{T}, \mathcal{A})$ consists of an implication list $\mathcal{T}$ and a binary power context pair $\mathcal{A}$, for the same signature. A knowledge base $(\mathcal{T}, \mathcal{A})$ is **consistent** if it has a **model**, by which we mean a binary context power pair that contains $\mathcal{A}$ and in which all implications from $\mathcal{T}$ hold. For an object $g \in G$ and a terminological attribute $m \in M_{\mathrm{DL}}$, we write

$$(\mathcal{T}, \mathcal{A}) \models g\ I\ m$$

if $(\mathcal{T}, \mathcal{A})$ extended by $\neg(g\ I\ m)$ is inconsistent.

The "open world" interpretation of $\mathcal{A}$ is expressed by a modified definition for the derived context:

Definition 25 The **partial terminological context** derived from a terminological knowledge base $(\mathcal{T}, \mathcal{A})$ is the pair $(\mathbb{K}_+, \mathbb{K}_?)$, where

$$\mathbb{K}_+ := (G, M_{\mathrm{DL}}, I_+), \quad \mathbb{K}_? := (G, M_{\mathrm{DL}}, I_?),$$

with incidence relations given by

$$\begin{aligned} g\ I_+\ m \ &:\Longleftrightarrow \quad (\mathcal{T}, \mathcal{A}) \models g\ I_{\mathrm{DL}}\ m \\ g\ I_?\ m \ &:\Longleftrightarrow \quad (\mathcal{T}, \mathcal{A}) \not\models \neg(g\ I_{\mathrm{DL}}\ m). \end{aligned}$$

◇

It can easily be checked that if $(\mathcal{T}, \mathcal{A})$ is consistent, then $(\mathbb{K}_+, \mathbb{K}_?)$ is indeed a partial context in the sense of Section 4.3.4. This partial context is used to testify that certain implications do *not* hold in the universe. Recall from Proposition 31 (p. 171) that an implication $A \to B$ is *refuted* by $(\mathbb{K}_+, \mathbb{K}_?)$ iff there is an object $g \in G$ such that $A \subseteq g^+$ and $B \not\subseteq g^?$. We say that an implication is refuted by $(\mathcal{T}, \mathcal{A})$ iff it is refuted by its derived partial context.

The goal of an exploration is to enrich the knowledge base so that it eventually becomes complete. This requires a definition:

Definition 26 Let $(\mathcal{T}, \mathcal{A})$ be a terminological knowledge base, which has $(\mathbb{U}, \mathbb{U}_2)$ (the "universe") as a model, and let N be a set of attributes. We say that $(\mathcal{T}, \mathcal{A})$ is N-complete with respect to $(\mathbb{U}, \mathbb{U}_2)$ iff the following three conditions are equivalent for all implications $A \to B$ over N:

1. $A \to B$ holds in the universe $(\mathbb{U}, \mathbb{U}_2)$,
2. $A \to B$ follows from $\mathcal{T}$, and
3. $A \to B$ is not refuted by $(\mathcal{T}, \mathcal{A})$.

[10] Some authors use the name **ontology**.

$(\mathcal{T}, \mathcal{A})$ *completes* a knowledge base $(\mathcal{T}_0, \mathcal{A}_0)$ (with respect to $(\mathbb{U}, \mathbb{U}_2)$ and N) iff it is complete and contains $(\mathcal{T}_0, \mathcal{A}_0)$. ◊

Attribute exploration cannot directly be applied to the derived partial context because of its infinitely many attributes. But any finite selection N of its attributes can be explored, and the knowledge base can be completed with respect to this finite choice.

Barış Sertkaya has implemented this method into his ontology completion tool OntoComP (see [195]), which is a plugin for the Protégé 4 ontology editor. His open-source libraries FCALib and OntoComPLib are available from his web site `https://www.frankfurt-university.de/~sertkaya/software.html`.

6.3 Concept Exploration [11]

> ...in many situations one has only a vague knowledge of a context although many of its concepts are fairly clear. For instance, it seems to be impossible to write down a comprehensive context of musical instruments, but everyone uses concepts as violin, trumpet, guitar, string instrument, etc. and meaningful sentences such as "a violin is a string instrument". In such cases of concepts of a vague context we have a modified determination problem: *How can one extend a partial lattice of concepts to a complete concept lattice of some context?*
>
> R. Wille [215]

Concept exploration aims at answering this question by exploring a "vaguely known" concept lattice generated by some set S of "basic concepts". The algorithm generates questions about the lattice of the form "Is $\mathfrak{c}_1$ a subconcept of $\mathfrak{c}_2$?" in order to obtain local insight into the structure of the lattice.

Already in the first publication on formal concept analysis, the basic conception of concept exploration is mentioned (Wille [215]). Its scheme is demonstrated by examples in [217] and [219]. U. Klotz and A. Mann further elaborated the method in their master thesis [135]. Unfortunately, no easy to use implementation of the method has come to our knowledge.

By developing such an exploration tool, one encounters two difficulties. The first is that, even with only three basic concepts, the resulting lattice may be infinite. The second difficulty results from the fact that, in many applications, the meaning of the supremum (i. e., the least common superconcept) of some given concepts is less intuitive to the investigator than the meaning of their infimum (i. e., the greatest common subconcept). While the infimum always corresponds to the conjunction of the attributes describing the concepts, the supremum, in general, does not correspond to their disjunction. However, if this

[11] Parts of this section are from a manuscript by Gerd Stumme [205] with his kind permission.

correspondence holds, then also the first difficulty is solved, since the distributivity of the lattice, which is enforced by conjunction and disjunction, guarantees the finiteness of the results. For such applications, *distributive concept exploration* is designed [204]. The algorithm performs a "depth-first exploration", adding the basic concepts one by one, each after having determined the sublattice generated by the previously added basic concepts. It will be discussed in the fourth subsection of this chapter. The third one contains a detailed example of the exploration of four simple mathematical concepts. We begin with the detailed presentation of the algorithms, followed by two examples.

6.3.1 Sublattice generation via double exploration

We first delineate concept exploration as a mathematical procedure for generating a finite sublattice of a (complete) lattice $(L, \leq)$ from a subset $S \subseteq L$. The conceptual interpretation will be discussed later. The lattice $(L, \leq)$ is thought to be unknown and accessible only by questioning a domain expert.

The method follows the generation process of the sublattice. First it computes the set $S^{(1)} := S^{\wedge}$ consisting of all infima of the elements of $S^0 := S$. Then $S^{(2)} := S^{\wedge\vee}$ is computed, the set of all suprema of $S^{\wedge}$. The next step provides $S^{(3)} := S^{\wedge\vee\wedge}$ and so on.

The elements of $S^{(k)}$ will be recorded in the form of lattice terms expressing how they are iteratively obtained from the "basic" elements in S by forming infima and suprema. It naturally happens that different subsets have the same suprema or infima, and the domain expert must be consulted in order to avoid doublings. Since there are exponentially many subsets, a parsimonious strategy of questioning the expert is inevitable. Several authors have investigated the algorithmics of sublattice generation, see e.g. Kauer & Krupka [130] for a recent contribution. We use a double exploration technique.

The construction produces a sequence of formal contexts

$$(S^{(1)}, S^{(0)}, \leq), \quad (S^{(1)}, S^{(2)}, \leq), \quad (S^{(3)}, S^{(2)}, \leq), \quad (S^{(3)}, S^{(4)}, \leq), \quad \dots$$

It terminates as soon as $S^{(k-1)} = S^{(k)}$. All sets $S^{(k)}$ are subsets of $(L, \leq)$, and the incidence relation of the formal contexts is always the order relation of the lattice, restricted to the respective sets. Moreover, a list of implications is maintained.

The next formal context in the above-mentioned sequence is obtained from the previous one by an attribute exploration or an object exploration of a slightly modified nature, called **$\bigwedge$-exploration** and **$\bigvee$-exploration** (Stumme [203]). Recall from the Basic Theorem 1 (p. 18) that $(L, \leq) \cong \mathfrak{B}(L, L, \leq)$ holds for every complete lattice $(L, \leq)$. This immediately provides us with two families of implications over L: the attribute implications (called **$\bigwedge$-implications**) and the implications between (sets of) objects (which we call **$\bigvee$-implications**). We write

$A \xrightarrow{\wedge} B$ for an implication between the attribute sets A and B and $A \xrightarrow{\vee} B$ if A and B are considered as sets of objects. Of course, objects and attributes are the same in the case of $(L, L, \leq)$, but the interpretation differs. Note that

- $A \xrightarrow{\wedge} B$ holds in $(L, L, \leq)$ iff $\bigwedge A \leq \bigwedge B$, and
- $A \xrightarrow{\vee} B$ holds in $(L, L, \leq)$ iff $\bigvee A \geq \bigvee B$.

$\bigwedge$- and $\bigvee$-exploration are, up to a modification, just the usual exploration technique applied to these two types of implications. The modification is that when the domain expert rejects a suggested implication $A \to A''$, no counterexample is asked for, but the set of all $a \in A''$ is requested for which $A \to a$ holds in the domain $(L, \leq)$. This set is then entered in the formal context as the object intent of the object $\bigwedge A$ in the case of $\bigwedge$-exploration; for $\bigvee$-exploration, the set will be the extent of the new attribute $\bigvee A$.

More explicitly: when a pseudo-intent A is found, the query proposed to the domain expert in the case of $\bigwedge$-exploration is

which elements of A'' are greater than or equal to $\bigwedge A$?

and in the case of $\bigvee$-exploration it is

which elements of A'' are less than or equal to $\bigvee A$?

The general schema of the concept exploration algorithm is shown in Algorithm 31. Since the situation is perfectly symmetric, we focus on $\bigwedge$-exploration described in Algorithm 32.

Algorithm 32 has two parts. The first is the interactive $\bigwedge$-exploration. It results in a formal context (G, M, I) the intents of which are precisely the subsets A of M that satisfy

$$A = \{m \in M \mid m \geq \bigwedge A\}.$$

Each such intent A corresponds to a unique element $\bigwedge A$ of $M^\wedge$, and this is bijective: for an arbitrary $X \subset M$, the set $A := \{m \in M \mid m \geq \bigwedge X\}$ is an intent satisfying $\bigwedge A = \bigwedge X$. The second part of Algorithm 32 introduces, for every intent A of (G, M, I), an object named $\bigwedge A$ to $(M^\wedge, M, \leq)$ with intent $\{m \in M \mid m \geq \bigwedge A\}$. For convenience, we identify the terms $\bigwedge\{m\}$ and m. When available, concept names are used instead of terms. This will be the case for the second example.

There is an extra line in Algorithm 32, which is needed when it is repeatedly invoked during concept exploration. It must be avoided that the domain expert is asked unnecessary questions. The algorithm therefore must record all available information. Hence, whenever a new object $\bigwedge A$ is inserted into (H, M, J), the implication $A \xrightarrow{\wedge} \bigwedge A$ should be added to $\mathcal{L}_\wedge$. Moreover, the

Algorithm 31 Sketch of the Concept Exploration algorithm

Input: A finite subset S of a complete lattice $(L, \leq)$.
Interactive input: Required for $\bigwedge$- and $\bigvee$-exploration via Algorithm 32.
Output: The sublattice of $(L, \leq)$ generated by S, if finite.

$k := 0$
$(S^{(-1)}, S^{(0)}, \leq_0) := (\emptyset, S, \emptyset)$
$\mathcal{L}_\wedge := \mathcal{L}_\vee := \emptyset$
repeat
  $k := k + 1$
  if k is odd **then**
    compute $(S^{(k)}, S^{(k-1)}, \leq_k)$, $\mathcal{L}_\wedge$, and $\mathcal{L}_\vee$ {Algorithm 32}
  else
    compute $(S^{(k-1)}, S^{(k)}, \leq_k)$, $\mathcal{L}_\wedge$, and $\mathcal{L}_\vee$ {dual of Algorithm 32}
until $S^{(k)} = S^{(k-1)}$
return $(S^{(k)}, \leq_k)$

order relation also gives rise to background implications, because, for single objects and attributes, we have

$$a \xrightarrow{\wedge} b \iff a \leq b \iff b \xrightarrow{\vee} a.$$

Since these implications are needed only later, it may be more efficient to store them separately until Algorithm 32 finishes. Moreover, since M is a subset of $M^\wedge$, information about the order of M may be reused if present.

6.3.2 Two examples of concept exploration

Concept exploration is illustrated in this book by three examples, two of them in this subsection. The first focuses on the technicalities of the algorithm and is purely mathematical. It shows in detail how a sublattice of a small lattice is obtained from a generating set. All elements of this sublattice are produced after one $\bigwedge$-exploration and a subsequent $\bigvee$-exploration. A second $\bigwedge$-exploration then shows that no further elements are produced and that the concept exploration is thus finished. This last step produces no queries to the domain expert, since all implications can be concluded from the background knowledge. The fact that this exploration terminates after one double exploration step depends on a special property of the lattice (distributivity). In general, there is no bound on the number of steps. FL(3), the *free lattice on three generators*, is infinite. Thus the concept exploration algorithm, when applied to its generators, will not ter-

Algorithm 32 $\bigwedge$-Exploration within a complete lattice $(L, \leq)$

Input: A context $(G, M, \leq)$, where both G and M are finite subsets of a complete lattice $(L, \leq)$; sets $\mathcal{L}_\wedge$ and $\mathcal{L}_\vee$ of valid $\bigwedge$- and $\bigvee$-implications over M.

Interactive input: $\{*\}$ Upon request, decide which elements $a \in A'' \setminus A$ satisfy $a \geq \bigwedge A$.

Output: The context $(M^\wedge, M, \leq)$ and enlarged sets $\mathcal{L}_\wedge$ and $\mathcal{L}_\vee$ of valid $\bigwedge$- and $\bigvee$-implications.

```
A := ∅
while A ≠ M do
    if A ≠ A'' then
        B := {a ∈ A'' \ A | a ≥ ⋀A}                                    {*}
        if B ≠ ∅ then
            L∧ := L∧ ∪ {A → B}
        if B ≠ A'' \ A then
            extend (G, M, I) by a new object with intent A ∪ B
    A := NEXT CLOSURE(A, M, L∧)                          {see Algorithm 3}

(H, M, J) := (∅, M, ∅)
A := FIRST INTENT(G, M, I)        {can be computed with Algorithm 2}
while A ≠ ⊥ do
    extend H by the lattice term ⋀A
    extend J by {⋀A} × A
    (for concept exploration, extend L∧ and L∨ as described in text)
    A := NEXT INTENT(G, M, I)    {can be computed with Algorithm 3}
return (H, M, J), L∧, L∨
```

minate. But it will produce after each $\bigwedge$- or $\bigvee$-exploration step an intermediate lattice, and it can be shown that these lattices cannot be obtained by fewer steps.

The setting for our first example is displayed in Figure 6.16, which shows a 20-element lattice. Four of its elements are highlighted and labelled **A**, **B**, **C**, and **D**, and the sublattice generated by these elements is to be determined. Figure 6.17 shows what is generated after the first $\bigwedge$-exploration. It is the eight-element $\bigwedge$-subsemilattice formed by all infima of the four generators. The resulting formal context, as shown in Figure 6.17, is then subject to a $\bigvee$-exploration, which generates five more elements, shown in Figure 6.18.

The second $\bigwedge$-exploration produces no further elements, and the concept exploration algorithm is finished. Two diagrams, both in Figure 6.19, document the sublattice generated by $\{\mathbf{A}, \mathbf{B}, \mathbf{C}, \mathbf{D}\}$.

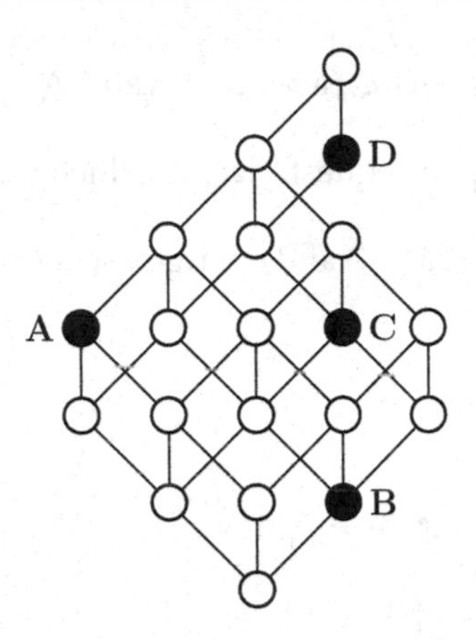

Figure 6.16: A small lattice, in which a subset $S := \{\mathbf{A}, \mathbf{B}, \mathbf{C}, \mathbf{D}\}$ is highlighted. Figures 6.17 to 6.19 show step by step how Algorithm 31 determines the sublattice generated by S

$\leq$	A	B	C	D
$\bigwedge\emptyset$				
D				×
C ∧ D			×	×
B ∧ C ∧ D		×	×	×
A	×			
A ∧ D	×			×
A ∧ C ∧ D	×		×	×
A ∧ B ∧ C ∧ D	×	×	×	×

$\mathbf{C} \xrightarrow{\wedge} \mathbf{D}$
$\mathbf{B} \xrightarrow{\wedge} \mathbf{C}, \mathbf{D}$
$\mathbf{A}, \mathbf{D} \xrightarrow{\wedge} \mathbf{A} \wedge \mathbf{D}$
$\mathbf{A}, \mathbf{C}, \mathbf{D} \xrightarrow{\wedge} \mathbf{A} \wedge \mathbf{C} \wedge \mathbf{D}$
$\mathbf{A}, \mathbf{B}, \mathbf{C}, \mathbf{D} \xrightarrow{\wedge} \mathbf{A} \wedge \mathbf{B} \wedge \mathbf{C} \wedge \mathbf{D}$

Figure 6.17: The results of the first $\bigwedge$-exploration. The first seven lines of the formal context are the result of the exploration, and the first two implications are the ones that were confirmed. All intents except for the full one are already present. This one is added (last line) to obtain $(S^\wedge, S, \leq)$. Implications 3–5 are recorded for later use

≤	A	B	C	D	(A∧C)∨B	(A∧D)∨B	A∨B	(A∧D)∨C	A∨C	A∧B	A∧C	A∧D	A∨D
⋀∅													×
D				×									×
C ∧ D			×	×				×	×				×
B ∧ C ∧ D		×	×	×	×	×	×	×	×				×
A	×						×		×				×
A ∧ D	×			×		×	×	×	×			×	×
A ∧ C ∧ D	×		×	×	×	×	×	×	×		×	×	×
A ∧ B ∧ C ∧ D	×	×	×	×	×	×	×	×	×	×	×	×	×

Figure 6.18: The result of the first ⋁-exploration, with simplified terms. The first nine columns come from the exploration; the last four are ∩-reducible extents. No further implications had to be confirmed

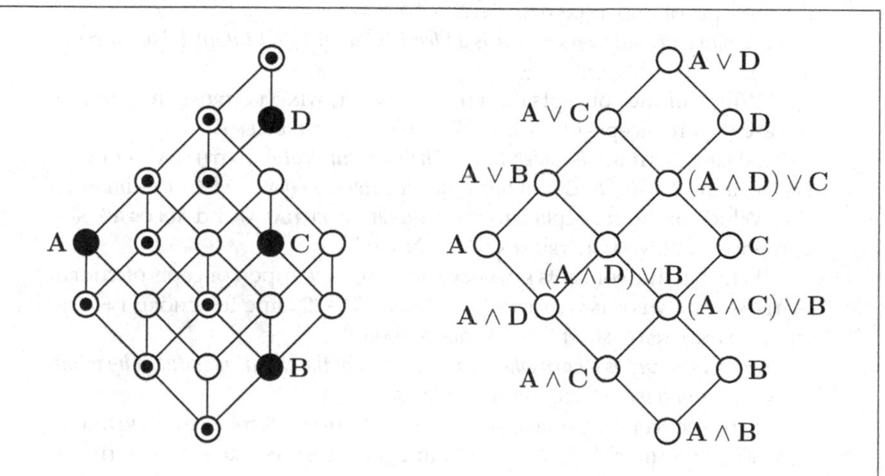

Figure 6.19: The second ⋀-exploration yields no further lattice elements, since all requested implications follow from the background knowledge derived from the order. The concept exploration terminates here. The lattice elements generated by S are indicated by the black dots in the left diagram. The right one displays the sublattice labeled with simplified lattice terms

The second concept exploration example, which is taken from Stumme [202], sheds light on the "conceptual" interpretation of the algorithm. Imagine a knowledge engineer who wants to model knowledge about ancient Greek musical instruments by conceptual graphs. She uses concept exploration for supporting the creation of an adequate type lattice. For instance, she might want to start with the following four basic types: CHORD INSTRUMENT, KITHARA, WIND INSTRUMENT, and AULOS. A kithara is a harp which, in Greek mythology, is the symbol of Apollo, while an aulos is an oboe-like instrument associated with Dionysus. The newly generated types are named using the naming mechanism described at the end of the previous section. The dialogue of the first $\bigwedge$-exploration (i. e., the first part of Algorithm 32) consists of eight questions:

> "Which of the concepts CHORD INSTRUMENT, KITHARA, WIND INSTRUMENT, and AULOS are superconcepts of $\bigwedge\emptyset$?" – "None!" – "Name for $\bigwedge\emptyset$?" – "INSTRUMENT."
>
> *The first $\mathcal{L}_\wedge$-pseudo-intent is the empty set. The infimum $\bigwedge\emptyset$ of the empty set is always the largest element of the lattice and can thus be understood as the concept of "everything" comprising all objects of the field of interest, which, in this example, are* INSTRUMENTS. *It is the first object added to G. The relation I remains empty.*
>
> "Which of the concepts KITHARA, WIND INSTRUMENT, and AULOS are superconcepts of CHORD INSTRUMENT?" – "None!"
>
> *The name* CHORD INSTRUMENT *is added to G with object intent {chord instrument }.*
>
> "Which of the concepts CHORD INSTRUMENT, WIND INSTRUMENT, and AULOS are superconcepts of KITHARA?" – "CHORD INSTRUMENT!"
>
> *The name* KITHARA *is added to G. The first implication,* KITHARA $\to$ CHORD INSTRUMENT, *is added to $\mathcal{L}_\wedge$. In this way, the extension of G and $\mathcal{L}_\wedge$ continues...*
>
> "Which of the concepts CHORD INSTRUMENT, KITHARA, and AULOS are superconcepts of WIND INSTRUMENT?" – "None!"
>
> "Which of the concepts KITHARA and AULOS are superconcepts of CHORD INSTRUMENT $\wedge$ WIND INSTRUMENT?" – "None!" – "Name for CHORD INSTRUMENT $\wedge$ WIND INSTRUMENT?" – "AEOLIAN HARP."
>
> *An aeolian harp is a harp where the vibration of the chords is induced by wind (either natural wind or generated by bellows).*
>
> "Is the concept AULOS a superconcept of CHORD INSTRUMENT $\wedge$ KITHARA $\wedge$ WIND INSTRUMENT?" – "Yes!" – "Name for CHORD INSTRUMENT $\wedge$ KITHARA $\wedge$ WIND INSTRUMENT?" – "NOTHING."
>
> *There is no kithara being a wind instrument. We obtain the "absurd type".*
>
> "Is the concept KITHARA a superconcept of CHORD INSTRUMENT $\wedge$ WIND INSTRUMENT $\wedge$ AULOS?" – "Yes!"
>
> *We obtain again the concept* NOTHING. *Since there is no lattice term added to G, we are not asked to provide a name.*

The result of the first $\bigwedge$-exploration is shown in Figure 6.20. Observe that the diagram has to be read as $\bigwedge$-semilattice, since only all infima are defined, but no suprema. In the following $\bigvee$-exploration, for instance, the supremum of KITHARA and AEOLIAN HARP will not be identified with CHORD INSTRUMENT, but with a proper subconcept of it, namely, with HARP.

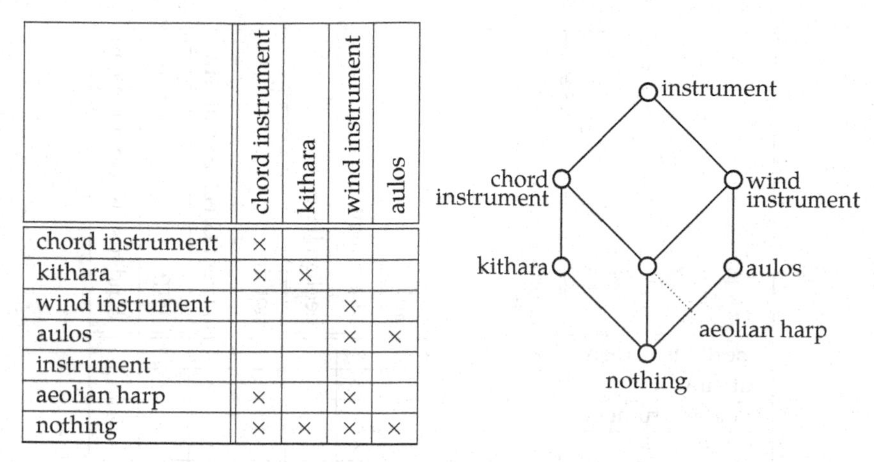

	chord instrument	kithara	wind instrument	aulos
chord instrument	×			
kithara	×	×		
wind instrument			×	
aulos			×	×
instrument				
aeolian harp	×		×	
nothing	×	×	×	×

Figure 6.20: The result of the first $\bigwedge$-exploration of ancient Greek musical instruments

The list $\mathcal{L}_\vee$ of $\bigvee$-implications now contains for every object g the implication $g \xrightarrow{\vee} \{h \mid g < h\}$: for example, WIND INSTRUMENT $\xrightarrow{\vee}$ AULOS, AEOLIAN HARP, NOTHING. These are used for the first $\bigvee$-exploration:

"Which of the concepts CHORD INSTRUMENT, KITHARA, WIND INSTRUMENT, AULOS, INSTRUMENT, AEOLIAN HARP and NOTHING are subconcepts of $\bigvee\emptyset$?" – "NOTHING!"

This justifies the choice of the name NOTHING *in the former $\bigwedge$-exploration, since $\bigvee\emptyset$ is always the smallest element of a lattice.*

"Which of the concepts CHORD INSTRUMENT, WIND INSTRUMENT, INSTRUMENT, and AEOLIAN HARP are subconcepts of KITHARA $\vee$ AULOS?" – "None!" – "Name for KITHARA $\vee$ AULOS?" – "ANCIENT GREEK INSTRUMENT."

Here and in the sequel, the concept NOTHING *is omitted in the lattice terms, since it is not relevant for the supremum.*

"Is the concept CHORD INSTRUMENT a subconcept of KITHARA $\vee$ AEOLIAN HARP?" – "No!" – "Name for KITHARA $\vee$ AEOLIAN HARP?" – "HARP."

"Is the concept WIND INSTRUMENT a subconcept of AULOS $\vee$ AEOLIAN HARP?" – "No!" – "Name for AULOS $\vee$ AEOLIAN HARP?" – "WIND INSTRUMENT WITH VIBRATING PART."

"Which of the concepts CHORD INSTRUMENT, WIND INSTRUMENT, and INSTRUMENT are subconcepts of KITHARA $\vee$ AULOS $\vee$ AEOLIAN HARP?" – "All!"

Here we decide to identify the concept KITHARA $\vee$ AULOS $\vee$ AEOLIAN HARP *with* INSTRUMENT.

The result of this $\bigvee$-exploration is shown in Figures 6.21 and 6.22. In the resulting so-called "weak partial lattice", all suprema are defined. Infima are only defined for its $\bigwedge$-subsemilattice shown in Figure 6.20.

	chord instrument	kithara	wind instrument	aulos	instrument	aeolian harp	nothing	ancient Greek instrument	harp	wind inst. w. vibrating part
chord instrument	×				×					
kithara	×	×			×			×	×	
wind instrument			×		×					
aulos			×	×	×			×		×
instrument					×					
aeolian harp	×		×		×	×			×	×
nothing	×	×	×	×	×	×	×	×	×	×

Figure 6.21: The formal context produced by the first $\bigvee$-exploration of ancient Greek musical instruments

In the next $\bigwedge$-exploration, the names ANCIENT GREEK CHORD INSTRUMENT, ANCIENT GREEK HARP, ANCIENT GREEK WIND INSTRUMENT, and ANCIENT GREEK WIND INSTRUMENT WITH VIBRATING PART are introduced. The following $\bigvee$-exploration provides only one new name, CHORD INSTRUMENT WITHOUT BOW. Finally, the fifth exploration (which is a $\bigwedge$-exploration again) runs through without generating any question. Hence, the concept exploration is terminated. The resulting lattice is shown in Figure 6.23. Of course, all infima and suprema are defined now.

The "degree of exactitude", i. e., the decision whether or not to identify two concepts, essentially depends on the purpose the exploration was done for. For instance, a very pedantic user might only accept the two implications KITHARA $\rightarrow$ CHORD INSTRUMENT and AULOS $\rightarrow$ WIND INSTRUMENT, and reject all others. Then every question generates a new concept. This process can be continued ad infinitum, converging towards the (infinite) free lattice generated by two two-element chains.

In general, the knowledge engineer can greatly benefit from the line diagram of the lattice freely generated by the ordered set $(S, \leq)$. In fact, the line diagram of $\mathrm{FL}(\underline{2} + \underline{2})$ was used for the exploration of instruments. Unfortu-

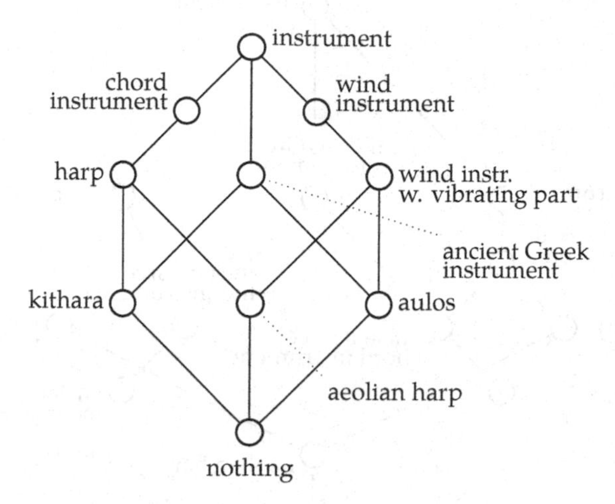

Figure 6.22: The $\bigvee$-subsemilattice obtained from the first $\bigvee$-exploration of ancient Greek musical instruments

nately, there are only very few free lattices over partially ordered sets which can be "drawn" (see Rival & Wille [185]).

6.3.3 Manual concept exploration

One of the earliest examples of concept exploration appeared in R. Wille's paper *"Bedeutungen von Begriffsverbänden"* (*Meanings of concept lattices*) [217]. Wille had no computer implementation available and performed the process manually on paper. We include his example here, because it reveals the graphical intuition on which the algorithm in Section 6.3.1 is based. Moreover, Wille has a slightly different strategy. He uses objects and attributes in his *four quadrants schema* and asks the domain expert to give evidence for the decisions. Instead of defining an abstract lattice $(L, \leq)$ as the exploration domain, Wille specifies a formal context. Typically, this context may have infinitely many objects and attributes, and it is only "vaguely known". It can be accessed through the domain expert.

Wille's example starts with four basic concepts of numbers:

A	even number,
B	square number,
C	sum of two squares,
D	sum of three squares.

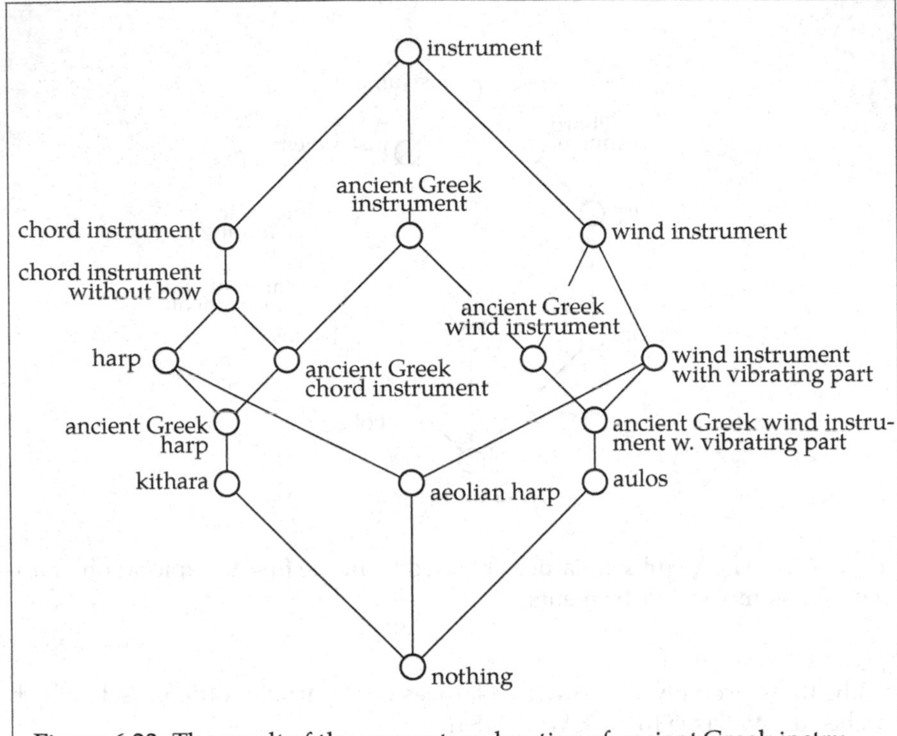

Figure 6.23: The result of the concept exploration of ancient Greek instruments

He first delimits the exploration by defining an object domain, the non-negative integers, and the set of attributes to be used. These are of the form

$$\text{congruent to } k_1 \text{ or } k_2 \text{ or} \ldots \text{or } k_j \text{ modulo } n,$$

abbreviated as "$\equiv k_1, \ldots, k_j \pmod{n}$". It will be sometimes shorter to list the remaining elements and to write "$\not\equiv l_1, \ldots, l_j \pmod{n}$" for "*not* congruent $\pmod{n}$ to any of $l_1, \ldots, l_j$".

As in Section 6.3.1, the general type of a concept exploration query is

$$\text{"Is } T_1 \text{ a subconcept of } T_2\text{?"},$$

where T_1 and T_2 are lattice terms in the basic concepts, i.e., expressions obtained from the basic concepts using infima and suprema.

Such a query may be answered either positively or by means of a counterexample. The latter must be given in form of an object-attribute pair (g, m), where g does *not* have m and, moreover, g is an object of T_1 and m is an attribute of T_2.

Then it must also be said which of the already considered concepts the object g and the attribute m belong to, except for the cases where this can be concluded from previously collected information. Wille records his findings in a growing quadratic schema organized as shown in Figure 6.24. The exploration starts at the center of the coordinate system.

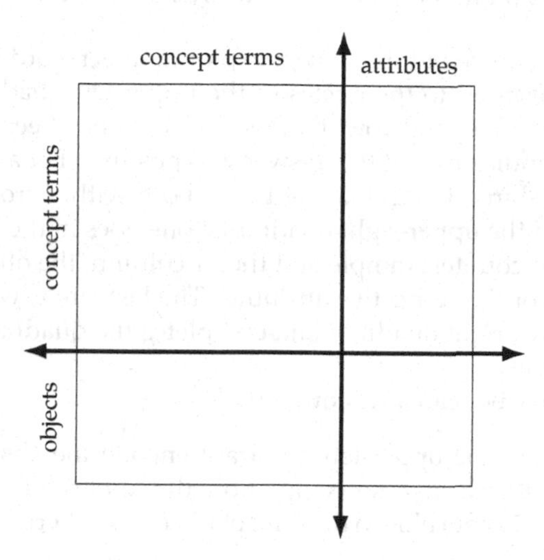

Figure 6.24: Wille's schema for a concept exploration, see Figure 6.25 for a concrete example

The order of the queries is not completely determined in Wille's approach, but the first question in the present example could be

Is **A** a subconcept of **B**?

A possible answer is "No, because 2 is an object of **A** and $\equiv 0, 1 \pmod 4$ is an attribute of **B**, but $2 \not\equiv 0, 1 \pmod 4$. Answers to subsequent queries could be

- 1 belongs to **B** and $\equiv 0 \pmod 2$ to **A**, but $1 \not\equiv 0 \pmod 2$,
- 6 belongs to **A** and $\not\equiv 3, 6, 7 \pmod 8$ to **C**, but $6 \equiv 3, 6, 7 \pmod 8$,
- 28 belongs to **A** and $\not\equiv 7, 15, 23, 28, 31 \pmod{32}$ to **D**, but $28 \equiv 7, 15, 23, 28, 31 \pmod{32}$,
- **B** is a subconcept of **C**,
- **B** is a subconcept of **D**,
- **C** is a subconcept of **D**.

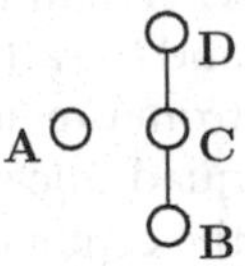

Now it is clear that the basic concepts B, C, and D form a three-element chain and that A is incomparable to the other three basic concepts. The largest possible lattice generated by such a configuration (the *free lattice* generated by this ordered set) is known to have 20 elements (Grätzer [118], p. 266). This shows that the concept exploration will terminate after only a few further questions.

The complete course of this concept exploration is recorded in Figure 6.25. The queries correspond to the circles in the upper-left quadrant. So if there is a circle in column T_1 and row T_2, then $T_1 \leq T_2$ has been asked. A cross within the circle indicates that the answer was positive. For a negative answer, three more circles are placed. Two of them, both with a cross, are placed in the lower-left and the upper-right quadrants: one goes in the row of the object that is used in the counterexample and the T_1-column, the other in the T_2-row and the column for the respective attribute. The last circle, without a cross, is placed in the lower-right quadrant and completes the quadrangle spanned by the first three circles.

Figure 6.25 may be read as follows:

- The crosses in the upper-left quadrant encode the abstraction order of the concepts. Indeed, after completion, the upper-left quadrant is, up to clarification, isomorphic to the dual of $(T, T, \leq)$, where $(T, \leq)$ is the lattice generated by S.
- The crosses in the lower-left quadrant show which of the objects belong to which concepts.
- The upper-right quadrant shows the relationship between concepts and attributes.
- The lower-right quadrant contains the incidence relation between the objects and attributes introduced in the exploration process.

The exploration starts at the center of the coordinate system. Newly formed terms are added to the left and the top of the schema, objects at the bottom, and attributes at the right. Cells in the upper-left quadrant that give rise to queries are marked by a circle; other crosses in that quadrant could be deduced.

The resulting lattice is uniquely determined, but the circles in Figure 6.25 are not, because the order of the queries is not fully prescribed. For example, the circle in the B-column and D-row of the schema comes from the fact that $B \leq D$ was asked before $C \leq D$. For the reverse order, that cross would not be encircled.

It happens that a new term is both $\leq$ and $\geq$ when compared with an already present one. This is the case for the last three terms in Figure 6.25. When no term for a new concept can be constructed, the exploration finishes.

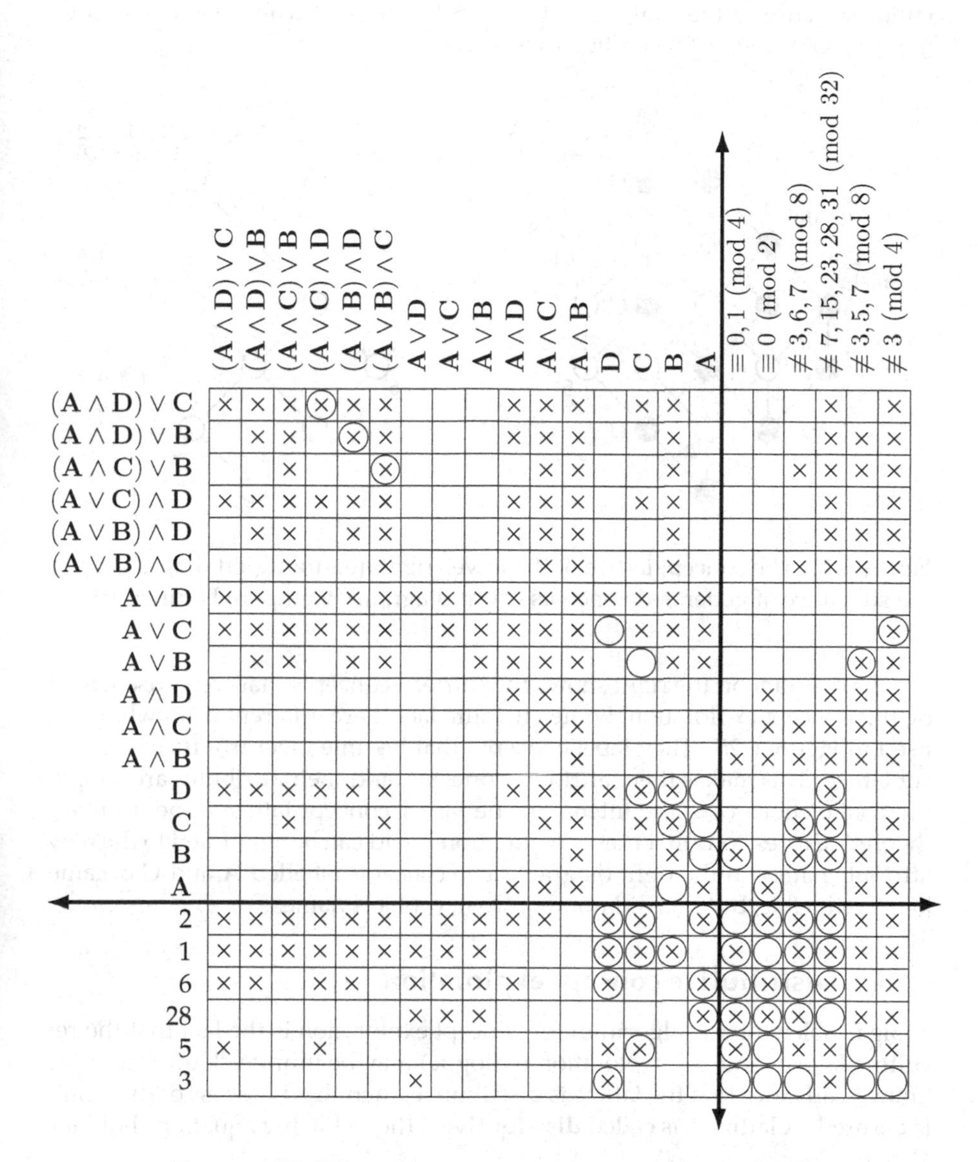

Figure 6.25: When the concept exploration is completed

The final result of this concept exploration is shown in Figure 6.26. The lattice structure is the same as in Figure 6.19, simply because we had chosen the very same lattice for our first example.

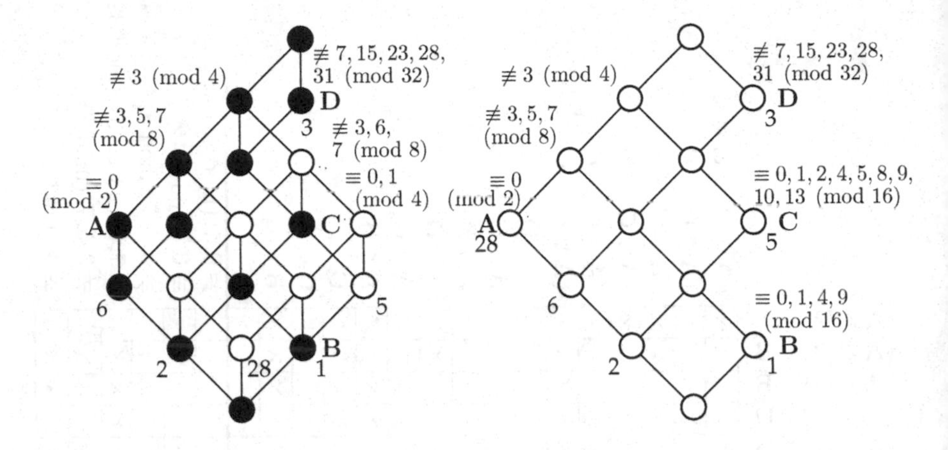

Figure 6.26: The concept lattice of the lower-right quadrant (left diagram), and the sublattice of generated concepts (right diagram). Compare Figure 6.19

The diagram on the right shows the thirteen concepts that were discovered by this concept exploration. Some attribute labels are different from what was used in Figure 6.25. They label concepts that became meet-irreducible in the sublattice. This makes them attribute concepts, and their attributes are simply the conjunctions over the intents in the larger concept lattice. The number-theoretic expressions that these conjunctions yield can be simplified to the new attribute names in the right diagram. The concepts labelled **A** and **C** became join-irreducible, but this did not require new object names.

6.3.4 Distributive concept exploration

A big problem of any algorithm for concept exploration is the fact that the resulting lattice (and the exploration dialogue) may be infinite. However, if we know in advance that the lattice is distributive, then the finiteness of the result is ensured. A lattice L is called **distributive** if the following equations hold for all $x, y, z \in L$:

$$x \wedge (y \vee z) = (x \wedge y) \vee (x \wedge z)$$
$$x \vee (y \wedge z) = (x \vee y) \wedge (x \vee z)$$

What additional assumptions imply the distributivity of the lattice? An important case is when the attribute set M that governs the conceptual hierarchy is closed under disjunction, because then

$$A' \cup B' = \{m \vee n \mid m \in A, n \in B\}'$$

holds for arbitrary sets $A, B \subseteq M$. As a consequence, the family of all concept extents is closed both under union and under intersection, and thus it is a sublattice of the distributive power set lattice. One interesting application is for *Description Logics* (see Section 6.2.3), where both conjunction and disjunction are often used as constructors for terminological attributes.

Stumme [204] has investigated distributive concept exploration in detail. He makes use of the *tensor product* of concept lattices, which is the co-product in the category of completely distributive complete lattices. Stumme explains how this fact can be utilized for a step-by-step construction, adding a single basic concept at a time, by forming the tensor product and then using the answers of the domain expert to define the congruence relation which factors this tensor product appropriately.

His detailed example of a concept exploration of zinks (a family of musical instruments) not only shows that his algorithm is of practical value. It also makes clear that the use of the tensor product is technically rather simple to handle. This is due to the fact that the tensor product of concept lattices corresponds to the direct product of their formal contexts, see [110]. The use of objects and attributes during such an exploration is similar to that in the previous subsection.

6.4 Exploration galore!

It appears that the possibilities of applying and generalizing attribute exploration are far from being exhausted. Sometimes it is difficult to keep up with the many new ideas that appear in the literature. The last section of this book therefore is devoted to promising new developments in the field. Three of them are sketched in Sections 6.4.1 to 6.4.3. Section 6.4.4 discusses a related framework from machine learning. The very last subsection contains a collection of hints to further interesting contributions.

6.4.1 Exploring faulty data

The main information source of an attribute exploration is the reliable domain expert. Answering queries may be tedious and expensive. It is therefore reasonable to ease the expert's working load whenever possible.

Borchmann [50, 53] has an approach allowing an additional information source, which is not required to be fully reliable. This is useful in practice, because it allows one to build an exploration on top of a large collection of observations, even if these – as always – contain a few errors. Simply including such data into the exploration knowledge base would lead to erroneous results. Borchmann's method shows that the acquisition of reliable knowledge may be supported by slightly faulty data.

Borchmann starts with a modest generalization of attribute exploration using a modified definition of the exploration base: he uses two implication lists $\mathcal{K}$ and $\mathcal{L}$ (over the same finite attribute set) instead of one. As usual, $\mathcal{L}$ is the list of confirmed implications and is considered not to be faulty. This requirement is dropped for the second list, $\mathcal{K}$. A typical choice for $\mathcal{K}$ could be the set of all association rules of high confidence in observed data, see Section 3.4.3 (p. 115).

For a better conformity with the standard version of attribute exploration, Borchmann assumes $\mathcal{L} \subseteq \mathcal{K}$, a condition which can easily be fulfilled by adding $\mathcal{L}$ to $\mathcal{K}$. Borchmann's algorithm does maintain a list of counterexamples for later, but does not refer to it. The basic version of attribute exploration is obtained when $\mathcal{K}$ generates precisely the implications that hold in the context of counterexamples.

The exploration algorithm is generalized in the most natural manner. It tries to construct an implication that follows from $\mathcal{K}$, but not from $\mathcal{L}$. If no such implication exists, then $\mathcal{K}$ and $\mathcal{L}$ generate the same implications. If such an implication exists, then the domain expert decides about the implication. If it is confirmed, it is added to the list $\mathcal{L}$. If a counterexample is given refuting the implication, then all implications in $\mathcal{K}$ that are not respected by the counterexample are modified. If C is the intent of the counterexample and $A \to B$ is an implication in $\mathcal{K}$ with $A \subseteq C$, then $A \to B$ is replaced by $A \to A \cup (B \cap C)$.

It is quite clear that the algorithm terminates (for finite M). Note that the termination of this algorithm does not usually conclude the exploration of the domain; it only finishes the extraction of valid implications from $\mathcal{K}$. Borchmann analyzes this in more detail. The following proposition summarizes some of his findings.

Proposition 43 *Let $\mathcal{L}_0 \subseteq \mathcal{K}_0$ be the two implication lists at the begin of the exploration and $\mathcal{L}_n \subseteq \mathcal{K}_n$ the lists when the algorithm terminates after n steps. Then it holds that*

- *every valid implication in $\mathcal{K}_0$ can be inferred from $\mathcal{L}_n$, and*
- *every implication in $\mathcal{L}_n$ is valid and can be inferred from $\mathcal{K}_0$.*

Moreover, if the queries are generated as usual, i.e., so that the premises are in lectic order and the conclusions are the $\mathcal{K}$-closures of the premises, then $\mathcal{L}_n \setminus \mathcal{L}_0$ is the relative canonical basis of the corresponding closure systems.

It is intuitively plausible, but not so easy to substantiate, that exploiting high-confidence association rules can support the domain expert. Indeed, the number of queries that the domain expert must answer is likely to rise (recall that the number of implications to verify is optimal for "unsupported" attribute exploration). But queries derived from high-confidence association rules may be easier to answer, and the remaining "difficult" questions may be better focused when a substantial amount of background knowledge has been accumulated. It would however be desirable to have more precise arguments for this, accompanied by convincing use cases.

6.4.2 Exploration in a fuzzy setting

There is a fuzzy-set version of Formal Concept Analysis, and it is well investigated. In particular, the Olomouc school of R. Bělohlávek and his co-workers has contributed a broad spectrum of results on this topic. Early publications are, e.g., by Burusco and Fuentes-Gonzáles [60] and, notably, the book by Pollandt [175], which is based on her Ph.D. thesis of 1995.

A fuzzy formal context still consists of a set G of "objects" and a set M of "attributes". But the incidence is no longer simply a binary relation encoding that an object "has" an attribute. Rather, it expresses the *degree* to which an object has an attribute. Such degrees (or *truth values*) may be numbers from $[0,1]$, but, more generally, can be chosen from any suitable ordered set $(L, \leq)$.

No generalization of Formal Concept Analysis is required when one uses the notion of a *many-valued context*, as in Definition 18 (p. 222), with $(L, L, \geq)$ as the conceptual scale for all attributes. Attribute exploration for such **fuzzy-valued contexts**[12], as Pollandt calls them, can be handled along the lines of Chapter 5. It is actually a particularly easy instance, because the scaling-induced background knowledge is harmless. The formal concepts of the derived context have (crisp) subsets of G as extents and order ideals of $(L, \leq)^M$ as (fuzzy) intents.

Most authors prefer a different notion of a fuzzy formal context, where the incidence relation is replaced by a fuzzy relation and both extents and intents may be proper fuzzy sets. This better reflects the idea of fuzziness and leads to a richer lattice structure. A more specific choice then is made for the domain of truth values: $(L, \leq)$ is assumed to be a *residuated* (complete) lattice, see [37] for a definition. Often an additional unary operation, called a *truth-stressing hedge*, is added for greater flexibility. A **fuzzy formal context** (G, M, I) connects sets G and M through a *fuzzy incidence relation* $I \in L^{G \times M}$.

Many notions of Formal Concept Analysis can be lifted to this setting, see Bělohlávek [37], but also Pollandt [175], who builds a bridge to non-fuzzy concept analysis through a double scaling process. Fuzzy derivation operators define fuzzy formal concepts in close analogy to the original definitions, and fuzzy implications can be introduced. Bělohlávek and Vychodil [39, 38] have shown that many of the algorithms discussed in this book can be adapted to the fuzzy setting.

[12]But note that the values themselves are not fuzzy, they are elements of L.

So what about fuzzy attribute exploration? This was considered by Glodeanu [113]. She finds that, under a strong assumption on the hedge (that it is the "globalization" hedge), exploration generalizes seamlessly. For other hedges, it is known that there is no analogue to the canonical basis of implications, and so technical difficulties arise. It should be pointed out that queries asked in a fuzzy exploration of this kind become more complicated. They include attributes with truth degrees, and the domain expert will have to provide fuzzy counterexamples to such queries. A convincing use-case study for such an exploration would be welcome, but a good understanding of fuzzy logic may be needed to come up with a practicable method.

6.4.3 Triadic data

The basic data type of Formal Concept Analysis, the formal context (G, M, I), represents a binary relation $I \subseteq G \times M$ between two sets G of "objects" and M of "attributes". This notion can be generalized to n-ary relations, see e.g. [214]. But then things get much more difficult already for $n = 3$. Many mathematical aspects of **Triadic Concept Analysis** (Wille & Lehmann [151]) are still waiting to be investigated.

The basic definitions are straightforward: a **triadic context**[13] is a quadruple (G, M, B, Y), where G, M, and B are sets (of "objects", "attributes", and "**conditions**", respectively), and $Y \subseteq G \times M \times B$ is a ternary relation. When $(g, m, b) \in Y$, this is interpreted as "the object g has the attribute m under the condition b". In analogy to what we have discussed in Section 1.3 (p. 13), **triadic concepts** can be defined. However, the *complete trilattices* that are formed by the triadic concepts of triadic contexts are rather exotic structures (so far). Their mathematical theory is still in its early stages (see, e.g., Biedermann [44, 45], Wille and Zickwolff [225] for first results). It is not obvious what generalization of attribute exploration to the triadic case should be like. There is not even a commonly accepted notion of a **triadic implication**.

Ganter and Obiedkov [104] have suggested a rather simple approach. It follows the intuition that a triadic context can be viewed as a family of (dyadic) formal contexts. If a triadic context is visualized as a three-dimensional box-shaped table, then such a family is obtained as the parallel "slices" in some direction. More precisely, we obtain from a triadic context $\mathbb{K} = (G, M, B, Y)$, for every condition $b \in B$, a formal context $\mathbb{K}_b := (G, M, I_b)$ by defining

$$(g, m) \in I_b :\iff (g, m, b) \in Y.$$

Actually, when writing down a triadic context, it is convenient to write this family of dyadic contexts instead. We do so in Figure 6.27. Unfortunately, we have no experience with use cases of reasonable size yet. Our example therefore

[13] For better distinction, will shall refer to the usual formal contexts as **dyadic** contexts.

is a toy example, but at least based on real-world data, copied from the tram schedule at a tram stop near the home of one of the authors. Four tram lines stop there, all coming from the city center, with line numbers 1, 6, 7, and 8. These are the objects of our triadic context. The attributes say at which time of the day these trams arrive. The times have been aggregated into four slots, "very early" (between 3 and 6 a.m.), "working hours" (between 7 a.m. and 7 p.m., including Sundays), "evening" (between 8 p.m. and 11 p.m.), and "late night" (after 11 p.m. until 2 a.m. of the next day). A cross indicates that at least one tram arrives in the respective time slot. The conditions are "Monday–Friday", "Saturday", and "Sunday".

Mo - Fr	very early	working hours	evening	late night
1		×	×	×
6	×	×		
7	×	×	×	×
8	×		×	×

Sat	very early	working hours	evening	late night
1	×			×
6				
7	×	×		×
8	×	×	×	×

Sun	very early	working hours	evening	late night
1			×	
6				
7	×	×		×
8		×	×	

Figure 6.27: Four tram lines stopping at "Kirche"

There are other ways of slicing the triadic context. Figure 6.28 displays the dyadic context of conditions and attributes for tram line 1 only.

Tram line 1	very early	working hours	evening	late night
Monday-Friday		×	×	×
Saturday	×			×
Sunday			×	

Figure 6.28: Conditions vs. attributes for tram line no. 1

In this example, there is nothing to explore. The data is complete. But one can imagine similar situations where an attribute exploration may be useful. Note that the three formal contexts in Figure 6.27 are completely independent of each other and therefore have implication systems that are independent as well. Since these implications live on the same set of attributes, we need to indicate the conditions under which an implication holds, and this is done by condition labels on top of the implication arrows. The following are examples of **conditional attribute implications** which hold in Figure 6.27:

$$\begin{aligned} \text{late night} &\xrightarrow{\text{Mo-Fr}} \text{evening}, \\ \text{working hours} &\xrightarrow{\text{Sat}} \text{very early}, \\ \text{very early} &\xrightarrow{\text{Sat, Sun}} \text{late night}. \end{aligned}$$

In principle, the implication systems for the different conditions could be explored independently, but two aspects suggest a slightly different strategy:

1. In order to stay within the triadic setting, counterexamples to queries should not be given in form of an object with its attributes, but rather, like in Figure 6.28, as an object together with a formal context showing its attributes according to all conditions.

2. Some implications, like the third one above, hold under several conditions, and the domain expert may prefer to confirm them after a single query.

This second aspect can be resolved by organizing the queries along the *lattice of conditional attribute implications*, which is the concept lattice of the formal context

$$(\text{Imp}(M), B, I),$$

where $\text{Imp}(M)$ is the set of all implications[14] over M, B the set of conditions and

$$(X \to Y)\ I\ b :\Longleftrightarrow X \xrightarrow{b} Y \text{ holds.}$$

Figure 6.29 shows this lattice for our example, however with drastically simplified labels: the implications holding under *no* condition are not listed at the top node; and, at any other node, the relative canonical basis with respect to the set of all implications that are listed below is given.

The suggested query strategy is to explore these implication systems node by node, beginning at the bottom element and then going upwards in the sequence, in which the lattice of conditional attribute implications would be generated by the NEXT EXTENT algorithm from Section 2.2 (p. 44).

[14] It suffices to use implications $A \to \{m\}$ with singleton conclusions

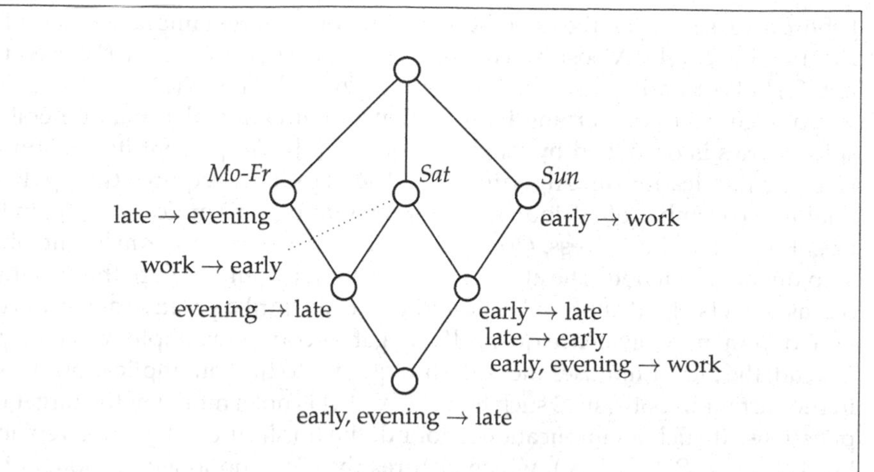

Figure 6.29: The lattice of conditional attribute implications with simplified labeling

6.4.4 Exploration with different queries

With attribute exploration, we learn the implication theory of a domain by posing queries about the validity of implications. This can be viewed as an instance of a general framework known in computational learning theory as **learning with queries** [7]. In this framework, rather than learning from a training dataset, the learning algorithm has access to a teacher, or an oracle, or – in our terms – a domain expert, which it can address with certain predefined types of queries. Probably, the most typical are equivalence and membership queries.

In a **membership query**, the learner asks whether a certain instance is an example of the notion being learnt. In fact, the standard term here is "concept" rather than "notion", but we will use the latter to avoid confusion with formal concepts. For the problem of learning implications, the membership query allows the learning algorithm to find out whether a particular attribute set satisfies the target implications, i.e., whether it is a member of their set of models.

An **equivalence query** is parameterized with a hypothesis describing the notion being learnt. If the hypothesis matches the notion, the answer is positive and learning may be terminated. Otherwise, the teacher must provide a counterexample covered by the hypothesis, but not by the target notion (*negative counterexample*) or vice versa (*positive counterexample*). In our case, the target notion and the hypothesis are implication sets $\mathcal{L}_*$ and $\mathcal{L}$, respectively. We need not guess $\mathcal{L}_*$ precisely: the learning is complete if $\mathcal{L}$ is equivalent to $\mathcal{L}_*$. If not, a counterexample is returned. A negative counterexample is an attribute set A satisfying the hypothesis $\mathcal{L}$, but not the target implications $\mathcal{L}_*$; i.e., $\mathcal{L}(A) = A \neq \mathcal{L}_*(A)$. A positive counterexample is an attribute set B closed under $\mathcal{L}_*$, but contradicting some implications from $\mathcal{L}$; thus, $\mathcal{L}(B) \neq B = \mathcal{L}_*(B)$.

An algorithm for learning Horn sentences with equivalence and membership queries is described by Angluin *et al.* in [9]. We present its version for learning implications in Algorithm 33. The algorithm requires time polynomial in the number of attributes, n, and the number of implications, m, in the target set $\mathcal{L}_*$; in the process, $O(mn)$ equivalence queries and $O(m^2n)$ membership queries are made. The algorithm starts with the empty hypothesis, which has as models all attribute subsets, and proceeds until a positive answer is obtained from an equivalence query. If a negative counterexample X is received instead, the algorithm uses membership queries to find an implication $A \to B$ in the current hypothesis $\mathcal{L}$ such that $A \cap X \neq A$ is not a model of the target implications. If such an implication is found, the implication $A \to B$ is replaced by $A \cap X \to B \cup (A \setminus X)$, which ensures that X is no longer a model of $\mathcal{L}$. Otherwise, adding the implication $X \to M$ produces the same effect. When a positive counterexample X is obtained from an equivalence query, every implication $A \to B$ of which X is not a model is replaced by $A \to B \cap X$. We refer the reader to [9] for further details including the proof of correctness and complexity analysis.

Arias and Balcázar show that Algorithm 33 always produces the canonical basis of the target implication set no matter what examples are received from the equivalence queries [12]. This is quite remarkable, since it gives us an output-polynomial algorithm for computing the canonical basis, something that no other algorithm we have seen before achieves. The catch is, of course, the superhuman power of the oracle answering equivalence queries: while attribute exploration asks only for positive counterexamples to implications, which belong to the domain under investigation, Algorithm 33 also needs negative counterexamples, which may be arbitrary attribute combinations.

In fact, even membership queries may present a problem for any realistic domain expert (whether it is a human or a computer program), unless we work with Horn domains, that is, domains where the set of object intents is closed under intersection. In this case, when asked a membership query about an attribute set C, the domain expert will only need to check if an object with intent C exists. However, in the general case, the expert will have to establish whether C is an intersection of object intents, which is a potentially nontrivial task.

The reason is that, as discussed in Section 4.1.3, the set of models of $\mathcal{L}_*$ can be much larger than the set of attribute combinations feasible in the domain, and the oracle of Algorithm 33 must be able to answer membership and equivalence queries with respect to this larger set of models. In contrast, Algorithm 19 ATTRIBUTE EXPLORATION expects the expert to answer queries with respect to the domain.

Algorithm 33 ATTRIBUTE EXPLORATION W. equivalence and membership queries

Input: An attribute set M over which a target set $\mathcal{L}^*$ of implications is defined.
Interactive input: {Equivalence oracle} Upon request, confirm that an implication set $\mathcal{L}$ over M is equivalent to $\mathcal{L}^*$ or give a counterexample $X \subseteq M$ such that $\mathcal{L}(X) = X \neq \mathcal{L}^*(X)$ or $\mathcal{L}(X) \neq X = \mathcal{L}^*(X)$.
{Membership oracle} Upon request, check if $X = \mathcal{L}^*(X)$ for $X \subseteq M$.
Output: The canonical basis of $\mathcal{L}_*$.

$\mathcal{L} := \varnothing$
while there is a counterexample X to $\mathcal{L}$ **do** {Equivalence oracle}
  if $\mathcal{L}(X) = X$ **then** {negative counterexample}
    $found :=$ **false**
    for all $A \to B \in \mathcal{L}$ **do**
      $C := A \cap X$
      if $A \neq C$ **and** $C \neq \mathcal{L}^*(X)$ **then** {Membership oracle}
        $\mathcal{L} := \mathcal{L} \setminus \{A \to B\}$
        $\mathcal{L} := \mathcal{L} \cup \{C \to B\}$
        $found :=$ **true**
        exit for
    if not $found$ **then**
      $\mathcal{L} := \mathcal{L} \cup \{X \to M\}$
  else {positive counterexample}
    for all $A \to B \in \mathcal{L}$ such that $A \subseteq X$ and $B \nsubseteq X$ **do**
      $\mathcal{L} := \mathcal{L} \setminus \{A \to B\}$
      $\mathcal{L} := \mathcal{L} \cup \{A \to B \cap X\}$

This is, of course, not at all to criticize Algorithm 33, but to explain why this polynomial-time algorithm is not ideally suited for our purposes and why we often resort to exponential-time procedures instead. Algorithm 33 is developed for Horn domains, but our setting is different: we want to learn the implication theory of an arbitrary domain or, to use the terminology from the artificial intelligence community, compute the **Horn approximation** of the domain [131].

It is interesting to see how Algorithm 19, the basic version of attribute exploration, fits the framework of learning with queries. Angluin [7] defines several common types of queries in addition to membership and equivalence queries. The query to check the validity of an implication $A \to B$ and to provide a positive counterexample if $A \to B$ is invalid is a special case of a **superset query**, which asks whether the set of models of the hypothesis includes all the instances of the target notion. Here, the hypothesis is $A \to B$, and we ask if its models include all object intents of the context we explore.

Algorithm 19 takes exponential time in the worst case, because it enumerates all the models of $\mathcal{L}_*$ and we know that the set of models of $\mathcal{L}_*$ can be exponentially larger than $\mathcal{L}_*$ itself, an extreme case being when $\mathcal{L}_*$ is empty and has 2^n models. However, the number of queries asked is linear in the combined size of the domain and its canonical basis, since each query results in a previously unseen object from the domain or in a new implication from the canonical basis of $\mathcal{L}_*$. This may seem a trivial bound, but it is not automatically achieved by algorithms requiring negative counterexamples from outside of the domain.

Arias *et al.* survey other types of queries proposed in the literature for learning Horn formulas; some of them are reducible to others [13].

6.4.5 Other contributions

- **Meschke's double exploration:** *"Concept Approximations"* is the main title of C. Meschke's Ph.D. dissertation [163], in which he thoroughly investigates approximative notions for concept lattices. His setting is as follows: he assumes that there is a **big world context** $\mathbb{K} := (G, M, I)$, which is too large to be completely studied. Instead one sees a small (in particular finite) subcontext $\mathbb{S} := (H, N, I_{H,N} := I \cap H \times N)$, the **scope**. The

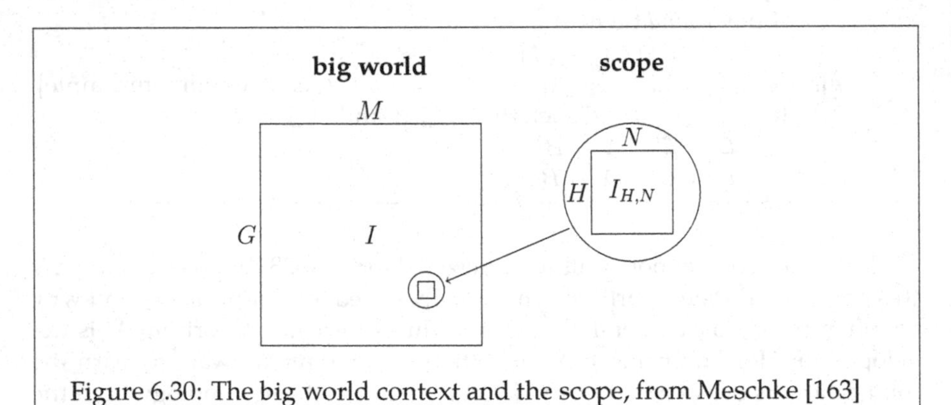

Figure 6.30: The big world context and the scope, from Meschke [163]

  concepts of the big world context can be studied through the traces which they leave on the scope in the following way: if (A, B) is a big world concept, then $(A \cap H, B \cap N)$ is the corresponding **conceptual trace**. However, these traces are usually not formal concepts of the scope; instead, they are only *preconcepts*, i.e., rectangular parts of the incidence relation, but not necessarily maximal ones.

  Meschke shows that the conceptual traces in a very natural manner generate a complete lattice $\mathfrak{S}(\mathbb{S}, \mathbb{K})$. He also shows how a formal context for

$\mathfrak{S}(\mathbb{S}, \mathbb{K})$ can be constructed: by a double exploration. He performs an attribute exploration on the scope (in the big world context) and obtains as the result a finite subcontext $(X, N, I \cap X \times N)$ of $\mathbb{K}$. An object exploration of the scope leads to a finite subcontext $(H, Y, I \cap H \times Y)$. The combined formal context $(X, Y, I \cap X \times Y)$ suffices for the determination of $\mathfrak{S}(\mathbb{S}, \mathbb{K})$, unless $\underline{\mathfrak{B}}(G, M, I)$ has nontrivial tolerances.

- **Reppe's attribute exploration with proper premises:** H. Reppe has investigated the use of proper premises for attribute exploration in [181] and his Ph.D. thesis [182]. His motivation was to find a basis for the implications with small premises, but his results are of interest for other questions as well.

- **Jäschke and Rudolph** [128] discuss the possibility of using web resources for a domain expert. They find "that there is a potentially wide range of options to employ the web for attribute exploration: querying search engines, asking questions on social question answering or crowdsourcing platforms, or retrieving counterexamples from the linked open data cloud." A prototype of their JAVA implementation is freely available.

- **Wollbold** in his Ph.D. thesis [227] uses attribute exploration with temporal logic for the study of gene regulatory networks. In [226] he studies temporal transitions. These are some of the few examples of extending the method to *processes*, using attribute exploration on temporal data. Another early example is due to S. Rudolph [106], who develops Formal Concept Analysis methods for Dynamic Conceptual Graphs. Temporal exploration would be very promising. However, these contributions show that the technical difficulties are severe and that more research is needed before an easily applicable version is achieved.

- **Wollbold, Köhling and Borchmann** [228] apply attribute exploration to the free radical theory of ageing. Their approach is particularly interesting, because they combine attribute exploration with proper premises, partially given counterexamples, and "ripple-down rules". Such rules differ from implications, which are "if—then" expressions, since they allow for "if—then—except if" statements.

6.5 Further reading

Rule exploration was the topic of M. Zickwolff's doctoral dissertation [230].

Applications of Formal Concept Analysis to Description Logics are discussed by Baader and Sertkaya [22].

Even for lightweight Description Logics, it is not obvious that finite models have finite implication bases. Positive results can be found in the work of Baader and Distel [19, 20], but see also [75].

R. Wille's project to combine Formal Concept Analysis with Conceptual Graphs as parts of a "Contextual Logic" ([221], but see also [68]) has recently been revived by S. Ferré [86], who proposes to extend FCA to G-FCA by introducing *graph contexts*. A graph context (G, M, I) has objects and attributes as usual for a formal context, but the incidence relation $I \subseteq G^* \times M$ connects object *tuples* with attributes.

Jäschke *et al.* propose an algorithm, called TRIAS, for building frequent triadic concepts from triadic data [127]. Cerf *et al.* go further and describe an algorithm DATA-PEELER that mines n-ary relations for polyadic concepts satisfying constraints from a certain class [62].

Learning with queries [7] has been applied to learning, for example, deterministic finite automata [6], description logic theories [91, 136], restricted first-order Horn theories [134, 14], preferences [137], and the structure of social networks [8]. There are variations of the learning model, in particular, those that account for possible errors in oracle answers [11, 10]. Kautz *et al.* have transformed the algorithm for learning propositional Horn theories from [9] into an output-polynomial randomized algorithm (with controllable error) for computing essentially the implication basis of a formal context [131].

Bibliography

Numbers at the end of a citation refer back to the citing page.

[1] Agrawal, R., Imielinski, T., Swami, A.: Mining association rules between sets of items in large databases. In: Proceedings SIGMOD Conf., pp. 207–216 (1993) 56, 82, 115

[2] Agrawal, R., Srikant, R.: Fast algorithms for mining association rules in large databases. In: J.B. Bocca, M. Jarke, C. Zaniolo (eds.) Proceedings VLDB'94, pp. 487–499. Morgan Kaufmann (1994) 62, 82

[3] Albano, A., Chornomaz, B.: Why concept lattices are large. In: [40], pp. 73–86 (2015) 49

[4] Alekseev, V.B.: The number of families of subsets that are closed with respect to intersections. Diskretnaya Matematika **1**(2), 129–136 (1989) 82

[5] Alonso-Jiménez, J.A., Aranda-Corral, G.A., Borrego-Dıaz, J., Fernández-Lebrón, M.M., Hidalgo-Doblado, M.J.: Extending attribute exploration by means of boolean derivatives. In: R. Bělohlávek, S.O. Kuznetsov (eds.) Concept Lattices and Their Applications. Proceedings CLA 2008, *CEUR WS*, vol. 433, pp. 121–132 (2008) 207

[6] Angluin, D.: Learning regular sets from queries and counterexamples. Information and Computation **75**(2), 87 – 106 (1987) 292

[7] Angluin, D.: Queries and concept learning. Machine Learning **2**, 319–342 (1988) 287, 289, 292

[8] Angluin, D., Aspnes, J., Reyzin, L.: Optimally learning social networks with activations and suppressions. Theoretical Computer Science **411**(29–30), 2729 – 2740 (2010) 292

[9] Angluin, D., Frazier, M., Pitt, L.: Learning conjunctions of Horn clauses. Machine Learning **9**, 147–164 (1992) 288, 292

[10] Angluin, D., Kriķis, M., Sloan, R.H., Turán, G.: Malicious omissions and errors in answers to membership queries. Machine Learning **28**(2), 211–255 (1997) 292

[11] Angluin, D., Slonim, D.K.: Randomly fallible teachers: Learning monotone DNF with an incomplete membership oracle. Machine Learning **14**(1), 7–26 (1994) 292

[12] Arias, M., Balcázar, J.L.: Construction and learnability of canonical Horn formulas. Machine Learning **85**(3), 273–297 (2011) 288

[13] Arias, M., Balcázar, J.L., Tîrnăucă, C.: Learning definite Horn formulas from closure queries. Theoretical Computer Science (to appear) 290

[14] Arias, M., Khardon, R.: Learning closed Horn expressions. Information and Computation **178**(1), 214 – 240 (2002) 292

[15] Armstrong, W.W.: Dependency structures of data base relationships. Proc. IFIP Congress pp. 580–583 (1974) 91

[16] Arnauld, A., Nicole, P.: La logique ou l'art de penser – contenant, outre les règles communes, plusieurs observations nouvelles, propres à former le jugement. Ch. Saveux (1668) 10

[17] Baader, F.: Computing a minimal representation of the subsumption lattice of all conjunctions of concepts defined in a terminology. In: Proc. Intl. KRUSE Symposium, August, pp. 11–13 (1995) 254

[18] Baader, F., Calvanese, D., McGuinness, D., Nardi, D., Patel-Schneider, P. (eds.): The Description Logic Handbook: Theory, Implementation, and Applications, second edn. Cambridge University Press (2007) 253, 258, 259

[19] Baader, F., Distel, F.: A finite basis for the set of $\mathcal{EL}$-implications holding in a finite model. In: [161], pp. 46–61 (2008) 292

[20] Baader, F., Distel, F.: Exploring finite models in the description logic $\mathcal{EL}_{\text{gfp}}$. In: [88], pp. 146–161 (2009) 292

[21] Baader, F., Ganter, B., Sertkaya, B., Sattler, U.: Completing description logic knowledge bases using formal concept analysis. In: M.M. Veloso (ed.) Proceedings IJCAI'07, pp. 230–235. AAAI Press (2007) 170, 261

[22] Baader, F., Sertkaya, B.: Applying formal concept analysis to description logics. In: [80], pp. 261–286 (2004) 291

[23] Babin, M.A., Kuznetsov, S.O.: Recognizing pseudo-intents is co$\mathcal{NP}$-complete. In: M. Kryszkiewicz, S. Obiedkov (eds.) Concept Lattices and Their Applications. Proceedings CLA 2010, *CEUR WS*, vol. 672, pp. 294–301. University of Sevilla, Spain (2010) 99

[24] Babin, M.A., Kuznetsov, S.O.: Computing premises of a minimal cover of functional dependencies is intractable. Discrete Applied Mathematics **161(6)**, 742–749 (2013) 99

[25] Baixeries, J., Kaytoue, M., Napoli, A.: Characterizing functional dependencies in formal concept analysis with pattern structures. Annals of mathematics and artificial intelligence **72**(1-2), 129–149 (2014) 234

[26] Baixeries, J., Sacarea, C., Ojeda-Aciego, M. (eds.): Formal Concept Analysis: Proceedings ICFCA 2015, *LNCS*, vol. 9113. Springer (2015) 297, 299

[27] Bartel, H.: Formale Begriffsanalyse und Materialkunde: Zur Archäometrie alt-ägyptischer glasartiger Produkte. MATCH Commun. Math. Comput. Chem **32**, 27–46 (1995) 234

[28] Bartel, H.G.: Mathematische Methoden in der Chemie. Spektrum Akad. Verlag (1996) 234, 235

[29] Bartel, H.G.: Über Möglichkeiten der Formalen Begriffsanalyse in der Mathematischen Archäochemie. In: [207] pp. 368–389 (2000) 218, 220

[30] Bartel, H.G., Nofz, M.: Exploration of NMR data of glasses by means of formal concept analysis. Chemometrics and Intelligent Laboratory Systems **36**(1), 53–63 (1997) 234

[31] Bastide, Y., Taouil, R., Pasquier, N., Stumme, G., Lakhal, L.: Mining frequent patterns with counting inference. SIGKDD Explorations, Special Issue on Scalable Algorithms **2**(2), 71–80 (2000) 116

[32] Bazhanov, K., Obiedkov, S.: Optimizations in computing the Duquenne–Guigues basis of implications. Annals of Mathematics and Artificial Intelligence **70**(1), 5–24 (2014) 121

[33] Becker, P.: ToscanaJ: Conceptual scaling of many-valued contexts. In: Engelbert Mephu Nguifo et al. (ed.) Supplementary Proc. of the 3rd Int. Conference on Formal Concept Analysis. Univ. of Artois (2005) 30

[34] Becker, P., Hereth, J.: The ToscanaJ suite for implementing conceptual information systems. In [107], pp. 324–348 (2005) 30

[35] Beeri, C., Bernstein, P.: Computational problems related to the design of normal form relational schemas. ACM TODS **4**(1), 30–59 (1979) 93

[36] Bell, D., LaPadula, L.: Secure computer systems: Mathematical foundations and model. Report M74-244, Mitre Corporation, Bedford, Mass. (1973) 147

[37] Bělohlávek, R.: Fuzzy relational systems: foundations and principles, vol. 20. Springer Science & Business Media (2012) 283

[38] Bělohlávek, R., Vychodil, V.: Attribute implications in a fuzzy setting. In: [164], pp. 45–60 (2006) 283

[39] Bělohlávek, R., Vychodil, V.: Fuzzy attribute logic over complete residuated lattices. Journal of Experimental & Theoretical Artificial Intelligence **18**(4), 471–480 (2006) 283

[40] Ben Yahia, S., Konecny, J. (eds.): Concept Lattices and Their Applications. Proceedings CLA 2015, *CEUR WS*, vol. 1466 (2015) 293, 302

[41] Bertet, K., Monjardet, B.: The multiple facets of the canonical direct unit implicational basis. Theoretical Computer Science **411**(22), 2155–2166 (2010) 107

[42] Bertet, K., Nebut, M.: Efficient algorithms on the Moore family associated to an implicational system. Discrete Mathematics and Theoretical Computer Science **6**, 315–338 (2004) 107

[43] Biba, K.: Integrity considerations for secure computer systems. Report TR-3153, Mitre Corporation, Bedford, Mass. (1977) 147, 155

[44] Biedermann, K.: How triadic diagrams represent conceptual structures. In: [156], pp. 304–317 (1997) 284

[45] Biedermann, K.: A foundation of the theory of trilattices. Ph.D. thesis, TU Darmstadt (1998) 284

[46] Birkhoff, G.: Lattice theory (first edition), *Colloquium Publications*, vol. 25. American Mathematical Soc. (1940) 30

[47] Birkhoff, G.: Lattice theory (third edition), *Colloquium Publications*, vol. 25. American Mathematical Society (1967) 13

[48] Blockeel, H.: Data mining: From procedural to declarative approaches. New Generation Comput. **33**(2), 115–135 (2015) 115

[49] Borchmann, D.: Context orbifolds. Diploma thesis, TU-Dresden (2009) 169

[50] Borchmann, D.: Learning Terminological Knowledge with High Confidence from Erroneous Data. Ph.D. thesis, Technische Universität Dresden (2011) 261, 281

[51] Borchmann, D.: Experience based nonmonotonic reasoning. In: P. Cabalar, T.C. Son (eds.) Logic Programming and Nonmonotonic Reasoning: 12th International Conference, LPNMR 2013, Corunna, Spain, September 15-19, 2013. Proceedings, pp. 200–205. Springer Berlin Heidelberg, Berlin, Heidelberg (2013) 121

[52] Borchmann, D.: A general form of attribute exploration. Tech. Rep. LTCS-Report 13-02, Technische Universität Dresden (2013) 184

[53] Borchmann, D.: Exploring faulty data. In: [26], pp. 219–235 (2015) 281

[54] Borchmann, D., Ganter, B.: Concept lattice orbifolds – first steps. In: [88], pp. 22–37 (2009) 169

[55] Bordat, J.P.: Calcul pratique du treillis de Galois d'une correspondance. Mathématiques et sciences humaines **96**, 31–47 (1986) 81

[56] Burmeister, P.: Merkmalimplikationen bei unvollständigem Wissen. In: W. Lex (ed.) Proceedings: Arbeitstagung Begriffsanalyse und Künstliche Intelligenz, October 6-8, 1988, pp. 15–46. Technische Universitat Clausthal, Clausthal-Zellerfeld (1991) 169

[57] Burmeister, P.: ConImp: Ein Programm zur Formalen Begriffsanalyse. In [207], pp. 25–56 (2000). English version `http://www.mathematik.tu-darmstadt.de/~burmeister/ConImpIntro.pdf` 30

[58] Burmeister, P., Holzer, R.: Treating incomplete knowledge in formal concept analysis. In [107], pp. 317–331 (2005) 169

[59] Burosch, G., Demetrovics, J., Katona, G.O.H., Kleitman, D.J., Sapozhenko, A.A.: On the number of databases and closure operations. Theoretical Computer Science **78**(2), 377–381 (1991) 82

[60] Burusco Juandeaburre, A., Fuentes-González, R.: The study of the L-fuzzy concept lattice. Mathware & soft computing. 1994 Vol. 1 Núm. 3 p. 209-218 (1994) 283

[61] Carpineto, C., Romano, G.: Concept Data Analysis. Wiley (2004) 31

[62] Cerf, L., Besson, J., Robardet, C., Boulicaut, J.F.: Closed patterns meet n-ary relations. ACM Trans. Knowl. Discov. Data **3**(1), 3:1–3:36 (2009) 292

[63] Cole, R.J., Ducrou, J., Eklund, P.W.: Automated layout of small lattices using layer diagrams. In: [164] (2006) 31

[64] Colomb, P., Irlande, A., Raynaud, O.: Counting of Moore families for n=7. In [149], pp. 72–87 (2010) 43, 82

[65] Colomb, P., Irlande, A., Raynaud, O., Renaud, Y.: About the recursive decomposition of the lattice of Moore co-families. presented at ICFCA 2011, Nicosia, Cyprus (2011) 82

[66] Colonius, H., Schulze, H.H.: Tree structures for proximity data. British Journal of Mathematical and Statistical Psychology **34**(2), 167–180 (1981) 253

[67] Conway, J.H., Burgiel, H., Goodman-Strauss, C.: The symmetries of things. CRC Press (2008) xvii

[68] Dau, F., Klinger, J.: From formal concept analysis to contextual logic. In: [107], pp. 81–100 (2005) 292

[69] Dau, F., Knechtel, M.: Access policy design supported by FCA methods. In: [189], pp. 141–154 (2009) 184

[70] Davey, B.A., Priestley, H.A.: Introduction to lattices and order. Cambridge University Press (2002). Second edition. 13

[71] Day, A.: The lattice theory of functional dependencies and normal decompositions. Int. J. Algebra Comput. **2**, 409–431 (1992) 107

[72] Delugach, H.S., Stumme, G. (eds.): Conceptual Structures: Broadening the Base. 9th International Conference on Conceptual Structures, ICCS, *LNCS*, vol. 2120. Springer (2001) 300, 301, 309

[73] Denning, D.: A lattice model of secure information flow. Comm. ACM **19**(5), 236–243 (1976) 147

[74] Distel, F.: Hardness of enumerating pseudo-intents in the lectic order. In [149] pp. 124–137 (2010) 99

[75] Distel, F.: Learning Description Logic Knowledge Bases from Data Using Methods from Formal Concept Analysis. Ph.D. thesis, Technische Universität Dresden (2011) 258, 260, 261, 263, 292

[76] Distel, F., Sertkaya, B.: On the complexity of enumerating pseudo-intents. Discrete Applied Mathematics **159**(6), 450–466 (2011) 121

[77] Duquenne, V.: GLAD (General Lattice Analysis & Design): a Fortran program for a glad user. MSH-Maison Suger, Paris (1992) 30

[78] Duquenne, V.: Latticial structures in data analysis. Theoretical Computer Science **217**(2), 407–436 (1999) 10

[79] Eiter, T., Gottlob, G.: Hypergraph transversal computation and related problems in logic and AI. In: S. Flesca, S. Greco, G. Ianni, N. Leone (eds.) Logics in Artificial Intelligence, *Lecture Notes in Computer Science*, vol. 2424, pp. 549–564. Springer Berlin Heidelberg (2002) 111

[80] Eklund, P. (ed.): Concept Lattices. Proceedings of the 2nd Int. Conference on Formal Concept Analysis, ICFCA, *LNAI*, vol. 2691. Springer (2004) 294, 299, 305

[81] Eklund, P.W., Ducrou, J., Brawn, P.: Information visualization using concept lattices: Can novices read line diagrams? In: [80], pp. 57–72 (2004) 31

[82] Ellson, J., Gansner, E., Koutsofios, L., North, S., Woodhull, G.: Graphviz – open source graph drawing tools. In: P. Mutzel, M. Jünger, S. Leipert (eds.) Graph Drawing, *Lecture Notes in Computer Science*, vol. 2265, pp. 594–597. Springer Berlin / Heidelberg (2002) 30

[83] Eschenfelder, D., Kollewe, W., Skorsky, M., Wille, R.: Ein Erkundungssystem zum Baurecht: Methoden der Entwicklung eines TOSCANA-Systems. In [207], pp. 254–272 (2000) 2

[84] Fajtlowicz, S.: On conjectures of Graffiti. Discrete mathematics **72**(1-3), 113–118 (1988) 184

[85] Ferré, S.: Incremental concept formation made more efficient by the use of associative concepts. Tech. rep., Institut National de Recherche en Informatique et en Automatique (INRIA) (2002). URL `http://www.inria.fr/RRRT/RR-4569.html` 81

[86] Ferré, S.: A proposal for extending formal concept analysis to knowledge graphs. In: [26], pp. 271–286 (2015) 292

[87] Ferré, S., Ridoux, O.: A logical generalization of formal concept analysis. In: G. Mineau, B. Ganter (eds.) Int. Conf. Conceptual Structures, LNCS 1867, pp. 371–384. Springer (2000) 234

[88] Ferré, S., Rudolph, S. (eds.): Formal Concept Analysis. Proceedings ICFCA 2009, *LNCS*, vol. 5548. Springer (2009) 294, 297, 308

[89] Ferré, S., Ridoux, O.: A file system based on concept analysis. In: Y. Sagiv (ed.) International Conference on Rules and Objects in Databases, no. 1861 in LNCS, pp. 1033–1047. Springer (2000) 156

[90] Fitz, S.: Die Farbglasuren spätbabylonischer Wandverkleidungen. Berichte der Deutschen Keram. Gesellschaft **59**(3), 179–185 (1982) 219

[91] Frazier, M., Pitt, L.: Classic learning. Machine Learning **25**(2), 151–193 (1996) 292

[92] Ganter, B.: Two basic algorithms in concept analysis. FB4–Preprint 831, TH Darmstadt (1984). Republished in [149] 82, 102, 184

[93] Ganter, B.: Algorithmen zur Formalen Begriffsanalyse. In [111] pp. 241–254 (1987) 184

[94] Ganter, B.: Finding closed sets under symmetry. FB4–Preprint 1307, TH Darmstadt (1990) 82

[95] Ganter, B.: Attribute exploration with background knowledge. Theoretical Computer Science **217**(2), 215–233 (1999) 169

[96] Ganter, B.: Begriffe und Implikationen. In [207] pp. 1–24 (2000) 234

[97] Ganter, B.: Conflict avoidance in additive order diagrams. Journal of Universal Computer Science **10**(8), 955–966 (2004) 31

[98] Ganter, B.: Qualitative Methoden beim Sprachvergleich (2004). Lecture notes 249, 250

[99] Ganter, B.: Relational Galois connections. In: [147], pp. 1–17 (2007) 157

[100] Ganter, B., Godin, R. (eds.): Formal Concept Analysis: Proceedings ICFCA 2005, *Lecture Notes in Computer Science*, vol. 3403. Springer Berlin Heidelberg (2005) 305, 306

[101] Ganter, B., Krauße, R.: Pseudo-models and propositional Horn inference. Discrete Applied Mathematics **147**(1), 43–55 (2005) 206

[102] Ganter, B., Kuznetsov, S.O.: Pattern structures and their projections. In: [72], pp. 129–142 (2001) 234

[103] Ganter, B., Kuznetsov, S.O.: Hypotheses and version spaces. In: A. de Moor, W. Lex, B. Ganter (eds.) Conceptual Structures for Knowledge Creation and Communication. Proceedings ICCS 2003, *LNCS*, vol. 2746, pp. 83–95. Springer (2003) 82

[104] Ganter, B., Obiedkov, S.: Implications in triadic contexts. In: K.E. Wolff, H.D. Pfeiffer, H.S. Delugach (eds.) Conceptual Structures at Work. Proceedings ICCS 2004, *LNCS*, vol. 3127, pp. 186–195. Springer (2004) 284

[105] Ganter, B., Reuter, K.: Finding all closed sets : A general approach. Order **8**, 283–290 (1991) 82

[106] Ganter, B., Rudolph, S.: Formal concept analysis methods for dynamic conceptual graphs. In: [72], pp. 143–156 (2001) 291

[107] Ganter, B., Stumme, G., Wille, R. (eds.): Formal Concept Analysis, Foundations and Applications, *LNAI*, vol. 3626. Springer (2005) 31, 295, 297, 298, 310

[108] Ganter, B., Wille, R.: Implikationen und Abhängigkeiten zwischen Merkmalen. In: P.O. Degens, H.J. Hermes, O. Opitz (eds.) Die Klassifikation und ihr Umfeld, pp. 171–185. Indeks–Verlag, Frankfurt (1986) 107, 111

[109] Ganter, B., Wille, R.: Contextual attribute logic. In: W.M. Tepfenhart, W.R. Cyre (eds.) Conceptual Structures: Standards and Practices. Proceedings ICCS 1999, *LNCS*, vol. 1640, pp. 377–388. Springer (1999) 192

[110] Ganter, B., Wille, R.: Formal Concept Analysis: Mathematical Foundations. Springer, Berlin/Heidelberg (1999) 3, 10, 30, 31, 76, 87, 93, 107, 111, 168, 175, 220, 281

[111] Ganter, B., Wille, R., Wolff, K.E. (eds.): Beiträge zur Begriffsanalyse. B.I.–Wissenschaftsverlag, Mannheim (1987) 31, 300, 309

[112] Glatz, V.: Quantorenelimination, ω-Kategorizität, Ultrahomogenität, Modellvollständigkeit – Eine Merkmalexploration –. Master's thesis, TU Dresden (2014) 175

[113] Glodeanu, C.V.: Attribute exploration with fuzzy attributes and background knowledge. In: M. Ojeda-Aciego, J. Outrata (eds.) Concept Lattices and Their Applications. Proceedings CLA 2013, *CEUR WS*, vol. 1062, pp. 69–80 (2013) 284

[114] Godin, R., Missaoui, R.: An incremental concept formation approach for learning from databases. Theoretical Computer Science **133**(2), 387 – 419 (1994) 121

[115] Godin, R., Missaoui, R., Alaoui, H.: Incremental concept formation algorithms based on Galois (concept) lattices. Computational Intelligence **11**(2), 246–267 (1995) 80, 81

[116] Goethals, B.: Frequent itemset mining implementations repository. `http://fimi.ua.ac.be/` 82

[117] Gollmann, D.: Computer Security. John Wiley & Sons Ltd, Chichester (1999) 151

[118] Grätzer, G.: General Lattice Theory. Birkhäuser Verlag (2003) 13, 278

[119] Guénoche, A.: Construction du treillis de Galois d'une relation binaire. Mathématiques et Sciences Humaines **109**, 41–53 (1990) 80

[120] Guigues, J.L., Duquenne, V.: Famille minimale d'implications informatives résultant d'un tableau de données binaires. Mathématiques et Sciences Humaines **24**(95), 5–18 (1986) 95

[121] Habib, M., Nourine, L.: The number of Moore families on $n = 6$. Discrete Mathematics **294**(3), 291–296 (2005) 43, 52

[122] Hammer, P.L., Kogan, A.: Optimal compression of propositional Horn knowledge bases: complexity and approximation. Artificial Intelligence **64**(1), 131 – 145 (1993) 121

[123] Hereth, J.: Relational scaling and databases. In: [179], pp. 62–76 (2002) 234

[124] Holzer, R.: Knowledge acquisition under incomplete knowledge using methods from formal concept analysis: Parts I and II. Fundam. Inform. **63**(1), 17–39, 41–63 (2004) 169

[125] Horn, A.: On sentences which are true of direct unions of algebras. Journal of Symbolic Logic **16**, 14–21 (1951) 121

[126] Jaenicke, J. (ed.): Evolution, *Materialienhandbuch Kursunterricht Biologie*, vol. 6. Aulis-Verlag Köln (1997) 224

[127] Jäschke, R., Hotho, A., Schmitz, C., Ganter, B., Stumme, G.: Trias – an algorithm for mining iceberg tri-lattices. In: Proceedings of the 6th IEEE International Conference on Data Mining (ICDM 06), pp. 907–911. IEEE Computer Society, Hong Kong (2006) 82, 292

[128] Jäschke, R., Rudolph, S.: Attribute exploration on the web. Preprint, available from `www.qucosa.de` (2013) 291

[129] Johnson, D.S., Yannakakis, M., Papadimitriou, C.H.: On generating all maximal independent sets. Information Processing Letters **27**(3), 119 – 123 (1988) 52

[130] Kauer, M., Krupka, M.: Subset-generated complete sublattices as concept lattices. In: [40], pp. 11–21 (2015) 266

[131] Kautz, H., Kearns, M., Selman, B.: Horn approximations of empirical data. Artificial Intelligence **74**(1), 129 – 145 (1995) 98, 121, 123, 289, 292

[132] Kestler, P.: Strukturelle Untersuchungen eines Varietätenverbandes von Gruppoiden mit unärer Operation und ausgezeichnetem Element. Ph.D. thesis, TU Bergakademie, Freiberg (2013) 174

[133] Khardon, R.: Translating between Horn representations and their characteristic models. J. Artif. Intell. Res. (JAIR) **3**, 349–372 (1995) 121

[134] Khardon, R.: Learning function-free Horn expressions. Machine Learning **37**(3), 241–275 (1999) 292

[135] Klotz, U., Mann, A.: Begriffexploration. Master's thesis, TH Darmstadt (1988) 265

[136] Konev, B., Lutz, C., Ozaki, A., Wolter, F.: Exact learning of lightweight description logic ontologies. In: Principles of Knowledge Representation and Reasoning: Proceedings KR 2014 (2014) 292

[137] Koriche, F., Zanuttini, B.: Learning conditional preference networks. Artificial Intelligence **174**(11), 685 – 703 (2010) 292

[138] Krauße, R.: Kumulierte Klauseln als aussagenlogische Sprachmittel für die Formale Begriffsanalyse. Master's thesis, TU Dresden (1998) 234

[139] Krolak-Schwerdt, S., Orlik, P., Ganter, B.: TRIPAT: a model for analyzing three-mode binary data. In: H.H. Bock, W. Lenski, M.M. Richter (eds.) Studies in Classification, Data Analysis, and Knowledge Organization, *Information systems and data analysis*, vol. 4, pp. 298–307. Springer (1994) 82

[140] Kuznetsov, S.O.: Interpretation on graphs and complexity characteristics of a search for specific patterns. Nauchno-Tekhnicheskaya Informatsiya Seriya 2 **1**, 23–27 (1989) 49

[141] Kuznetsov, S.O.: A fast algorithm for computing all intersections of objects in a finite semilattice. Nauchno-Tekhnicheskaya Informatika, Ser. 2 **1**, 17–22 (1993) 80

[142] Kuznetsov, S.O.: On computing the size of a lattice and related decision problems. Order **18**(4), 313–321 (2001) 49

[143] Kuznetsov, S.O.: On the intractability of computing the Duquenne-Guigues base. Journal of Universal Computer Science **10**(8), 927–933 (2004) 98, 99

[144] Kuznetsov, S.O., Obiedkov, S.: Comparing performance of algorithms for generating concept lattices. J. Exp. Theor. Artif. Intell. **14**(2-3), 189–216 (2002) 80, 81

[145] Kuznetsov, S.O., Obiedkov, S.: Counting pseudo-intents and #P-completeness. In: [164], pp. 306–308 (2006) 99

[146] Kuznetsov, S.O., Obiedkov, S.: Some decision and counting problems of the Duquenne-Guigues basis of implications. Discrete Appl. Math. **156**(11), 1994–2003 (2008) 99

[147] Kuznetsov, S.O., Schmidt, S. (eds.): Formal Concept Analysis. Proceedings ICFCA 2007, *Lecture Notes in Computer Science*, vol. 4390. Springer (2007) 300, 307

[148] Kwuida, L., Pech, C., Reppe, H.: Generalizations of Boolean algebras. An attribute exploration. Mathematica Slovaca **56**(2), 145–165 (2006) 175

[149] Kwuida, L., Sertkaya, B. (eds.): Formal Concept Analysis. Proceedings ICFCA 2010, *Lecture Notes in Computer Science*, vol. 5986. Springer (2010) 298, 300

[150] Langsdorf, R., Skorsky, M., Wille, R., Wolf, A.: An approach to automated drawing of concept lattices. In: K. Denecke, O. Lüders (eds.) General algebra and applications in discrete mathematics, pp. 125–136. Shaker Verlag (1997) 31

[151] Lehmann, F., Wille, R.: A triadic approach to formal concept analysis. In: G. Ellis, R. Levinson, W. Rich, J.F. Sowa (eds.) Conceptual Structures: Applications, Implementation and Theory. Proceedings ICCS 1995, *LNAI*, vol. 954, pp. 32–43. Springer (1995) 284

[152] Li, J., Mei, C., Lv, Y.: Incomplete decision contexts: approximate concept construction, rule acquisition and knowledge reduction. International Journal of Approximate Reasoning **54**(1), 149–165 (2013) 184

[153] Lichman, M.: UCI machine learning repository (2013). URL `http://archive.ics.uci.edu/ml` 114

[154] Lipner, S.: Nondiscretionary controls for commercial applications. In: Proc. IEEE Symp. Security and Privacy, pp. 2–10. IEEE CS Press (1982) 152

[155] Logan, W.B., Ochshorn, S., Place, C.: The Smithonian Guide to Historic America: The Pacific States. Stewart, Tabori & Chang Inc., New York (1989) 254

[156] Lukose, D., Delugach, H.S., Keeler, M., Searle, L., Sowa, J.F. (eds.): Conceptual Structures: Fulfilling Peirce's Dream, no. 1257 in LNAI. Springer, Heidelberg (1997) 296, 306, 308, 310

[157] Luksch, P., Skorsky, M., Wille, R.: On drawing concept lattices with a computer. In: W. Gaul, M. Schader (eds.) Classification as a tool of research, pp. 269–274. North–Holland (1986) 31

[158] Luxenburger, M.: Implications partielles dans un contexte. Mathématiques, Informatique et Sciences Humaines **29**(113), 35–55 (1991) 115, 116

[159] Maier, D.: The theory of relational databases. Computer software engineering series. Computer Science Press (1983) 93, 107, 121

[160] McCune, W.: Prover9 and Mace4 (2005–2010). `http://www.cs.unm.edu/~mccune/prover9/` 175

[161] Medina, R., Obiedkov, S. (eds.): Formal Concept Analysis: 6th International Conference, ICFCA 2008, Montreal, Canada, February 25-28, 2008. Proceedings, *LNCS*, vol. 4933. Springer Berlin Heidelberg (2008) 294, 309

[162] van der Merwe, D., Obiedkov, S., Kourie, D.: AddIntent: A new incremental algorithm for constructing concept lattices. In: [80], pp. 372–385 (2004) 81

[163] Meschke, C.: Concept approximations – approximative notions for concept lattices. Ph.D. thesis, TU Dresden (2011). Available from `www.qucosa.de` 290

[164] Missaoui, R., Schmidt, J. (eds.): Formal Concept Analysis, *Lecture Notes in Computer Science*, vol. 3874. Springer (2006) 296, 298, 304

[165] Moore, E.H.: Introduction to a form of general analysis, vol. 2. Yale University Press (1910) 82

[166] Nehmé, K., Valtchev, P., Rouane, M.H., Godin, R.: On computing the minimal generator family for concept lattices and icebergs. In: [100], pp. 192–207 (2005) 62

[167] Norris, E.M.: Maximal rectangular relations. In: M. Karpinski (ed.) Fundamentals of Computation Theory, *LNCS*, vol. 56, pp. 476–481. Springer (1977) 80

[168] Norris, E.M.: An algorithm for computing the maximal rectangles in a binary relation. Revue Roumaine de Mathématiques Pures et Appliquées **23**(2), 243–250 (1978) 80

[169] Nourine, L., Raynaud, O.: A fast algorithm for building lattices. Information Processing Letters **71**, 199–214 (1999) 81

[170] Obiedkov, S.: Modal logic for evaluating formulas in incomplete contexts. In: [179], pp. 314–325 (2002) 169

[171] Obiedkov, S., Duquenne, V.: Attribute-incremental construction of the canonical implication basis. Annals of Mathematics and Artificial Intelligence **49**(1), 77–99 (2007) 121

[172] Obiedkov, S., Kourie, D.G., Eloff, J.H.P.: On lattices in access control models. In: H. Schärfe, P. Hitzler, P. Øhrstrøm (eds.) Conceptual Structures: Inspiration and Application. Proceedings ICCS 2006, *LNCS*, vol. 4068, pp. 374–387. Springer (2006) 155

[173] Obiedkov, S., Kourie, D.G., Eloff, J.H.P.: Building access control models with attribute exploration. Computers & Security **28**(1-2), 2–7 (2009) 152, 184

[174] Outrata, J., Vychodil, V.: Fast algorithm for computing fixpoints of Galois connections induced by object-attribute relational data. Information Sciences **185**(1), 114 – 127 (2012) 80

[175] Pollandt, S.: Fuzzy–Begriffe. Springer (1997) 283

[176] Prediger, S.: Logical scaling in formal concept analysis. In: [156], pp. 332–341 (1997) 234

[177] Prediger, S.: Terminologische Merkmalslogik in der Formalen Begriffsanalyse. In [207], pp. 99–124 (2000) 258

[178] Prediger, S., Stumme, G.: Theory-driven logical scaling. In: E. Franconi, M. Kifer (eds.) Proc. 6th Intl. Workshop Knowledge Representation Meets Databases (KRDB'99), vol. CEUR Workshop Proc. 21 (1999) 234

[179] Priss, U., Corbett, D., Angelova, G. (eds.): Conceptual Structures: Integration and Interfaces, 10th International Conference on Conceptual Structures, ICCS 2002, *Lecture Notes in Computer Science*, vol. 2393. Springer (2002) 302, 306

[180] Priss, U., Old, L.J.: Conceptual exploration of semantic mirrors. In [100], pp. 21–32 (2005) 2

[181] Reppe, H.: Attribute exploration using implications with proper premises. In: P.W. Eklund, O. Haemmerlé (eds.) Conceptual Structures: Knowledge Visualization and Reasoning. Proceedings ICCS 2008, *LNCS*, vol. 5113, pp. 161–174. Springer (2008) 110, 111, 291

[182] Reppe, H.: Three Generalisations of Lattice Distributivity: An FCA Perspective. Shaker (2011) 291

[183] Revenko, A.: Automatic construction of implicative theories for mathematical domains. Ph.D. thesis, TU Dresden (2015) 175

[184] Revenko, A.V., Kuznetsov, S.O.: Attribute exploration of properties of functions on ordered sets. Fundamenta Informaticae **115**(4), 377–394 (2012) 175

[185] Rival, I., Wille, R.: Lattices freely generated by partially ordered sets: which can be "drawn"? Journal für Reine und Angewandte Mathematik (Crelles Journal) p. 56–80 (1979) 275

[186] Rudolph, S.: Relational exploration: combining description logics and formal concept analysis for knowledge specification. Ph.D. thesis, Technische Universität Dresden (2006) 261

[187] Rudolph, S.: Some notes on pseudo-closed sets. In: S. Kuznetsov, S. Schmidt (eds.) [147], pp. 151–165 (2007) 99

[188] Rudolph, S.: Foundations of description logics. In: Reasoning Web. Semantic Technologies for the Web of Data, pp. 76–136. Springer (2011) 253

[189] Rudolph, S., Dau, F., Kuznetsov, S.O. (eds.): Conceptual Structures: Leveraging Semantic Technologies: 17th International Conference on Conceptual Structures, ICCS 2009. Springer (2009) 298, 308

[190] Ryssel, U., Distel, F., Borchmann, D.: Fast computation of proper premises. In: A. Napoli, V. Vychodil (eds.) Concept Lattices and Their Applications. Proceedings CLA 2011, *CEUR WS*, vol. 959, pp. 101–113. INRIA Nancy–Grand Est and LORIA, France (2011) 107

[191] Ryssel, U., Distel, F., Borchmann, D.: Fast algorithms for implication bases and attribute exploration using proper premises. Annals of Mathematics and Artificial Intelligence **70**(1), 25–53 (2013) 107

[192] Sandhu, R.: Lattice-based access control models. IEEE Computer **26**(11), 9–19 (1993) 147

[193] SAS Institute Inc.: SAS/OR ® 9.2 User's Guide: Project Management. SAS Institute Inc., Cary, NC (2008). URL `http://support.sas.com/documentation/cdl/en/orpmug/59678/PDF/default/orpmug.pdf` 54

[194] Sertkaya, B.: Formal concept analysis methods for description logics. Ph.D. thesis, Technische Universität Dresden (2007) 261

[195] Sertkaya, B.: Ontocomp: A Protégé plugin for completing OWL ontologies. In: L. Aroyo, P. Traverso, F. Ciravegna, P. Cimiano, T. Heath,

E. Hyvönen, R. Mizoguchi, E. Oren, M. Sabou, E. Simperl (eds.) Proceedings of the 6th European Semantic Web Conference, (ESWC 2009), *LNCS*, vol. 5554, pp. 898–902. Springer Berlin Heidelberg (2009) 265

[196] Sertkaya, B.: Some computational problems related to pseudo-intents. In: [88], pp. 130–145 (2009) 99

[197] Sertkaya, B.: Towards the complexity of recognizing pseudo-intents. In: [189], pp. 284–292 (2009) 99

[198] Skorsky, M.: How to draw a concept lattice with parallelograms. In: R. Wille (ed.) Klassifikation und Ordnung, pp. 191–196. Indeks–Verlag (1989) 31

[199] Smith, G.: The modeling and representation of security semantics for database applications. PhD thesis, George Mason Univ., Fairfax, Va. (1990) 151

[200] Sowa, J.F.: Conceptual structures: information processing in mind and machine. Addison-Wesley Pub., Reading, MA (1983) 255

[201] Stumme, G.: Attribute exploration with background implications and exceptions. In: H.H. Bock, W. Polasek (eds.) Data Analysis and Information Systems. Statistical and Conceptual approaches. Proc. GfKl'95. Studies in Classification, Data Analysis, and Knowledge Organization 7, pp. 457–469. Springer (1996) 159, 160

[202] Stumme, G.: Concept Exploration – Knowledge Acquisition in Knowledge Systems. Shaker, Aachen (1997) 272

[203] Stumme, G.: Concept exploration — a tool for creating and exploring conceptual hierarchies. In: [156], pp. 318–331 (1997) 266

[204] Stumme, G.: Distributive concept exploration – a knowledge acquisition tool in formal concept analysis. In: O. Herzog, A. Günter (eds.) KI-98: Advances in Artificial Intelligence. Proc. 22. Jahrestagung, *LNAI*, vol. 1504, pp. 117–128. Springer, Heidelberg (1998) 266, 281

[205] Stumme, G.: Formal concept analysis: Methods and applications in computer science (2003). Lecture notes 114, 265

[206] Stumme, G., Taouil, R., Bastide, Y., Pasqier, N., Lakhal, L.: Computing iceberg concept lattices with Titanic. J. Data and Knowledge Engineering (DKE) **42**(2), 189–222 (2002) 62, 66, 82

[207] Stumme, G., Wille, R. (eds.): Begriffliche Wissensverarbeitung – Methoden und Anwendungen. Springer, Heidelberg (2000) 31, 295, 297, 299, 300, 306, 310

[208] Taouil, R., Bastide, Y.: Computing proper implications. In: E.M. Nguifo, M. Liquière, V. Duquenne (eds.) Proceedings of the CLKDD'01 Workshop on Concept Lattices-based Theory, Methods and Tools for Knowledge Discovery in Databases, pp. 49–61. Stanford University (2001) 107

[209] Valtchev, P., Duquenne, V.: On the merge of factor canonical bases. In: [161], pp. 182–198 (2008) 121

[210] Valtchev, P., Grosser, D., Roume, C., Hacene, M.R.: Galicia: An open platform for lattices. In: B. Ganter, A. de Moor (eds.) In Using Conceptual Structures: Contributions to ICCS 2003, pp. 241–254. Shaker Verlag (2003) 30

[211] Valtchev, P., Missaoui, R.: Building concept (Galois) lattices from parts: Generalizing the incremental methods. In: [72], pp. 290–303 (2001) 82

[212] Valtchev, P., Missaoui, R., Lebrun, P.: A partition-based approach towards constructing Galois (concept) lattices. Discrete Mathematics **256**(3), 801–829 (2002) 82

[213] Vanderveken, D.: Meaning and Speech Acts, vol. 1: Principles of Language Use, chap. Semantic Analysis of English Performative Verbs (with K. MacQueen). Cambridge University Press (2009) 157, 158

[214] Voutsadakis, G.: Polyadic concept analysis. Order **19**(3), 295–304 (2002) 284

[215] Wille, R.: Restructuring lattice theory: An approach based on hierarchies of concepts. In: I. Rival (ed.) Ordered Sets, pp. 445–470. Reidel, Dordrecht-Boston (1982) 18, 30, 265

[216] Wille, R.: Liniendiagramme hierarchischer Begriffssysteme. In: H.H. Bock (ed.) Anwendungen der Klassifikation: Datenanalyse und numerische Klassifikation, pp. 32–51. Indeks–Verlag, Frankfurt (1984). Line diagrams of hierarchical concept systems (English translation). *Int. Classif.* **11** (1984), 77–86 31, 33, 59

[217] Wille, R.: Bedeutungen von Begriffsverbänden. In [111], pp. 161–211 (1987) 265, 275

[218] Wille, R.: Dependencies of many-valued attributes. In: H.H. Bock (ed.) Classification and related methods of data analysis, pp. 581–586. North–Holland (1988) 234

[219] Wille, R.: Knowledge acquisition by methods of formal concept analysis. In: E. Diday (ed.) Data analysis, learning symbolic and numeric knowledge, pp. 365–380. Nova Science Publishers (1989) 265

[220] Wille, R.: Conceptual graphs and formal concept analysis. In: [156], pp. 290–303 (1997) 254

[221] Wille, R.: Contextual logic summary. In: G. Stumme (ed.) Working with conceptual stuctures. Contributions to ICFCA 2000. Shaker (2000) 292

[222] Wille, R.: Formal concept analysis as mathematical theory of concepts and concept hierarchies. In [107], pp. 1–33 (2005) 147, 149

[223] Wille, R., Wille-Henning, R.: Towards a semantology of music. In: U. Priss, S. Polovina, R. Hill (eds.) Conceptual Structures: Knowledge Architecture for Smart Applications. Proceedings ICCS 2007, *LNCS*, vol. 4604, pp. 269–282. Springer (2007) 184

[224] Wille, R., Zickwolff, M. (eds.): Begriffliche Wissensverarbeitung – Grundfragen und Aufgaben. B.I.-Wissenschaftsverlag (1994) 31

[225] Wille, R., Zickwolff, M.: Grundlagen einer Triadischen Begriffsanalyse. in [207] (2000) 284

[226] Wollbold, J.: Attribute exploration of discrete temporal transitions. arXiv preprint q-bio/0701009 (2007) 291

[227] Wollbold, J.: Attribute exploration of gene regulatory processes. Ph.D. thesis, Friedrich-Schiller-Universität Jena (2011) 291

[228] Wollbold, J., Köhling, R., Borchmann, D.: Attribute exploration with proper premises and incomplete knowledge applied to the free radical theory of ageing. In: C.V. Glodeanu, M. Kaytoue, C. Sacarea (eds.) Formal Concept Analysis. Proceedings ICFCA 2014, *LNAI*, vol. 8478, pp. 268–283. Springer (2014) 291

[229] Yevtushenko, S.A.: System of data analysis "Concept Explorer" (in Russian). In: Proceedings of the 7th national conference on Artificial Intelligence KII-2000, pp. 127–134 (2000). URL `http://conexp.sourceforge.net/` xvii, 30, 130

[230] Zickwolff, M.: Rule exploration: first order logic in formal concept analysis. Ph.D. thesis, TH Darmstadt (1991) 247, 291

[231] Zschalig, C.: An FDP-algorithm for drawing lattices. In: J. Diatta, P. Eklund, M. Liquière (eds.) Proceedings of CLA 2007, *CEUR-WS*, vol. 331 (2007) 31

List of Algorithms

Index

Printed in the United States
By Bookmasters